Lecture Notes of the Institute for Computer Sciences, Social Informatics and Telecommunications Engineering   417

More information about this series at https://link.springer.com/bookseries/8197

Shuai Liu · Xuefei Ma (Eds.)

Advanced Hybrid Information Processing

5th EAI International Conference, ADHIP 2021
Virtual Event, October 22–24, 2021
Proceedings, Part II

 Springer

Editors
Shuai Liu (iD)
Hunan Normal University
Changsha, China

Xuefei Ma
Harbin Engineering University
Harbin, China

ISSN 1867-8211 ISSN 1867-822X (electronic)
Lecture Notes of the Institute for Computer Sciences, Social Informatics
and Telecommunications Engineering
ISBN 978-3-030-94553-4 ISBN 978-3-030-94554-1 (eBook)
https://doi.org/10.1007/978-3-030-94554-1

This Springer imprint is published by the registered company Springer Nature Switzerland AG
The registered company address is: Gewerbestrasse 11, 6330 Cham, Switzerland

Preface

We are delighted to introduce the proceedings of the second edition of the European Alliance for Innovation (EAI) International Conference on Advanced Hybrid Information Processing (ADHIP 2021), which took place online during October 22–24, 2021. This conference brought together researchers, developers, and practitioners around the world who are leveraging and developing hybrid information processing technology for smarter and more effective research and application. The theme of ADHIP 2021 was "social hybrid data processing".

The technical program of ADHIP 2021 consisted 94 full papers selected out of 254 submissions, with an acceptance rate of approximately 37.0%. The conference tracks were as follows: Track 1—Industrial Application of Multi-modal Information Processing; Track 2—Industrialized big data processing; Track 3—Industrial automation and intelligent control; and Track 4—Visual information processing. Aside from the high-quality technical paper presentations, the technical program also featured two keynote speeches. The two keynote speakers were David Camacho from the Universidad Politécnica de Madrid (UPM), Spain, who is currently a Full Professor and the head of the Applied Intelligence and Data Analysis (AIDA: https://aida.etsisi.uam.es) research group, and also acts as an Associate Editor of several journals including Information Fusion, Ambient Intelligence and Humanized Computing, Expert Systems, and Cognitive Computation, etc., and Muhammad Sajjad, who is an ERCIM Research Fellow with Norges Teknisk-Naturvitenskapelige Universitet (NTNU), Norway, the Head of the Digital Image Processing Laboratory at Islamia College University Peshawar, Pakistan, and an editor of a number of journals, such as IEEE Access and IEEE Trans ITS.

Coordination with the steering committee, comprising Imrich Chlamtac and Guanglu Sun, was essential for the success of the conference. We sincerely appreciate their constant support and guidance. It was also a great pleasure to work with such an excellent organizing committee team for their hard work in organizing and supporting the conference. In particular, we are grateful to the Technical Program Committee who completed the peer-review process for technical papers and helped to put together a high-quality technical program. We are also grateful to Conference Manager Natasha Onofrei for her support and all the authors who submitted their papers to the ADHIP 2021 conference and workshops.

We strongly believe that the ADHIP conference provides a good forum for all researchers, developers, and practitioners to discuss all science and technology aspects that are relevant to hybrid information processing. We also expect that the future ADHIP conferences will be as successful and stimulating as this year's, as indicated by the contributions presented in this volume.

December 2021

Shuai Liu
Yun Lin
Chen Cen
Danda Rawat
Zheng Ma

Organization

Steering Committee

Imrich Chlamtac (Chair)	University of Trento, Italy
Guanglu Sun	Harbin University of Science and Technology, China

Organizing Committee

General Chair

Zhongxun Wang	Yantai University, China

General Co-chairs

Wei Xue	Harbin Engineering University, China
Joey Tianyi Zhou	Institute of High Performance Computing, A*STAR, Singapore
Guan Gui	Nanjing University of Posts and Telecommunications, China
Shifeng Ou	Yantai University, China
Bo Nørregaard Jørgensen	University of Southern Denmark, Denmark

Technical Program Committee Chair

Shuai Liu	Hunan Normal University, China

Technical Program Committee Co-chairs

Chen Cen	Institute for Infocomm Research A*STAR, Singapore
Danda Rawat	Howard University, USA
Yun Lin	Harbin Engineering University, China
Zheng Ma	University of Southern Denmark, Denmark

Sponsorship and Exhibit Chairs

Jinwei Wang	Yantai University, China
Ali Kashif	Manchester Metropolitan University, UK

Local Chairs

Zhuoran Cai	Yantai University, China
Yidong Xu	Harbin Engineering University, China

Workshops Chairs

Zhuoran Cai	Yantai University, China
Ming Yan	Agency for Science, Technology and Research (A*STAR), Singapore
Ting Wu	Beihang University, China
Ya Tu	Harbin Engineering University, China

Publicity and Social Media Chairs

Pengfei He	Yantai University, China
Lei Chen	Georgia Southern University, USA

Publications Chairs

Weina Fu	Hunan Normal University, China
Wenjia Li	New York Institute of Technology, USA
Meiyu Wang	Harbin Engineering University, China

Web Chairs

Gang Jin	Yantai University, China
Ruolin Zhou	University of Massachusetts Dartmouth, USA

Posters and PhD Track Chair

Liyun Xia	Hunan Normal University, China

Panels Chair

Peng Gao	Hunan Normal University, China

Demos Chair

Jingyi Li	Hunan Normal University, China

Technical Program Committee

Adam Zielonka	Silesian University of Technology, Poland
Amin Taheri-Garavand	Lorestan University, Iran

Yong Jun Qin   Guilin Normal College, China
Yun Lin   Harbin Engineering University, China
Zheng Ma   University of Southern Denmark, Denmark

Contents – Part II

Research and Analysis with Intelligent Education

Contents – Part I

AI System Research and Model Design

Method Research on Internet of Things Technology

Design of Regional Sharing Model of Enterprise Accounting Information Based on Blockchain Technology

Lin Chen[✉] and Yan Yu

Wuhan QingChuan University, Wuhan 430204, China

Abstract. In practice, there are too many sharing channels in the regional network, which leads to the low practical performance of the traditional enterprise accounting information regional sharing model. In order to solve this problem, this paper proposes a regional sharing model of enterprise accounting information based on block chain technology. Building a shared model framework based on block chain technology, carrying accounting model in the framework, and sorting out accounting information. Then build the sharing channel for different types of accounting information and deploy the sharing protection mechanism to further protect the security of enterprise information. The experimental results show that compared with the traditional shared model, the proposed model has better fault tolerance and higher throughput, and its overall application practicality is superior to the traditional shared model.

Keywords: Blockchain technology · Enterprise accounting information · Regional sharing · Sharing protection mechanism

1 Introduction

At present, blockchain has attracted the attention of all walks of life in the world, and there are different opinions on blockchain technology. Blockchain is applied in various industries, but the research depth is not enough [1]. Some foreign financial institutions are studying the blockchain, while the domestic ones are studying the blockchain by people in various industries and at all stages, resulting in a flood of blockchain development, which inevitably leads to insufficient development depth and uneven results [2].

Blockchain has been applied in various industries, such as the following fields: in the financial field, blockchain technology can make use of the coherence and cohesion between nodes to reduce the participation of third-party intermediaries, so as to realize the direct contact between nodes, reduce costs and improve the efficiency of transactions; The application in the field of the Internet of Things mainly makes it possible to reduce the cost of logistics through some characteristics of the blockchain technology, record the production and transportation process of products, and form a comprehensive supply chain management mode. Applications in the field of public service, chain block

S. Liu and X. Ma (Eds.): ADHIP 2021, LNICST 417, pp. 3–14, 2022.
https://doi.org/10.1007/978-3-030-94554-1_1

technology to the public management in the energy and transportation, etc. are inter-linked, let these technologies more close to the daily life, and be able to solve the quality problems in some areas, in the field of digital copyright applications, through regional chain technology, encryption feature to identify the true and false goods, to effectively lifecycle of digital rights management [3].

In enterprise management, there are great security risks in Regional Sharing of accounting information. In the process of information sharing, different regional channels or channel subnets may lead to the emergence of "information island" because of economic interests, data ownership and other factors. The private information owned by different subnets can not be circulated, which affects the social production efficiency and consumes repeated human and economic costs. In addition, it is contrary to the original intention of decentralization [4]. Usually, in an alliance chain environment with multiple channel structures, different channels need to communicate with each other, and all the organizations in the two channels can be added to a new channel, but this will force the unrelated organizations in the original channels to join them, which will make the whole network too bloated [5].

With the innovative development of science and technology, information construction is becoming more and more perfect, and the construction of regional sharing model of accounting information of various enterprises is also gradually improved. Later because of domestic block chain technology development, combined with industry is not very widely used, the chain blocks the development of technology is limited in certain areas, and can be used to make it to get comprehensive development difficult, which requires more and more people actively research and promote the application of the block chain technology, for the enterprise accounting information sharing model construction to provide technical support, Make its development more and more high-end, more and more high technical content.

Therefore, based on a large number of existing literature, in order to improve the application advantages of the sharing model, this study designed a regional sharing model of enterprise aggregated information based on blockchain technology. First of all, the sharing model framework is built based on blockchain technology, and the accounting model is equipped in the framework, so as to realize the analysis and sorting of accounting information. Then, special sharing channels are built for different types of accounting information, and sharing protection mechanism is deployed to further protect the security of enterprise information. The simulation results also prove that the model has the advantages of better fault tolerance and higher throughput.

2   Model Design

2.1   Model Frame Design

The model adopts the basic architecture of blockchain model, including P2P network, message propagation mechanism, block data structure, etc. the specific architecture is shown in Fig. 1.

In addition to the basic architecture of blockchain technology, the basic framework of the model also includes: the interaction between the credit reference institution and

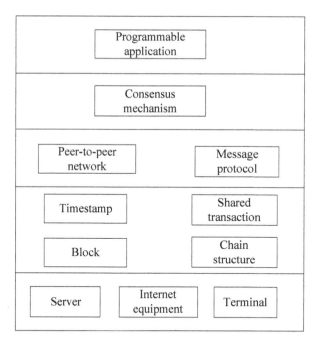

Fig. 1. Technical architecture of blockchain

the information subject. The credit reference institution needs to obtain the user's authorization for the requested data first, and then the credit reference institution can send a request to the credit information provider to obtain the corresponding data by virtue of the authorization information signed by the information subject [6]. In the interaction with the cloud server, after receiving the request, the credit information provider encrypts and uploads the relevant data to the cloud server according to certain data format requirements. In order to protect the rights and interests of the credit information provider, the credit information provider initiates the transaction containing the summary, request object, metadata and other information of the uploaded data to the blockchain node to obtain the upload voucher [7]. The request records of user authorization and credit reference institutions, as well as the data information uploaded by credit information providers are recorded on the blockchain. The blockchain has the characteristics of non tampering and traceability, and is jointly maintained by multiple nodes. Therefore, it can be used as the basis for conflict handling and provide a fair and reliable credit information sharing environment [8].

In the interaction with regulators, it is inevitable that there will be some disputes in the sharing of credit information. Therefore, the model includes regulators as the authority of dispute settlement to protect the rights and interests of all participants. The basic framework of enterprise credit information sharing model based on blockchain technology is shown in Fig. 2.

According to the actual needs of enterprises, the accounting model is carried on the shared model framework.

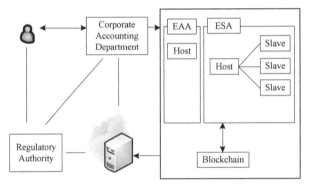

Fig. 2. Dasic framework of enterprise accounting information regional sharing model

2.2 Design of Accounting Model

Under blockchain technology, the accounting mode has changed some technical means in the basic organization and the original concrete operation method before the continuation. The latest financial software is used in the accounting of colleges and universities, which can not only improve the efficiency of accountants, but also make the accounting work more accurate, scientific and systematic, improve the quality of accounting information in Colleges and universities, and enhance the timeliness of accounting information [9]. The application of financial software can accelerate the processing and application of financial information data in colleges and universities, and provide comprehensive and accurate financial information analysis and financial report to university managers and information users.

Under blockchain technology, the connection among accounting vouchers, accounting books and financial reports can be better realized. The design of accounting method should consider the relationship between voucher and voucher, voucher and account book, account book and account book, account book and report, report and report [10]. Only by coordinating these relations can accounting work smoothly be carried out. Block chain is composed of each block, and the next block is formed on the basis of the previous block. The connection between blocks can ensure that each financial data can be traced. The accounting book block is formed on the basis of financial information provided by the accounting voucher block, thus providing information support for financial report.

The design of accounting mode should simplify the process of handling work as much as possible and reduce repetitive work, such as registering the accounting voucher according to the original voucher, which will lead to repetitive work and increase the workload. Under the premise of following the actual situation, we can use the original voucher to replace the bookkeeping voucher and improve the summary method of bookkeeping voucher. Under the blockchain accounting mode, under the coordination of smart contracts, the system can automatically execute programs and reduce workload after meeting the set conditions. The non tamperability of the blockchain can ensure that the content of the contract will not be changed, and the system will execute automatically only when the conditions are met [11]. The functions of decentralization and whole network backup can reduce audit workload and simplify audit procedures.

Blockchain technology can strengthen the internal control of accounting work. The design of accounting model should be combined with post responsibility system, clear division of labor and mutual restriction. Blockchain distributed ledger means that a transaction is completed by multiple nodes in multiple places, and each node records a complete account book, so they can supervise the legitimacy of the transaction with each other, and no node can record the account separately, so as to avoid the occurrence of a single bookkeeper being controlled or bribed to record false account. The independent completion and mutual supervision of each node ensure the internal control of accounting work. In addition, according to the characteristics of the blockchain smart contract, the setting rules of the blockchain will automatically determine whether its trading activities are effective [12]. Therefore, the trading party does not need to obtain the trust of the other party through the way of public identity, so as to avoid artificial collusion and make the transaction reasonable control.

2.3 Building a Shared Channel

In order to transmit data from the target subnet to the relay link ledger and from the relay link ledger to the request subnet, it is necessary to build a contact channel. You can call the API related to channel creation in SDK, or use configtxgen tool to generate offline configuration information related to channel creation and update. The new application channel mainly generates the following files: (1) generates the creation block that starts sorting service and supports checking the contents of the block; (2) Generate configuration trade files and support the inspection of trade contents; (3) Generate the configuration transaction file used to update the channel organization Anchor Peer.

Change the contents of the configTX.YAML configuration file. After adding the configuration template for the new channel, add an Application to specify the organization information of the Application channel and specify the name of the associated federation in the Profile section. Next, the channel configuration block is generated in turn, and the file to update the organization anchor node is generated. Create contact channel according to channel configuration transaction file, add organization nodes to contact channel in turn, update anchor node, install and instantiate chain code.

Considering the privacy of enterprise accounting information, the shared protection mechanism is loaded in the constructed shared link. For link separated working paths L_1 and L_2, end-to-end shared protection paths P_{1r} and P_{2r} are established respectively. In addition, P_{1e} and P_{2e} constitute the protection for some working paths $0 \leftrightarrow 1$ and $0 \leftrightarrow 6 \leftrightarrow 7$. Let η_i denote the O type of each link, then the reliability of the working circuit, the sectional working circuit and the corresponding part of the protection circuit are respectively expressed as:

$$\eta_L = \prod_{i \in lp} \eta_i \tag{1}$$

$$\eta_e = \prod_{i \in lw} \eta_i \tag{2}$$

$$\eta_r = \prod_{i \in sp} \eta_i \tag{3}$$

The comprehensive reliability of the established business connection is:

$$\eta_c = \frac{\eta_L(\eta_e + \eta_r(1 - \eta_e))}{\eta_e} \tag{4}$$

Assuming that the reliability of each link is 0.98, the reliability requirement of service connection is 0.95, and the reliability of working path is 0.94, it obviously can not meet the reliability requirement. It is necessary to establish a protection path. For the actual network, due to the limitation of network connectivity, sometimes some protection paths can not be found. At this time, in order to ensure the establishment of business connection request, the complete protection path between the source and destination nodes should also become a necessary alternative to further protect the Regional Sharing Security of enterprise accounting information. So far, the regional sharing model of enterprise accounting information based on blockchain technology has been designed.

3 Experimental Research on Regional Sharing Model of Enterprise Accounting Information Based on Blockchain Technology

3.1 Design Shared Data Items

The implementation of blockchain technology in the sharing module first requires enterprises to join the blockchain network as blockchain nodes. In the experimental case, the process was started on ports 5000 to 5005 of the local machine, and six blockchain nodes were successfully constructed through the flash framework to simulate the registration of six enterprises, with ports 5000 and 5001 as EAA nodes and ports 5002 to 5005 as ESA nodes Registration.

Enterprises sharing credit information can encrypt and upload the relevant credit information to the cloud server according to the request, upload the encrypted summary and metadata of credit information to the blockchain network in the form of transaction, or directly upload the credit information to the blockchain network in the form of transaction. For example, when a bank decides whether to issue a short-term loan to an enterprise, it should focus on the historical credit records and capital turnover of the enterprise. The data items to be collected are the basic situation, historical credit records and financial status of the enterprise. Therefore, the format standard for uploading various data should be formulated. The experimental case uses the sample data of Guiyang big data exchange's "other financial activity enterprise big data" to simulate the enterprise credit information to be shared, and assumes that these data are scattered in six enterprises. The rest of the reference data format is shown in Table 1.

When simulating the process of enterprise information sending, the main information is sent to the sending interface by post request in the experiment, which is received by two EAA nodes. The EAA node verifies the structure of the information and the accuracy of the uploaded information. After the verification is successful, the accounting information is signed and encapsulated, and then broadcast to the master node in ESA. The master node collects the transactions signed by EAA node, verifies the signature and packages the transactions to generate blocks to prevent the transaction data, time and order from being tampered.

Table 1. Reference format of shared data items

Sample field (parameter description)	Parameter name	Data type	Is it required?
Basic information of enterprise	Entall	String	Optional
Financial information	Financellist	String	Optional
Product competitors	Jzdslist	String	Optional
Related companies	Glgslist	String	Optional
Investment and financing events	Trzallist	String	Optional
Shareholder structure	Gdjglist	String	Optional
Trademark patent	Sbzllist	String	Optional
Customer relationship	Khgxlist	String	Optional
Negative news	Fmxwlist	String	Optional
Litigation record	Ssjllist	String	Optional
Foreign investment information	Dytz	String	Optional
Customs punishment	Cf_haiguan	String	Optional
Environmental punishment	Cf_huanjing	String	Optional
Industrial and commercial penalties	Cf_gongshang	String	Optional

The experiment is divided into fault tolerance test and throughput test, mainly based on comparative experiment. In the experiment, the traditional regional sharing model of enterprise accounting information is introduced. After the experiment, based on the experimental results, the practical application level of each sharing model is compared and analyzed.

3.2 Experimental Results and Analysis of Fault Tolerance

Fault tolerance is an important performance of shared model, which reflects the maximum number of node errors that shared model can tolerate. In order to test the fault-tolerant performance of the designed shared model, that is, whether the block chain module can recover itself without affecting the overall performance when the node fails. In the experimental study, the consensus process of shared model is simulated under the condition of 0, 1, 2 and 3 failure nodes. The experimental results are shown in Table 2.

According to the results in Table 2, under the sharing conditions of different number and different types of failure nodes, only the shared model in this paper can be identified correctly. Only the other two traditional shared models can ensure the recognition success without failure nodes and few failure nodes. In the other two cases, the recognition fails.

In conclusion, the proposed block chain based enterprise accounting information regional sharing model can effectively identify the transmission nodes and ensure the reliability of the sharing model.

Table 2. Experimental results of fault tolerance

	Number of failed nodes	Failure node type	Consensus results
Traditional sharing model 1	0	Nothing	Success
	1	Master node	Success
		Slave node 1	Success
		Slave node 2	Success
		Slave node 3	Success
	2	Master + Slave 1	Fail
		Slave node 1 + slave node 2	Fail
	3	Nodes other than master	Fail
		All nodes except one slave node	Fail
Traditional sharing model 2	0	Nothing	Success
	1	Master node	Success
		Slave node 1	Success
		Slave node 2	Success
		Slave node 3	Success
	2	Master + Slave 1	Fail
		Slave node 1 + slave node 2	Fail
	3	Nodes other than master	Fail
		All nodes except one slave node	Fail
Article sharing model	0	Nothing	Success
	1	Master node	Success
		Slave node 1	Success
		2-node slave	Success
		Slave node 3	Success
	2	Master + Slave 1	Success

(*continued*)

Table 2. (*continued*)

	Number of failed nodes	Failure node type	Consensus results
		Slave node 1 + slave node 2	Success
	3	Nodes other than master	Success
		All nodes except one slave node	Success

3.3 Experimental Results and Analysis of Shared Information Throughput

Throughput is a measure of the ability of a sharing model to process sharing requests per unit time. The higher the throughput, the more requests the model can process and the higher the performance of the model. The calculation formula of block throughput is as follows:

$$TPS_t = \frac{tras_t}{t} \tag{5}$$

In the formula, t represents the time interval from sending information sharing request to block confirmation, and $tras_t$ represents all information sharing requests contained in a block. In the experiment, the time interval for the master node to initiate a consensus is defined as t.

In order to simply test the ability of sharing model to process information sharing requests, JMeter performance testing tool is used to simulate 50 users to send transactions to a sharing process in a single machine environment. The shared content simulates the data information uploaded by enterprises. Four groups of experiments are conducted in 10s, 20s, 30s and 40s respectively. 20 experiments are repeated in each time interval, and 2 is selected The average of 0 experiments is used as the throughput of this time interval. The test results of four time intervals are drawn by the third-party software MATLAB. The experimental results are shown in Fig. 3.

The experimental results in Fig. 3 show that the throughput of different models gradually increases with the increase of the transaction time of shared information. However, the information throughput of the two traditional sharing models varies greatly in the process of sharing, and the throughput level is low. In the experiment, the throughput variation of the sharing model proposed in this paper is relatively stable and the throughput is relatively high, which makes the information sharing efficiency higher.

By comprehensive analysis of the results of fault-tolerant experiment and throughput experiment, it can be seen that the regional sharing model of enterprise accounting information designed in this paper based on block chain technology has high fault-tolerant rate and high sharing efficiency, which proves that its application performance is better.

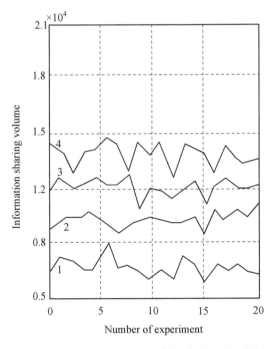

(a)Experimental results of traditional shared model 1

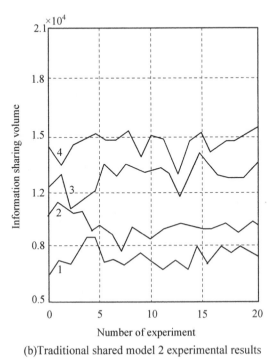

(b)Traditional shared model 2 experimental results

Fig. 3. Experimental results of information throughput of different sharing models

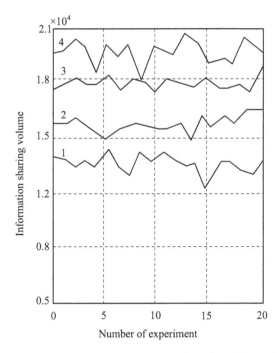

(c)This paper shares the experimental results of the model

Fig. 3. continued

4 Conclusion and Prospect

In this paper, the regional sharing model of enterprise accounting information is studied. On the basis of using block chain technology to build a sharing model framework, sharing channels are built for different types of accounting information, and sharing protection mechanisms are deployed to further protect the security of enterprise information. This study provides some referential significance for other institutions to build regional accounting information sharing model. However, from the perspective of the current development of blockchain technology, the research in this paper still has some deficiencies. In the design process, this paper mainly designs and adjusts the specific situation of regional sharing of accounting information and the development characteristics of block chain technology. However, in order to consider the timeliness of sharing, the process of sharing accounting information takes a long time. In the following research, the model of this paper will be further optimized from the perspective of improving the timeliness of sharing.

Fund Projects. Research on the Impact of Private Chain Based on Blockchain Technology on Enterprise Financial Sharing Center (B2019352) Hubei Provincial Department of Education.

References

1. Yang, Y., Shi, X.: Design of cross-border logistics operation model in "lanmei cooperation area" based on block chain. Sci. Technol. Manag. Res. **40**(05), 139–145 (2020)
2. Wu, G., Chen, Y., Zeng, X.: Design and analysis of token model based on blockchain technology. Comput. Sci. **47**(S1), 603–608 (2020)
3. Shen, K., Gao, S., Zhu, J.: LibRSM: digital library information resources security sharing model based on consortium blockchain. J. Natl. Libr. China **28**(02), 77–86 (2019)
4. Huang, K., Lian, Y., Feng, D., et al.: Cyber security threat intelligence sharing model based on blockchain. J. Comput. Res. Dev. **57**(04), 836–846 (2020)
5. Zhai, S., Wang, Y., Chen, S.: Research on the application of blockchain technology in the sharing of electronic medical records. J. Xidian Univ. (Nat. Sci.) **47**(05), 103–112 (2020)
6. She, W., Chen, J., Gu, Z., et al.: Location information protection model for IoT nodes based on blockchain. J. Appl. Sci. **38**(01), 139–151 (2020)
7. Yang, S.: Research on interorganizational relationship design of supply chain accounting information sharing. Commun. Finan. Account. **42**(15), 45–48 (2019)
8. Liu, S., Liu, G., Zhou, H.: A robust parallel object tracking method for illumination variations. Mob. Netw. Appl. **04**(01), 5–17 (2019)
9. Yu, J., Zhang, H., Li, S., et al.: Data sharing model for internet of things based on blockchain. J. Chin. Comput. Syst. **40**(11), 2324–2329 (2019)
10. Wang, Y.: Cloud data integrity verification algorithm for accounting informationization in sharing mode. Mod. Electron. Tech. **42**(05), 87–89 (2019)
11. Yuan, G., Guo, Y.: Application of blockchain technology in accounting field. Finan. Account. **15**(06), 73–74 (2019)
12. Jin, X., Zhang, X.: Research on information transfer mechanism of financial sharing center of enterprise group under structural hole theory. Finan. Account. Mon. **28**(17), 26–32 (2020)

Automatic Information Scheduling System of 5G Intelligent Terminal Based on Internet of Things Technology

Guo-yi Zhang$^{(\boxtimes)}$, Hai-long Zhu, Dan-ke Hong, Si-tuo Zhang, and Shan-ke Huang

Digital Grid Research Institute, China Southern Power Grid, Guangzhou 510670, China
zgy329@aliyun.com

Abstract. It is difficult to extract accurate feature data in conventional automatic scheduling system, which leads to large automatic scheduling error. Therefore, 5G intelligent terminal information automatic scheduling system is designed based on Internet of Things technology. In the design of hardware system, the bus circuit, 5G intelligent network management node system and 5G mobile communication chip based on Internet of Things are designed. In the design of software system, the initial solution of automatic scheduling is calculated, the automatic scheduling characteristics of 5G intelligent terminal based on Internet of Things are extracted. Experimental results show that the average scheduling error is 0.1604 and 0.2632 lower, the absolute error is 0.0503 and 0.0568 lower, and the data obtained by the two methods are smaller than those by the conventional methods. Therefore, the system has a more accurate 5G intelligent terminal information automatic scheduling capabilities.

Keywords: Internet of things technology · 5G · Intelligent terminal · Information automatic scheduling system

1 Introduction

With the rapid development of information technology in the 21st century, the information released through 5G intelligent terminal has become an important pillar of people's daily interaction, so the scheduling of such information has become an important method of production, operation and management of enterprises [1]. But in the actual process, some unexpected events often occur, this kind of event has great damage to the accuracy of program scheduling, which may cause great error. So this paper analyzes how to reduce the error of 5G intelligent terminal information scheduling system. Document [2] Coding of communication information by establishing a standardized unified system enhances the integration of the system and the sharing of information. Document [3] Obtaining the network server address for TCP/IP through an application in the client and controlling the security of the server using standard Internet technology [2, 3]. However, in practical application, the scheduling characteristics of 5G intelligent terminals are not accurately extracted, which leads to a large error in the automatic scheduling of 5G

© ICST Institute for Computer Sciences, Social Informatics and Telecommunications Engineering 2022
Published by Springer Nature Switzerland AG 2022. All Rights Reserved
S. Liu and X. Ma (Eds.): ADHIP 2021, LNICST 417, pp. 15–25, 2022.
https://doi.org/10.1007/978-3-030-94554-1_2

intelligent terminals. Therefore, this paper proposes to design an automatic scheduling system for 5G intelligent terminal information by using the Internet of Things technology. The bus circuit of the system is designed through zero group grounding, which makes the grounding line anti-interference. The parameters of bus and gateway nodes are designed, and the hardware part of the system is built with the core chip. The objective function is constructed, and standard frequency, initial frequency, optimal frequency and fundamental frequency are used to identify the automatic scheduling characteristics of 5 G intelligent terminal information samples.

2   Design of Hardware for 5G Intelligent Terminal Information Automatic Scheduling System

2.1   Design Bus Circuit

In the design of the bus circuit of the system, a portion of the power output condition is provided by the power division module shown in Fig. 1, where the output device uses a 3.0V standard voltage.

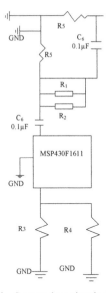

Fig. 1. System bus circuit design

However, after some time of testing, it is found that the overall design of the circuit as shown in Fig. 1 needs to achieve the purpose of grounding anti-interference through zero group, so different power line numbers are needed to make these grounding lines bypass interference and complete the automatic scheduling of the whole system module [4]. In this process, two 30Ω resistors can be used as the matching resistors of the smart terminal transmission module, and five grounded power supplies can be connected as the output of line impedance.

Controller STM32F207VE internal integration of the powerful CAN controller, CAN communication module design only need to design the interface circuit between the microprocessor and the driver [5]. In the system design, the master node and the child node adopt the same driving circuit, the driver adopts the whole encapsulation technology, the internal integration of the power isolation, electrical isolation and CAN transceiver circuit, isolation voltage up to DC2500V, simple interface, easy to use, can be a good CAN-BUS bus between the nodes of the complete isolation and independence of electricity. The TXD, RXD pins of the module are fully compatible with the CAN controllers of 3.3V and + 5V, and no external components are needed. Compared with the existing design scheme, the DC-DC power module, optocoupler module and ESD protection circuit are simplified, and the stability and security of the node are improved.

2.2 Design of 5G Intelligent Gateway Node System

The design of 5G intelligent terminal information gateway node is the key node of the whole automatic scheduling system. The gateway node realizes the automatic scheduling of information by setting the parameters of CAN bus and its sub-nodes [6]. Among them, the main function of each sub-node is to collect field information and transmit it to the information gateway node of 5G intelligent terminal through CAN bus. When there is any problem at the production site, the on-duty personnel shall be alerted by means of LCD display and sound and light alarm. The on-duty personnel may learn the information of each node in detail from the control platform, and issue control commands to the sub-node through the control platform or keyboard [7]. The 5G intelligent terminal information gateway node communicates with the control platform through TCP/IP and LTE to ensure the effective control of the scene. 5G intelligent terminal information gateway node circuit mainly includes CAN transceiver circuit, Ethernet interface circuit, LTE communication circuit, keyboard circuit, LCD display circuit and acoustooptic alarm circuit, the specific structure is shown in Fig. 2.

Considering that the data processing capacity of the 5G intelligent terminal information gateway node is required to be high, the 32-bit processor STM32F207VE based on the ARM Cotex-M3 core is selected. The chip uses the latest 90-nm process to produce a new generation of STM32 products, which can achieve the processing capacity of 150DMIPS at a high speed of 120 MHz, up to 512k bytes of on-chip flash memory and 128k bytes of SRAM, and the adaptive real-time flash memory accelerator enables the on-chip flash memory to execute the code with a high-speed zero wait of 120 MHz [8]. The chip carries 1 way 10/100M 5G intelligent terminal interface, 2 CAN channels, 6 UART interfaces, in addition to ADC, DAC, USB, PWM and other peripheral equipment. In this way, 5G intelligent terminal information automatic scheduling system can communicate with TCP/IP and CAN bus stably, and it can be ensured to be included in the running module of automatic scheduling system.

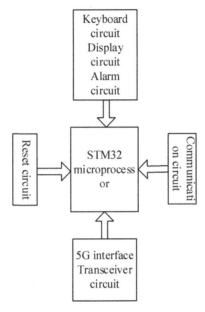

Fig. 2. 5G smart terminal gateway node system architecture

2.3 Design of 5G Mobile Communication Chip Based on Internet of Things

In the design of communication chip of 5G intelligent terminal information gateway node, in order to realize the remote transmission of data, expand the communication bandwidth and improve the communication compatibility. The MAC chip of STM32F207VE integrates special DMA inside and provides MII or RMII for Ethernet communication. The PHY physical layer control chip adopts DP83848CW to simplify the circuit design diagram of the hardware as shown in Fig. 3. The PB11 of STM32 is connected with TX _ EN to enable the sending of pins [9]. PB12, PB13 connect with TX _ DO, TXD _ 1, send data. PD9, PD10 connect with TX _ D0, TXD _ 1, and receive data. The X1 provides a clock signal for receiving and sending data. DP83848CW is connected with network transformer socket J0011D21BNL to realize the differential transmission of TCP/IP digital signal.

LTE5G intelligent terminal information gateway node communication chip USR-LTE-754 as the core, can easily achieve high-speed communication with the LTE network. The USR-LTE-754 is a compact, feature-rich M2M product that can be easily and quickly integrated into the design system. The chip software has complete functions, covers most common application scenarios, supports custom sign-up packets and heartbeat packets, supports 4-way Socket connections, and supports transparent cloud access with high speed and low latency [10]. The UARTTX and UART-RX pins of USR-LTE-754 are connected with PC11 and PC10 of STM32 respectively to realize the bidirectional data transparent transmission from serial port to LTE network. The chip has integrated the function of SIM card and connected with the corresponding pins of SIM card base in design. CSI _ CLK, CSI _ RST and CSI _ D pins are connected in parallel with the capacitors around 47pF to filter the interference of RF signal. The chip provides switch

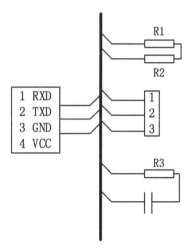

Fig. 3. 5G mobile communication chip

RES and POWER _ KEY pins to control the switch of STM32. Through the LED status display chip working state. The indicator pins of the chip are LinkA, LED _ W and LED _ N, which indicate the network connection, the chip operation and the specific network status respectively. The pin level is 18V, and the driving indicator needs to match the level and drive the indicator through the triode.

3   Design of 5G Intelligent Terminal Information Automatic Scheduling System Software Based on Internet of Things Technology

3.1   Initial Solution of Computational System Automatic Scheduling

Suppose an automated scheduling schedule has three machines in progress, m_1, m_2, m_3, respectively, generating, encapsulating, and transporting jobs. The number of methods of intelligent terminal information automatic scheduling is n_1, and the number of schemes formed by the method is n_2, and then the number of initial solutions is n_3 [11]. Where $m = m_1 + m_2 + m_3$ and $n = n_1 + n_2 + n_3$ for transshipment. In order to facilitate the calculation and reduce the computational efficiency problems caused by too many variables, the varieties of the three initial solutions are calculated separately and the following objective functions are established:

$$c = \sum_{i=1}^{m} \sum_{j=1}^{n} X_{mn} D_{mn} \tag{1}$$

In the expression, the constraint of the objective function, the number of intelligent terminal information automatic scheduling methods and the number of intelligent terminal information automatic scheduling schemes are expressed. In this process, the system calculates the number of methods and schemes [12], and calculates the implementation coefficient matrix of 5G intelligent terminal information automatic dispatching through ArcGIS:

$$X = \begin{pmatrix} 1\,1\,0\,0\,0 \\ 1\,1\,1\,0\,0 \\ 1\,1\,1\,1\,0 \\ 1\,1\,1\,1\,1 \\ 0\,1\,1\,1\,1 \\ 0\,0\,1\,1\,1 \\ 0\,0\,0\,1\,1 \end{pmatrix} \tag{2}$$

The objective value function obtained from matrix calculation is as follows:

$$X' = \left(2324\ 1956\ 3817 \right)^{T} \tag{3}$$

Using the program of simple form method and two-stage method, the scheduling value can be obtained after the calculation of standard type, initial type, optimal type, basic type, iterative type, etc.

3.2 Extracting Automatic Scheduling Characteristics of 5G Intelligent Terminal Based on Internet of Things

Before extracting the automatic dispatching features of 5G intelligent terminal, it is necessary to calculate the automatic dispatching features first, and to recognize the effects of various automatic dispatching features in the 5G communication process [13]. In this process, there are several parameters which are inevitably affected. This paper uses standard frequency, initial frequency, optimal frequency and fundamental frequency to identify the automatic scheduling characteristics of a 5G intelligent terminal information sample.

First is about the standard frequency change, in eliminates in the standard frequency the process, often appears the signal overload the question. In order to eliminate the influence of this kind of signal, filter eliminator is usually used to reduce the signal frequency. If the short-time framing coefficient of a signal is set to $f(n)$, the adaptive function of the signal can be obtained:

$$f(n) = \sum_{n=1}^{N-K} W_x(n)(n+k) \tag{4}$$

In the formula, N indicates the number of frames in the signal, K indicates the time parameter for signal elimination, and $0 \leq k \leq n$, usually the value of k is between 50 and 100 Hz, and $W_x(n)$ indicates the range of change in the window function of the signal.

The initial frequency is usually used to reflect the frame amplitude of a 5G smart terminal information sample, which plays a key role in the expression of 5G smart terminal information. When calculating the initial frequency, the weighted squared sum of the sampling points can be obtained by calculating the average energy of the information data of a specific 5G intelligent terminal [14]. The specific calculation formula is as follows:

$$E = \sum_{i=1}^{N} A_x^2(\alpha) \tag{5}$$

In the formula, E represents the numeric value of the initial frequency, with the unit of kJ; N represents the frame length of the information of a segment of 5G intelligent terminal; i represents the frame number of the information of the segment of 5G intelligent terminal; A represents the average value of the information of the segment of 5G intelligent terminal; and α represents the electric energy calculation parameters of the segment of the information of the segment of 5G intelligent terminal. Because the 5G intelligent terminal information will be averaged in the process of this calculation, the data are not less than zero integers. The specific parameters of energy summation can be obtained in the open form in daily calculation.

The optimal frequency is usually due to a short period of time in a 5G intelligent terminal information changes 5G intelligent terminal information, there is a short zero-energy phenomenon, such phenomenon can obviously lead to a large difference between 5G intelligent terminal information data, and this kind of phenomenon usually exists in the 5G intelligent terminal information with large emotional fluctuations. The method of extracting the optimal frequency is very simple, and can be determined directly by the change of the symbol, namely:

$$\theta = \frac{1}{2} \sum_{i=1}^{N} |\text{sgn}(\beta) - \text{sgn}(\beta - 1)| \tag{6}$$

In the formula, N represents the frame length of the information data of the 5G intelligent terminal with the phenomenon of short-time zero-crossing; i represents the frame number of the information of the 5G intelligent terminal; and β represents the frequency coefficient of the information of the 5G intelligent terminal. Of these, sgn is typically a symbolic function that has a value of 1 at $\beta > 0$ and a value of 0 at $\beta \leq 0$.

Most of the basic variable frequencies are based on the analysis of the functions of 5G intelligent terminals. If the computing capacity of 5G intelligent terminals is linearly and positively correlated with the information of 5G intelligent terminals, the information frequency of 5G intelligent terminals in this section is lower; if the computing capacity of 5G intelligent terminals is unable to be correlated with the information data of 5G intelligent terminals, the information frequency of 5G intelligent terminals in this section is higher. When a frequency filter is constructed based on this concept, it can be concluded that the frequency calculation formula for the information data of the 5G intelligent terminal is as follows (4):

$$f(x) = 3325 \ln(1 + \frac{f(v)}{700}) \tag{7}$$

In the formula, $f(x)$ represents the value of the frequency filter function, and $f(v)$ represents the calculation function of the frequency. The cosine transform parameters of the

5G intelligent terminal information data can be obtained by logarithmic analysis of the function value. By accurately extracting the information features of the four kinds of 5G intelligent terminals, the accuracy of the identification system can be directly optimized, so the accuracy and reliability of the parameters should be paid attention to. However, it is very difficult to determine these four kinds of parameters perfectly only by 5G intelligent terminal information, and its accuracy is very difficult to guarantee. Therefore, other algorithms can be introduced to test the results of these four kinds of parameters to ensure the accuracy of 5G intelligent terminal information automatic scheduling system.

3.3 Designing Optimization Algorithms for Automatic Scheduling Systems

In order to get more accurate 5G intelligent terminal information automatic scheduling system, the 5G intelligent recognition is classified as a calibration reference in the scheduling algorithm, which is shown in Fig. 4.

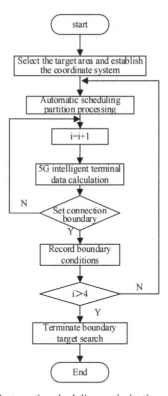

Fig. 4. Automatic scheduling optimization algorithm

As shown in Fig. 4, you first need to set four boundary conditions to schedule partitions through intelligent terminal computing devices. When all the four boundary conditions are satisfied, the boundary target can be terminated and the search can be stopped, so that the input data and parameter can be obtained in the automatic scheduling.

Then, according to the data obtained from the automatic scheduling search, converted to the above parameters, become another vector of the automatic scheduling features, so as to achieve the optimization of this algorithm.

4   Experimental Research

4.1   Experimental Preparation

This system mainly uses Eclipse as the JAVA language development tool; the database aspect uses PL/SQL to develop the SQL statement, the view, the stored procedure and so on; the development work mainly carries on under the Windows XP, the temporary application server chooses Tomcat 5.0; uses Microsoft Visual SourceSafe 6.0 as the version control tool. The software environment on the database server is mainly divided into two aspects, namely, the operating system software Solaris and the database software Oracle 9i. The software environment on the application server includes the operating system software UNIX and the application server software Websphere. The operating system software needs to choose the version of Win2000 or above and the IE browser is IE6 or above. SUNV4900 is used for the database server, SUNV490 for the application server and PC for the client machine in the hardware environment.

According to the above conditions, the data simulation test environment is built, and the same 5G intelligent terminal information database is tested by using this system and several traditional dispatching systems. Data monitoring network is the core of the database, and the best scheduling feature is selected by 5G intelligent terminal. The best component model is obtained by reducing the dimension of the original data. The final parameter of the database is 45 by default. When one sample is a training dataset, the other sample can be verified by cycling until all the data are traversed, and the best model parameters can be obtained. In general, the accuracy of the algorithm needs to be tested more than 5 times in order to get more accurate values. In order to determine the superiority and accuracy of the system design method in this paper, this method is tested separately with the conventional three methods, and its superiority is analyzed through the comparison of accuracy.

4.2   Scheduling Error Test

The validity of the 5G intelligent terminal information automatic scheduling system based on Internet of Things technology is verified and tested according to the above experimental environment and parameter settings. The average and absolute errors in the process of automatic dispatching shall be calculated separately, and the calculation formulas are as follows:

$$\begin{cases} \alpha = \frac{1}{|\delta|} \sum_{\tau_i \in R} \left| \tau_i - \tau_i' \right| \\ \beta = \sqrt{\frac{1}{\delta} \sum_{\tau_i \in R} \left(\tau_i - \tau_i' \right)^2} \end{cases} \tag{8}$$

In the formula: α represents the average error of the dispatching result; β represents the absolute error of the dispatching result; δ represents the actual number of dispatching scoring items; τ_i represents the true rating of the automatic dispatching of 5G intelligent terminal information; τ' represents the forecast rating of the system; R represents the scoring set. The smaller the values of α and β are, the smaller the error of the scheduling result is. Based on the above calculation, any 10 sets of data in the data set are selected, and the data set is trained and tested according to the ratio of 8:2. According to the above formulas, the performance evaluation results of different systems are obtained, as shown in Table 1 below.

Table 1. Comparison of scheduling errors

Test set	The system in this paper		Conventional system 1		Conventional system 2	
	α	β	α	β	α	β
1	0.1534	0.2267	0.3886	0.3794	0.2568	0.4596
2	0.2367	0.9657	0.4685	0.4752	0.7456	0.6935
3	0.4587	0.1145	0.1426	0.5544	0.1459	0.4756
4	0.1544	0.0112	0.4455	0.1452	0.6357	0.3247
5	0.5556	0.1214	0.4425	0.6895	0.4582	0.2476
6	0.1445	0.5214	0.1144	0.2448	0.9634	0.1475
7	0.0112	0.3145	0.3514	0.2569	0.7851	0.5475
8	0.1123	0.2552	0.5527	0.1456	0.2447	0.1474
9	0.1145	0.4425	0.4752	0.3697	0.1448	0.4256
10	0.0221	0.4425	0.1455	0.2578	0.2145	0.1145
Mean value	0.1963	0.3016	0.3527	0.3519	0.4595	0.3584

As shown in Table 1, the average error of the system used in this paper is 0.1963, 0.1604 and 0.2632 lower than the two conventional systems, 0.3016 in absolute error, 0.0503 and 0.0568 lower than the two conventional systems, both absolute error and relative error are lower than the two conventional systems. Therefore, the average error and absolute error of the system designed in this paper are lower, and it has more accurate effect for the automatic scheduling of 5G intelligent terminal information.

5   Concluding Remarks

In this paper, an automatic scheduling system of 5G intelligent terminal information is designed based on Internet of Things technology. The bus circuit of the system is designed through zero group grounding, which makes the grounding line anti-interference. Taking standard frequency, initial frequency, optimal frequency and fundamental frequency as objective functions, the scheduling characteristics of 5G intelligent terminal information samples are extracted, and the automatic scheduling of 5G intelligent terminal information is realized. Through the above three experiments, it can be found that the 5G intelligent terminal information automatic scheduling system based on Internet of Things technology has better performance, and can reduce errors in the automatic scheduling process. Due to the limited conditions, the method studied in this paper is verified in the simulation environment, but has not been applied in practice. In the future research, it needs to be applied in the actual environment to meet the actual needs and improve the application performance of the design system.

References

1. Li, N.: Design of material supply chain distribution intelligent scheduling system based on big data. Mod. Electron. Tech. **43**(22), 184–186 (2020)
2. Tang, W., Yang, C.: Design and implementation of power dispatching system based on data mining technology. Autom. Instrum. (5), 55–58 (2019)
3. Zhao, Y., Huang, L., Liu, H.: Design of power dispatching data automatic backup system based on deep learning. Mod. Electron. Tech. **43**(20), 42–45, 49 (2020)
4. Mo, X., Ding, X., Li, Q., et al.: Design and implementation of water resources dispatching system in Erhai Basin. Water Resour. Informatiz. (1), 59–63 (2020)
5. Wang, W., Li, Y., Chen, M., et al.: Design and implementation of online management system of power dispatching object with certificate. Electric Eng. (18), 78–79, 82 (2020)
6. Liu, J., Lu, H., Han, D., et al.: Design of cross-regional topology verification system for power grid dispatching data based on control cloud. Autom. Instrum. (4), 141–144 (2020)
7. Zhang, Y., Yu, L., Wu, Y., et al.: Design of grid dispatching instruction management system based on identification technology. Electron. Des. Eng. **28**(21), 33–37 (2020)
8. Guo, L., Yuan, Q., Li, Y., et al.: 8-element MIMO antenna design based on 5G intelligent terminal. J. Hangzhou Dianzi Univ. **39**(5), 8–12 (2019)
9. Song, Y., Song, S.: Design and development of a personal smart-device based distributed virtual reality field teaching system. Res. Explor. Lab. **39**(2), 227–232 (2020)
10. Huang, W.: Analysis of security threats and key technologies of the perception layer of internet of things based on smart home. J. Xi'an Univ. (Nat. Sci. Ed.) **24**(1), 59–64 (2021)
11. Yang, L., Ma, L., Cui, Y., et al.: Research on security transmission model of medical image data based on internet of things technology. China Med. Devices **36**(2), 54–57 (2021)
12. Song, Q., Jiang, F., Li, C., et al.: Research on the application of underground ditch and pipe environmental monitoring system based on Internet of Things technology in power system. Electron. Des. Eng. **29**(2), 94–98 (2021)
13. Liu, S., Liu, G., Zhou, H.: A robust parallel object tracking method for illumination variations. Mob. Netw. Appl. **24**(1), 5–17 (2018). https://doi.org/10.1007/s11036-018-1134-8
14. Liu, S., Bai, W., Liu, G., et al.: Parallel fractal compression method for big video data. Complexity **2018**, 2016976 (2018)

Data Access Control Method of Power Terminal Based on 5G Technology

Hai-long Zhu[✉], Guo-yi Zhang, Dan-ke Hong, Si-tuo Zhang, and Shan-ke Huang

Digital Grid Research Institute, China Southern Power Grid, Guangzhou 510670, China

Abstract. Due to the limitation of network environment, traditional terminal data access control methods can not establish a connection channel with fast access devices, which makes the data access response time longer. Therefore, the design of data access control method for power terminal based on 5G technology is proposed. According to the classification results, the data access requests are assigned to the business areas based on 5G slicing technology, so as to improve the efficiency of data access. At the same time, build a model to ensure that the power terminal data access can be processed on time, and finally realize the power terminal data access control, improve the response performance of group mobile phone access request, and improve the power terminal data access speed. The simulation results show that the data access control method designed in this paper is obviously better than the traditional control method. Under the same thread condition, the data access response time is almost the same, and the method in this paper has a slight advantage. Under different thread conditions, with the increase of the number of threads, the access response time of the control method in this paper is significantly lower than that of the traditional method, which effectively improves the data access speed.

Keywords: Data access · 5G technology · Power terminal · Access control

1 Introduction

With the rapid development of mobile communication technology and Internet technology, higher requirements are put forward for real-time access and data response speed of terminal equipment [1]. At present, there are some problems in power terminal identity authentication and data access control. Starting from the 2G era, the access of mobile terminal equipment to the power enterprise Intranet has gradually become a demand point. With the rapid development of 3G and 4G, the transmission speed of mobile data networks has improved a lot, making the transmission of images, voice and video possible, and the user experience has become better and better [2]. With the development and popularization of 5G, a large number of mobile terminal equipment, including Internet of things equipment, have been developed, access to the enterprise Intranet has become possible [3]. The characteristics of 5G provide many scenarios and requirements for mobile terminal devices to access the enterprise intranet. With the good development of

S. Liu and X. Ma (Eds.): ADHIP 2021, LNICST 417, pp. 26–39, 2022.
https://doi.org/10.1007/978-3-030-94554-1_3

the network environment, the power terminal equipment began to access the power grid safely, so as to facilitate people's daily life. In reference [4], the heterogeneous network (C/U) which separates the control plane from the user plane is an evolutionary method of the fifth generation (5G) network to achieve high system coverage and high capacity. In order to minimize the signaling load of the core network when a macro base station fails in the control plane, a scheme is proposed to transfer the control right of the cell covered by the faulty macro base station to the adjacent macro base station. Based on the analysis of the average handover rate between cells, the extended coverage rate of adjacent macro stations under the constraint of transmit power and the load balance index of the system, the maximum formula of macro blind source handover is constructed and further solved by convex optimization method. In reference [5] the rapid growth of the number of mobile devices, massive data and higher data rate are promoting the development of the fifth generation (5G) wireless communication. 5G network has three unique characteristics: ubiquitous connectivity, extremely low latency and high-speed data transmission through the use of new technologies to equip future nano millimeter wave wireless communication systems and large-scale multiple input multiple output (MIMO) systems with extreme base station and device density, as well as an unprecedented number of nano antennas. These 5G new technologies are introduced, especially for the nano materials used in antenna. Due to a large number of MIMO technology and ultra densification technology, traditional antennas can not provide new frequencies for smaller antennas, while nano antennas are used for 5G. Nano materials for broadband millimeter wave antennas are introduced. However, due to the limitation of the traditional network channel, it is unable to meet the centralized access of mass power Internet of things terminal equipment at the same time. Therefore, for the research of power terminal data access control, based on 5G technology, a power terminal data access control method is designed, which has a good protection and reference for the security of sensitive data of power system terminal.

2   Design of Power Terminal Data Access Control Based on 5G Technology

With the rapid development of Internet technology, the use of smart phones is becoming more and more popular [6]. Table 1 shows the survey of smart phone use rate from 2016 to 2020.

Table 1. 2016–2020 smartphone usage survey

Particular year	Utilization rate
2016	65.32%
2017	72.10%

(continued)

Table 1. (*continued*)

Particular year	Utilization rate
2018	90.06%
2019	97.26%
2020	98.01%

According to Table 1, more than 98% of smart phones are used, which makes more and more service projects begin to be carried out through mobile terminals, which facilitates people's lives [7]. The same is true of power data, and the data access control of power terminal has become a problem to be solved. In order to better ensure the speed of user data access, based on 5G slicing technology, the data access area of power terminal is divided, and then the data access control model is established to control the data access direction of power terminal, and finally the data access control is realized.

2.1 Development of Data Access Classification Schemes

The design of data access control method for power terminal promotes the normal operation of the method by formulating data access classification scheme in the model [8]. In the control process, how to classify the data access requests of power terminals is the main problem to be solved [9]. The classification scheme includes three main parts: user (U), cloud service provider (CSP) and master certification authority (MA). The relationship between them is shown in Fig. 1.

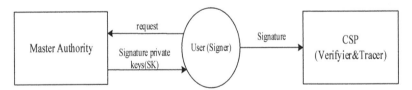

Fig. 1. U, CSP, MA relationship diagram

According to the content of Fig. 1, some problems are analyzed as follows: firstly, the existing access requests of users are classified, and the threshold access structure is a more efficient way, so this paper selects the threshold structure as the basis for data access classification [10]. Then, in cloud computing, users can protect the integrity of data in some ways [11], and a more important role is to authenticate users. The main starting point of this paper is how to make the cloud platform trace the user's identity while verifying the user's access request. This paper sets the symmetric traceability key TK to encrypt the user's identity, and only after the cloud platform is verified successfully can it pass the calculation The TK is obtained by calculation, and then the user identity is obtained by decryption. One is to ensure the security of the whole scheme, and the other is to improve the computational efficiency [12]. Finally, if the user's access request is revoked or changed, it should be updated and classified in time [13]. In the specific operation of the scheme, it is necessary to select two multiplication groups $Q1$ and $Q2$ of prime p for MA, and set e as bilinear mapping, then the formula (1) can be obtained.

$$e = Q1 \times Q1 \rightarrow Q2 \tag{1}$$

In the multiplication group $Q1$, $g1$ and $g2$ are randomly generated, and the main key $x \in Z$, $H1, H2, H3:\{0, 1\} \rightarrow g2$ are selected as three Hash functions that can resist collision. Then for any attribute i, the random number $ti \in Zp$ is selected, and the formula (2) can be obtained.

$$\begin{aligned} PK &= (g1, g2, H1, H2, H3, t_i)_{i \in A}, \\ MK &= x \end{aligned} \tag{2}$$

Since the threshold access structure is used for classification, the threshold access structure is defined as $T_{k,v}()$ and the relationship between this structure and attribute set ω is shown in formula (3).

$$T_{k,v}(\omega) = (v, k) = \begin{cases} 1, |\omega \cap v| \geq k \\ 0, other \end{cases} \tag{3}$$

In the formula (3), k is the threshold, ω is the attribute set required by the authentication request, and v is the attribute set of the access structure. The meaning of the above formula is to use (v, k) to represent the threshold policy access structure T. only when the attribute set of the user authentication request and the attribute set of the access structure T have the same attributes greater than k, then the attribute set ω satisfies the access structure T, and the access can be classified to the region, otherwise, the access can be classified not adopted [14]. In the attribute set v of access structure T, let the attribute set $\omega1$ be the same as the attribute set of access structure T and user signature, and let the attribute set $\omega2$ be different from the attribute set of access structure T and user signature. In other words, the intersection $\omega1$ of the signature attribute set and the attribute set of the access structure must be greater than the threshold k to pass the verification. Thus, the relationship among the attribute sets ω, $\omega1$, and $\omega2$ can be obtained, as shown in formula (4).

$$\begin{aligned} \omega1 &= \omega \cap v = A \cap v \\ \omega2 &= v - \omega = v - A \\ \omega1 &\cup \omega2 = v \\ \omega1 &\cap \omega2 = \sigma \end{aligned} \tag{4}$$

When the cloud platform receives the verification from user U, the cloud platform needs to verify whether the request classification is correct, that is, judge whether the request s meets the access structure $T_{k,v}()$, and the judgment equation is shown in (5).

$$\frac{\prod_{i=1}^{|\omega|} e(g, sig_2^{q(i)+t_i})^{\Delta_{i,\omega}(0)}}{e(g, s_2)e(H_2(m), s_1)} = Z \tag{5}$$

Judge whether it is true through formula 5. If it is true, it means that the intersection of attribute set ω and attribute set v of access structure T is greater than k, then accept m and request s, and the verification is passed. Otherwise, the next classification verification is refused. Lagrange interpolation theorem should be used in verification, as shown in formula (6).

$$q(x) = \sum_{i=1}^{|\omega|} q(i)\Delta_{i,\omega}(x)$$
$$\Rightarrow q(0) = \sum_{i=1}^{|\omega|} q(i)\Delta_{i,\omega}(x) = x$$
(6)

Finally, the classification of data access requests is determined by the formula calculation results.

2.2 Division of Business Area Based on 5G Technology

After the power terminal data access classification is completed, the terminal data area is divided based on 5G technology, so that the power terminal data access classification corresponds to the corresponding slice area, so as to promote the data access speed [15]. The so-called 5G is the abbreviation of the fifth generation mobile communication technology, as a new generation of mobile communication technology after 4G system [16]. Increasing data rate, decreasing energy demand, increasing system capacity and expanding equipment connection scale are the performance goals of 5G [17].

5G has three characteristics

(1) It is eMBB (enhanced mobile broadband), which provides ultra high speed broadband services;
(2) Ultra high reliable ultra low delay communication (uRLLC), data communication low delay, real-time is very strong;
(3) mMTC (massive machine communication) supports the connection of massive devices and provides the basic guarantee for the interconnection of everything.

In 5G technology, with the maturity of cloud computing and virtualization technology, network slicing mode becomes possible [18]. In 5G network environment, network slicing refers to dividing some network related resources. According to the actual needs, the network areas with different functions are set up. With 5G network segmentation technology as the core, a number of logical subnets with different characteristics are presented, and these logical subnets are isolated. Wireless network, transmission network and core network form the network slice combination. For the logical subnet formed by slicing, the slicing management system realizes the unified management. And according to the different characteristics of logical subnet, it meets the diversified needs [19]. For power terminal system, the biggest advantage of 5G network slicing technology is that it can freely choose slicing characteristics, set logical subnet to meet user access requirements, and realize data access control of power terminal through low delay, high throughput and high network efficiency. Reasonable data access control is conducive to improve products and services, and improve user experience satisfaction. In addition, the impact of network slicing on the rest of the network does not need to be considered by operators, which saves operation time and reduces operation cost. Through the application of network slicing technology to bring better economic benefits [20]. The schematic diagram of network slice is shown in Fig. 2.

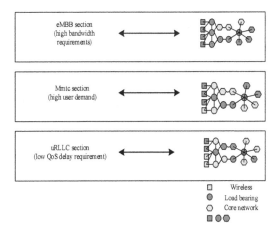

Fig. 2. Diagram of a network slice

Based on the characteristics of 5G such as high bandwidth, large connection and low latency, 5G application scenarios are mainly divided into application scenarios, application examples and network requirements, as shown in Table 2. According to Table 2, the first commercial scenario is mobile bandwidth, which is characterized by high speed. On this basis, services in the 4G era will be further improved.

Table 2. 5G application scene division

Application scenarios	Application cases	Network demand
Mobile broadband	4K/8K Ultra HD video, holography, AR/VR	High capacity, high storage
Massive internet of things	Mass sensors: related sensors such as power, agriculture, family, smart city, etc.	Large scale connection, mostly static
Mission critical internet of things	Unmanned driving, intelligent industry, smart grid	Low delay, high reliability

5G will bring great changes to the mobile terminal equipment access to the intranet of the enterprise. Through the previous dedicated physical network providing services for different vertical applications, it will be transformed into "network factory" [21–23].

The network slice can be customized by software to provide guaranteed service quality for different mobile terminal equipment and scenarios. It is based on the same set of physical infrastructure equipment (wireless network, bearer network and core network) Generate logical isolated, independent running network slices from different business requirements. 5G will create a new mobile Internet. Through 5G slicing technology, the access area of mobile terminal equipment is divided according to the diversified target. The main business characteristics of grid company include high real-time requirements, long data connection and high bandwidth. Therefore, the network of power grid needs to ensure absolute security, and mobile terminal equipment is connected to the internal network.

The realization of network slice is based on network virtualization, the essence of virtualization is mapping and the form is encapsulation. The problem of network slicing is to solve the mapping problem, which aims to allocate network resources. Data needs to be sealed to achieve the isolation between network slices, and the isolation mechanism will be the key to the implementation of network slicing. For the application of network slicing, whether it is a special line of enterprise, SLA with low delay service or independent control of vertical industry, it is essential that the network slicing be rigid or elastic isolated to a certain extent. Because 5G architecture is still developing, it brings new requirements and challenges to the isolation of network slice, especially in the case of end-to-end security, and the end-to-end 5C network Slicing Based on slicing isolation is the first step to realize user security. After studying the mechanism of network slice isolation, some isolation mechanisms and technologies are pointed out, and the latest trends and challenges of slice isolation are analyzed. Generally speaking, compared with the existing network isolation technology, 5G network needs to integrate and manage various isolation mechanisms on different levels of network, and provide end-to-end isolation that meets the requirements. These technologies may not be new, such as sandbox based isolation, virtual machine based isolation and VLAN. Generally speaking, there are several aspects of network slice isolation:

Traffic isolation: when multiple network slices share network resources, it is necessary to ensure that the data flow through one slice does not flow to other slices.

Bandwidth isolation: any given slice and data flow through a slice will be allocated a certain bandwidth resource, and other slice bandwidth resources should not be encroached on. CPU resources on the node are similar.

Processing isolation: when network resources are shared between network slices, the processing of packets on the slice shall be independent of other slices. If on the same physical server, packet processing for multiple slices must be run independently by virtualization, container environment separation.

Storage isolation: slice related data should be stored independently.

In order to provide guaranteed performance services for multi tenants on the general physical network, network slicing needs data/control/management/service plane isolation. From the effect of isolation, there are two layers of isolation: hard isolation and soft isolation.

Hard isolation is achieved by providing independent circuit switched connections, dedicated to a virtual network. Circuit switched connections can be provided by flex, dedicated wavelength or dedicated TDM slots.

Soft isolation is realized by using packet technology to multiplex traffic statistics from two or more virtual networks to common circuit switched connections. Nvo3, MPLS and deterministic networks are typical soft isolation technologies, but slices isolated by soft isolation technology may still compete for underlying resources in extreme cases.

Compared with the previous smart grid technology, it can provide more security and flexibility for the development of 5G wireless network. According to the research on network slicing, it is still at a higher level of architecture and requirements. The specific implementation of network slicing is mainly formulated by 3GPP. According to the definition of network slicing in 3GPP, network slicing is a logical network that provides specific network capabilities and network characteristics. Network slicing includes core network and access network slicing, covering slicing architecture, slicing selection, slicing roaming, etc. The research on network slicing in 3GPP is shown in Table 3.

Table 3. 3GPP research on network slice

Project	Function
SA1	The demand principle of network slicing is proposed
SA2	Research the architecture and scheme of network slicing
SA3	Study the security and isolation scheme of slice
SA5	This paper studies the arrangement management of network slice

In addition, based on the service gateway, combined with 5G slicing technology, 5G slicing isolation at the service level can be achieved. Different business scenarios use different slices. For example, powerful Internet of things devices use the characteristics of the massive Internet of things to divide slices and connect a large number of devices to the enterprise intranet. Each slice and network channel pass through the security access area. After passing the authentication, it can easily and freely access to the enterprise intranet. Effectively improve the power terminal data access speed.

2.3   Construction of Data Access Control Model for Power Terminal

Mobile terminals need to consider real-time and concurrency when accessing background data. In single thread data communication, there can only be one thread in a process, and the rest of the process must wait for the current thread to finish. As a result, it takes a long time for the system to complete a very small task, and this way is not strong enough to deal with emergencies. If a connection is different often, it will cause the whole program to crash. 5G slicing technology is used to make better use of system resources. In practical application, how to ensure that the user access requirements are correctly assigned to the sliced area has become a difficult problem. In order to realize the information exchange between mobile terminal and server, a data access control model based on socket communication mechanism and Java multithreading method is designed and implemented, as shown in Fig. 3. The data access control model runs on the server, and a main thread control program responds to the request of the mobile terminal in real time. The application of socket communication mechanism is the basis of the normal operation of the model. Socket is a socket specification based on transport layer protocol. It defines three types of sockets: streaming socket, datagram socket and original socket. They support TCP/IP and UDP protocols respectively. Socket is independent of the application layer protocol and consists of two main components: socket server and socket client 3. When the client establishes a connection with the remote host application layer server, the client first establishes a connection with the proxy server, and the address and port of both application layer servers will be passed to the proxy server. In the socket protocol system, when the client wants to establish a connection with the application layer server, it first establishes a connection with the socket proxy server, and the relevant address and port of the application layer server will be passed to the socket proxy server in this process. After authentication and negotiation between the client and the socket server, the socket server will establish the corresponding TCP or UDP connection with the remote server according to the request of the socket 'client, so as to realize the corresponding application protocol.

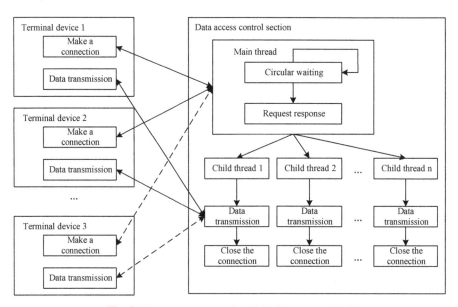

Fig. 3. Data access control model of power terminal

In the data access control model of power terminal, the server is always in the waiting state when no connection request comes, and is ready to provide services at any time. When a terminal sends a connection request to the server, the main process of the server immediately establishes a new server Socket connection, according to the type of request to find the corresponding 5G slice for sub area, at the same time, a sub thread in this area to process the request of the mobile terminal, when the server starts the sub thread, it immediately returns to the listening state, waiting for the next connection request. After each sub thread starts, it completes the data transmission with the terminal device independently. The design of data access control model uses concurrent service mode to work, and each mobile terminal is processed by an independent thread. Because the main thread hardly communicates with the sub thread, the system overhead is small, and the resource utilization efficiency is high. Because different threads are associated with different objects and handle their own abnormal events, the software has strong robustness. What's more, each sub thread receives data synchronously, which ensures the real-time requirement of data transmission.

2.4 Implementation of Data Access Control

According to the classification of data access of power terminal, the access request is classified, and the corresponding access area is replaced according to the classification results. Finally, the access control model is used to realize the overall control of access. When the server application starts, the monitoring status is triggered. There are two options for initialization and destruction of the monitoring interface at the server. Through monitoring the servlet container, call the star tpdaserver() method to start a main thread, which will be in the listening state after the main thread is created. At the same time, it has its own stack, and can derive multiple sub threads according to the needs. However, after listening and receiving the request, the dynamic call creates the sub thread function, derives independent sub threads, and processes the requests with different port numbers. After the main thread is sent, the sub thread is not controlled and scheduled. The sub thread connects with the client and handles the exception. After the data access control model is started, the request from the terminal is received by listening to the specified local port. Every time a request is received, a thread is used by implementing the Java runnable interface. A run() method is defined in the runnable interface. When instantiating a thread object, an object implementing the runnable interface can be passed in as the parameter. The thread class will call the runnable object The run() method then executes the contents in the run() method to start the new mobile terminal thread, and completes the request and response of the mobile terminal. After each sub thread is started, it can get the request content from the mobile terminal and submit it to the XML parsing class for processing. The XML document is parsed and encapsulated into Java object, passed to the background processing business logic and interacted with the database. After each sub thread is started, a deadlock detection thread is designed, which is a separate thread for handling deadlock.

3 Simulation Experiment

In order to verify that the terminal data access control method designed in this paper can effectively improve the user access speed in practical application, a simulation experiment was carried out. In order to ensure the scientificity of the experiment, the design method is taken as the experimental group, and two traditional terminal data access control methods, cloud based data resource context aware access control method in reference [7] (traditional method 1) and edge computing based access control method in reference [8] (traditional method 2), are selected as the control group. Under the same experimental conditions, three methods are applied to access control, and the experimental data are analyzed to verify the application effect of this method. Before the start of the experiment, the running environment required for the simulation experiment is set. The development machine is Mac OS X 10.11.4. The computer parameters used are shown in Table 4 below.

Table 4. Computer parameters used

Serial number	Name	Parameter
1	Computer name	Mac mini
2	Operating system	Windows 10
3	Processor	2.80 GHz
4	Monitor	21 in.
5	Memory	8 GB
6	Hard disk	1867 MHO DDR3

Several computers with the same parameters as those in Table 4 are selected to implement three kinds of terminal data access control methods respectively. Taking threads as experimental conditions, this paper analyzes from the same thread and different threads to clarify the effect of power terminal data access control methods.

3.1 Data Access Response of Same Thread Power Terminal

For the computer that implements the data access control method of power terminal, 300 threads are opened to simulate the data access of power terminal. Each thread automatically performs 15 million addition calculations by computer. Repeat the simulation experiment for 7 times, and finally make statistics on the access response time. The results are shown in Fig. 4.

According to Fig. 4, it can be found that under the application conditions of two traditional power terminal data access control methods, the response time of access is between 2 s–2.5 s. After the application of the design method in this paper, the longest data access response time is only 1 s, and the response time of most experimental results is less than 1 s. Through the analysis of the data access response results of the same thread power terminal, the conclusion is that although the response time of the three methods is almost the same, the method in this paper has a small advantage.

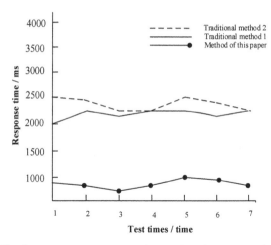

Fig. 4. Data access response of same thread power terminal

3.2 Data Access Response of Different Threaded Power Terminal

In order to show the effect of different power terminal data access control methods more clearly, 100, 200, 500, 1000, 2000, 5000 threads are opened respectively, each thread carries out 10 million times addition calculation, and finally, the average response time is counted. The statistical results are shown in Fig. 5.

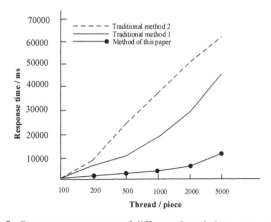

Fig. 5. Data access response of different threaded power terminal

Compared with the experimental results of the same thread, the data access response results of different threads shown in Fig. 5 show the effect of different data access control methods of different power terminals. The response time of the data in this paper increases with the increase of the number of threads, but the overall growth trend is relatively slow. When 2000 threads are opened at the same time, the response speed is about 3.12 s, which can meet the data access requirements of power terminals. The response time of the two traditional methods increases with the increase of threads. When 2000 threads are opened simultaneously, the response time of the traditional method reaches 30.08 s and 51.35 s, which greatly affects the data access effect. In conclusion, the method designed in this paper has good effect on data access control of power terminal. It can better meet the needs of users.

4 Conclusion

In this paper, 5G technology is the core, and a data access control method of power terminal is designed. Through the application of 5G chip technology and the establishment of data access control model of power terminal, the data access control of power terminal is realized finally. The speed and efficiency of power mobile terminal equipment accessing the intranet of enterprises under 5G environment are improved, and the response time of data access is reduced. Although the data access control method designed in this paper improves the response speed of data access effectively in practical application, the research results are not perfect because of the insufficient research time and the limited of the environment. In the future, this will be the focus of in-depth research.

References

1. Ahmad, W., Radzi, N., Samidi, F.S., et al.: 5G technology: towards dynamic spectrum sharing using cognitive radio networks. IEEE Access **8**, 14460–14488 (2020)
2. Ameen, H.A., Abdelmonem, K., Elgamal, M.A., et al.: A 28 GHz four-channel phased-array transceiver in 65-nm CMOS technology for 5G applications. AEU-Int. J. Electron. Commun. **98**, 19–28 (2019)
3. Roth, J.D., Tummala, M., Mceachen, J.C.: Fundamental implications for location accuracy in ultra-dense 5G cellular networks. IEEE Trans. Veh. Technol. **68**(2), 1784–1795 (2019)
4. Yang, R.: Transferring in control plane for 5G heterogeneous networks. High Technol. Lett. **25**(03), 43–48 (2019)
5. Hao, H., Hui, D., Lau, D.: Material advancement in technological development for the 5G wireless communications. Nanotechnol. Rev. **9**(1), 683–699 (2020)
6. Anagnostopoulos, N.A., Ahmad, S., Arul, T., et al.: Low-cost security for next-generation IoT networks. ACM Trans. Internet Technol. **20**(3), 1–31 (2020)
7. Kayes, A., Rahayu, W., Dillon, T., et al.: Context-aware access control with imprecise context characterization for cloud-based data resources. Future Gener. Comput. Syst. **93**, 237–255 (2019)
8. Ndikumana, A., Tran, N.H., Ho, T.M., et al.: Joint communication, computation, caching, and control in big data multi-access edge computing. IEEE Trans. Mob. Comput. **19**(6), 1359–1374 (2019)
9. Zhang, Q., Wang, S., Zhang, D., Wang, J., Zhang, Y.: Time and attribute based dual access control and data integrity verifiable scheme in cloud computing applications. IEEE Access **7**, 137594–137607 (2019)
10. Sn, A., Sk, A., Ds, A.: Modeling access control on streaming data in apache storm. Procedia Comput. Sci. **171**, 2734–2739 (2020)
11. Yang, S., Xu, Z., Hou, Y., et al.: Substrate integrated waveguide filter based on novel coupling-enhanced semicircle slots for 5G applications. IEICE Electron. Express **16**(8), 20190125 (2019)

12. Ghosh, A., Maeder, A., Baker, M., et al.: 5G evolution: a view on 5G cellular technology beyond 3GPP release 15. IEEE Access **7**(99), 127639–127651 (2019)

13. Pham, Q.V., Fang, F., Ha, V.N., et al.: A survey of multi-access edge computing in 5G and beyond: fundamentals, technology integration, and state-of-the-art. IEEE Access **8**, 116974–117017 (2020)

14. Zhou, L., Rodrigues, J.J., Wang, H., Martini, M., Leung, V.C.: 5G multimedia communications: theory, technology, and application. IEEE Multimedia **26**(1), 8–9 (2019)

15. Bailey, W.H., Cotts, B., Dopart, P.J.: Wireless 5G radiofrequency technology — an overview of small cell exposures, standards and science. IEEE Access **8**, 140792–140797 (2020)

16. Hao, J., Huang, C., Ni, J., et al.: Fine-grained data access control with attribute-hiding policy for cloud-based IoT. Comput. Netw. **153**, 1–10 (2019)

17. Li, H., Pei, L., Liao, D., Chen, S., Ming Zhang, D.: FADB: a fine-grained access control scheme for VANET data based on blockchain. IEEE Access **8**, 85190–85203 (2020)

18. Pallavi, K.N., Kumar, V.R.: Authentication-based access control and data exchanging mechanism of IoT devices in fog computing environment. Wireless Pers. Commun. **116**(4), 3039–3060 (2021)

19. Munoz-Arcentales, A., López-Pernas, S., Pozo, A., et al.: An architecture for providing data usage and access control in data sharing ecosystems. Procedia Comput. Sci. **160**, 590–597 (2019)

20. Shanthi, P., Umamakeswari, A.: Privacy preserving time efficient access control aware keyword search over encrypted data on cloud storage. Wireless Pers. Commun. **109**(4), 2133–2145 (2019)

21. Liu, S., Bai, W., Liu, G., et al.: Parallel fractal compression method for big video data. Complexity **2018**, 2016976 (2018)

22. Liu, S., He, T., Dai, J.: A survey of CRF algorithm based knowledge extraction of elementary mathematics in Chinese. Mob. Netw. Appl. **26**(5), 1891–1903 (2021). https://doi.org/10.1007/s11036-020-01725-x

23. Liu, S., Fu, W., He, L., et al.: Distribution of primary additional errors in fractal encoding method. Multimed. Tools Appl. **76**(4), 5787–5802 (2017)

Design of Urban Traffic Node Management Scheme Based on 5G Wireless Network Architecture

Xiao-hua Mo[✉]

Nanning University, Nanning 530200, China

Abstract. Effective management of urban traffic nodes can improve the order and efficiency of road traffic, and play a pivotal role in alleviating urban traffic congestion. To this end, a city traffic node management scheme based on 5G wireless network architecture is designed. This method first collects relevant traffic characteristic data, and then uses 5G cellular mobile communication technology to establish the connection between the control center and the on-board unit and roadside unit to realize real-time vehicle and road data transmission. Finally, it uses the algorithm based on dynamic programming to perform urban traffic node signal lights. Control and realize urban traffic flow scheduling. The results show that: under the application of the designed scheme, the number of vehicles in the queue is reduced, and the average delay time of vehicles is reduced, thus effectively improving the utilization efficiency of road traffic resources and alleviating the problem of urban traffic congestion.

Keywords: 5G wireless network architecture · Urban traffic node · Signal light management scheme

1 Introduction

Intelligent transportation system is the development direction of future transportation system. It effectively integrates advanced information technology, data communication transmission technology, electronic sensor technology, control technology and computer technology to the entire ground transportation management system, and finally establishes a It is a real-time, accurate and efficient comprehensive transportation management system that plays a role in a large-scale and all-round way [1]. The development of intelligent transportation is closely related to the development of the Internet of Things. Only when the Internet of Things technology continues to develop, the intelligent transportation system be more perfect. Intelligent transportation is a concrete manifestation of the high degree of materialization of road transportation.

As an important part of smart transportation, smart city traffic nodes can guide and divert traffic in time and space through the collection and transmission of traffic information, reduce congestion, and protect pedestrian safety. Because of the complicated traffic conditions of the traffic nodes, the large flow of people and vehicles, and the difficulty of

S. Liu and X. Ma (Eds.): ADHIP 2021, LNICST 417, pp. 40–50, 2022.
https://doi.org/10.1007/978-3-030-94554-1_4

optimizing control, it has become the primary problem to be solved in the establishment of an intelligent transportation system.

In this paper, under the support of 5G network, combined with the location, speed, route and other information of various target objects of traffic behavior, design an urban traffic node management plan. First collect traffic characteristic parameters, and then on the basis of traffic characteristic parameters, use part of the networked vehicle data to obtain the vehicle arrival matrix. At the same time, the NEMA dual-loop phase structure is introduced, and the phase duration and phase sequence are optimized by the dynamic programming method. The optimization model with the minimum delay as the goal realizes the real-time adaptive phase sequence traffic signal control at the traffic node. Through this research, it is hoped to provide reference and reference for the traffic scheduling problem of urban nodes and relieve the traffic pressure of traffic nodes.

2   5G Wireless Network and Urban Traffic

5G refers to the fifth-generation mobile phone mobile standard, also called the fifth-generation mobile communication technology. The 5G network has obvious advantages, which are mainly reflected in the transmission speed, bandwidth capacity, network delay and so on. According to related tests, the current fastest transmission speed of 5G network can reach 7.5–10.0 Gb/s, and according to the latest test of University of Salisbury, the fastest download speed of 5G network reaches 1 Tb/s. The 5G communication technology with super bearing capacity will surely open a new era of interconnection of all things [2]. The functional characteristics of the 5G network have a strong supporting role for the Internet of Things, making the high-quality and efficient networking of things and things, people and people, and things and people a reality, laying a good foundation for building a smart society with all things connected. The bandwidth capacity of the 5G network will be more than one hundred times that of the 4G network. The Internet of Things and mobile devices will no longer need to wait for access to the network, and fast and large amounts of information can be exchanged anytime and anywhere. The above-mentioned performance of 5G network is very beneficial to the operation of Internet of Things equipment, and will play a major role in traffic conditions, vehicle detection, unmanned driving, and intelligent transportation [3].

A new generation of intelligent transportation based on 5G wireless network consists of three parts: control center, vehicle terminal (vehicle user), and handheld terminal (pedestrian user). The control center is the core and brain of the entire intelligent intersection system, including 5G communication base stations, traffic lights, various sensors, cloud servers, etc. The 5G communication base station is used for large-capacity and high-speed information exchange between the control center and end users. The traffic signal receiving control center generates control instructions based on real-time traffic flow, and realizes the dynamic adjustment of the green light release time in each direction between the minimum green light time (to take care of pedestrians crossing the street) and the maximum green light time (considering traffic in other directions) to reduce green light loss Time, reducing red light waiting time, improving the efficiency of intersections, and reducing vehicle exhaust emissions [4]. Various types of sensors include video traffic cameras, radar speed measurement devices, illegal capture cameras, etc., which are used to obtain information on the number of vehicles and pedestrians in various directions and manage traffic violations such as speeding through red lights. The cloud server is the brain of the entire intersection. The number of users, location, speed, route and other information collected by various sensors are calculated and processed to issue corresponding control instructions to guide vehicles and pedestrians to efficiently and safely pass the intersection.

2.1 Collection of Relevant Traffic Data

OBU On-board Unit
The on-board unit includes on-board sensors, information processing modules, and communication modules. Vehicle-mounted sensors can collect vehicle information such as vehicle speed, position, lane change, etc., and then transmit it to other vehicles and RSU roadside units through 5G wireless network after processing by the vehicle-mounted information processing module [5, 6]. The structure diagram of the OBU on-board unit is shown in Fig. 1:

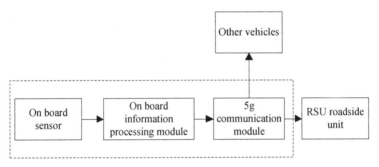

Fig. 1. OBU on-board unit

RSU Roadside Unit
The roadside unit includes a vehicle information receiving module, a data processing calculation module, and a communication module. The vehicle information receiving module receives the vehicle speed, position, status information and the current signal state of the road intersection sent by the vehicle unit through the 5G wireless network, and extracts the information for After calculating the real-time signal timing data, the data processing calculation module calculates the required timing plan according to the set calculation method and transmits it to the roadside signal machine for execution. The schematic diagram of the roadside unit structure is shown in Fig. 2.

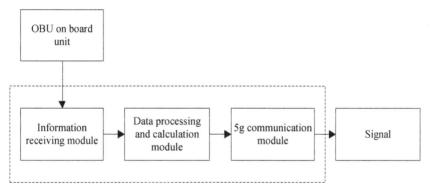

Fig. 2. RSU roadside unit

2.2 Data Transmission

The data transmission process is shown in Fig. 3, the system uses 5G cellular mobile communication technology to establish the connection between the vehicle and the roadside controller, and realizes the real-time vehicle road data exchange [7]. Data such as vehicle position and speed are directly applied to signal timing optimization control.

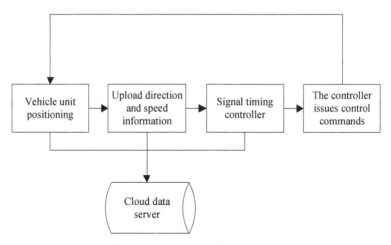

Fig. 3. Data transmission flow chart

2.3 Fuzzy Control of Traffic Node Signals

Traffic signal management of urban traffic nodes is the most basic form of urban traffic signal control, and it is also the basis of urban traffic signal management systems. Usually management methods are divided into timing control, induction control and adaptive signal control.

(1) Timing management is to formulate different timing plans according to several typical conditions based on historical traffic flow. This method is suitable for scenarios where the changing pattern of traffic flow is fixed, and the ability to adapt to changing traffic flow is relatively poor [8].

(2) Induction management detects the traffic demand arriving at the intersection through the vehicle detector, so that the signal timing can adapt to the control method of traffic changes. For intersections with large and irregular traffic flow, induction management has better results, but Induction management can only respond to the detected information of the arriving vehicles, and cannot truly respond to the traffic demand of each phase.

(3) Adaptive signal management uses the traffic flow information obtained by the detector to predict the arrival of vehicles and the length of the intersection queue in the short term, and find the optimal signal timing scheme according to the defined objective function, which is often compared to induction control Have better results. There are several popular adaptive signal management systems, including SCATS, SCOOT, OPAC, RHODES, etc.

In the current traffic signal management and control system, most of the data collection is provided by roadside inspection equipment, including coil detectors, video detectors, etc. There are two limitations to the use of these detectors. First of all, these detectors can only be detected when the vehicle passes the position of the detector. After the vehicle passes the detector, the state of the vehicle (including position, speed, acceleration, etc.) cannot be sensed; secondly, the roadside detection required by the system Equipment often requires high installation costs, and when one or more detectors have problems, the effect of the adaptive signal control system will be greatly reduced [9–11].

In the 5G network environment, the vehicle-road collaboration technology can obtain real-time networked vehicle information on the road, including the location, speed, acceleration, etc. of the connected vehicle, so as to accurately estimate and predict the arrival of vehicles at the intersection, which is real-time Adaptive signal control provides a good data basis. At the same time, based on the environment of vehicle-road collaboration, there is no need to spend a lot of money to arrange fixed-point detector equipment. Each connected vehicle is equivalent to a "mobile sensor", and when one or more connected vehicles have a communication failure, it is quite The coverage rate of connected vehicles is slightly reduced, and it will not seriously affect the effect of the adaptive signal control algorithm.

The research in this chapter is carried out in a 5G network environment. On the basis of 2.1 traffic characteristic parameter estimation, by estimating the information of unconnected vehicles, constructing a vehicle arrival matrix for a period of time in the future, and designing real-time adaptive signal management strategies. The adaptive signal management in this chapter draws on the COP algorithm in the RHODES system, and on this basis introduces the NEMA double-loop phase structure to increase the flexibility of phase sequence changes. This section specifically analyzes the signal control problems of the NEMA double-loop phase structure and uses dynamic programming methods to model and optimize [12–13].

The optimization proposition is to optimize the evaluation index in the prediction time window T by optimizing the duration of each stage. Suppose there are P phase groups for signal phase at a single intersection, and each phase group is taken as a phase. Under the NEMA dual-loop phase structure, the number of phase groups P is 2, j is the phase group number, and j0 is the starting point for optimization Phase group number, i is the sequence number of the dynamic planning decision-making phase. In order to unify the phase group sequence number and the phase sequence number, definition $Y(i)$ means the phase group sequence number corresponding to phase i, and the phase group sequence number j and the decision stage number i satisfy:

$$Y(i) = (i + j_0) \bmod P \tag{1}$$

Define state variable s_i and decision variable x_i, take the time s_i from the start of the optimization to the end of the i-th stage as the state variable, and the phase group duration x_i allocated by the i-th stage as the decision variable. The relationship between state variables and corresponding decision variables at each stage is shown in Fig. 4.

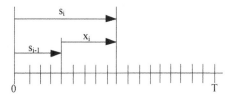

Fig. 4. The relationship between phases and states of the algorithm model

The state transition equation is:

$$s_{i-1} = s_i - x_i \tag{2}$$

When given a state variable s_i, there will be many possible decision values corresponding to it, and different state variable values will have different decision values. Define decision set $X_i(s_i)$:

$$X_i(s_i) = \begin{cases} 0, & \text{if } s_i < X_i^{\min} \text{ or } T - \sum_{k=1}^{i-1} X_k^{\min} < X_i^{\min} \\ 0, X_i^{\min}, \ldots, X_i^{\max}, & \text{if } T - \sum_{k=1}^{i-1} X_k^{\min} > X_i^{\max} \\ 0, X_i^{\min}, \ldots, T - \sum_{k=1}^{i-1} X_k^{\min}, & \text{if } T - \sum_{k=1}^{i-1} X_k^{\min} < X_i^{\max} \end{cases} \tag{3}$$

In the above formula, T represents the number of steps in the optimized time window; the unit time step is 2 s, $X_i(s_i)$ represents the decision set; $X_i^{\max}$ and $X_i^{\min}$ represent the maximum and minimum allowed by the i stage phase group, respectively.

The recursive formula and boundary conditions of the optimal index function are as follows:

$$\begin{cases} f_i(s_i) = \min_{x_i \in X_i(s_i)} [v_i(s_i, x_i) + f_{i-1}(s_{i-1})] \\ f_0(s_1) = 0 \end{cases} \tag{4}$$

In the above formula, $f_i(s_i)$ represents the optimal value function of a given state s_i, which represents the current state and the optimal index values of all previous stages; $v_i(s_i, x_i)$ represents the index function when the state variable of stage i is s_i and the decision is x_i, the specific evaluation index Function $v_i(s_i, x_i)$ is a sub-optimization problem, which will be expanded in the next section. The optimal value function here uses the forward recursion formula, after obtaining the optimal decision within the optimization step through the forward recursion, the optimal decision of each stage is recursively deduced in the reverse direction. The detailed forward recursion method is as follows:

Step 1: Initialization, $f_0 = 0$, $i = 1$;
Step 2: For $(s_i) = 0$, $X_i^{\min}$, ..., T, ask:

$$f_i(s_i) = \min_{x_i}\left[v_i(s_i, x_i) + f_{i-1}(s_{i-1})\right], x_i \in X_i(s_i) \tag{5}$$

The optimal decision of recording stage i in state s_i is $x_i'(s_i)$;
Step 3: If $i \geq P$ and there are $f_{i-k}(T) = f_i(T)$ for all $k < P$, then end the recursion, otherwise $i = i + 1$ go to step 2;

When considering the two adjacent phases $i - 1$ and i, in the same state, more phases will allow more phase group changes, making the evaluation function value smaller, namely $f_{i-k}(T) \geq f_i(T)$, when $f_{i-k}(T) = f_i(T)$, it proves that the latter phase The phase group does not have a better effect on the index function, so it is considered reasonable to use $f_{i-k}(T) = f_i(T)$ as the end condition in the state T.

After the forward recursion is over, assuming that stage I is the final decision-making stage, the optimal decision value in the predicted time window T is obtained, and then reverse recursion to obtain the optimal decision sequence in the time window T, the steps are as follows:

Step 1: Let $S'I - (P - 1) = T$;
Step 2: For $i = I - (P - 1)$, ..., 1, read the optimal decision value $x_i'(s_i)$ under the optimal state S_i' of this stage, and obtain the optimal state $s_{i-1}' = s_i' - x_i'(s_i)$ of the previous stage according to the state transition equation, and then obtain the optimal decision sequence by analogy.

3   Simulation Experiment Analysis

In order to verify the dynamic trunk line coordination control algorithm, this section builds a simulation road network on the SUMO simulation platform, uses C# secondary development to implement the algorithm, and accesses the GAMS/CPLEX solver to assist in the solution of the optimization model. The experimental diagram of the urban traffic road network simulation system is shown in Fig. 5.

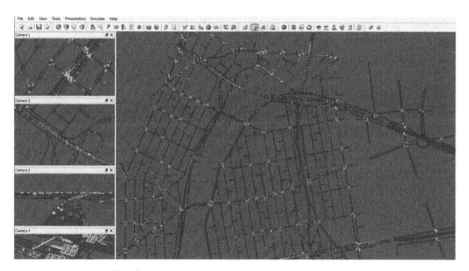

Fig. 5. Urban traffic road network simulation system

3.1 Simulation Tools

(1) NS-3 is a relatively novel simulation tool compared with GloMoSim, OPNET and other network simulation software. Its powerful functions bring great convenience and help to experimental research. In fact, NS3 is not the next generation software of NS2, but a brand new discrete event simulation software. In NS3, the simulator is all written in C++, with only selective Python language binding. Therefore, the simulation script in NS3 can be written in C++ or Python. Because NS3 can generate pcap package trace files, it can also be used in combination with other simulation tools to analyze the simulation process by importing trace files generated by other software.

(2) SUMO (Simulation of Urban Mobility) simulation platform is a microscopic, continuous road traffic simulation software, mainly developed by the German Aerospace Center. SUMO came out in 2000. The main purpose of SUMO as an open source, microscopic road traffic simulation is to provide traffic research organizations with a tool to implement and evaluate their own algorithms. There is also to achieve complete traffic simulation without involving all the necessary components and equipment, such as setting traffic network commands and realizing traffic control.

3.2 Simulated Road Network

Before the NS3 simulation simulation, the SUMO tool was used to generate an effective mobile mode. The simulation scene used a simulation area of 2000×2000 m^2 and used the main street layout of the urban area. The city simulation area includes 32 two-way roads and 17 road intersections, as shown in Fig. 6. Vehicle nodes are randomly released above the road and start movement in two directions at the same time. The intelligent driver model manages the movement of vehicle nodes on the road. Table 1 summarizes all the basic simulation parameters. The simulation results shown below are the average of several simulations.

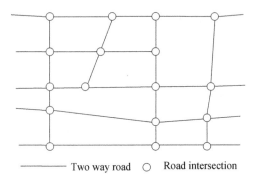

——— Two way road   ○   Road intersection

Fig. 6. Simplified model of urban traffic node routes

Table 1. Simulation parameter settings

Parameter name	Parameter settings
Simulation area	2000 m × 2000 m
Number of intersections	17
Transmission range	265 m
Number of vehicle nodes	75–250
Number of roads	32
Vehicle speed	30–60
MAC	Layer protocol
Channel capacity	2 Mbps
Business model	CBR
Packet size	128
Weight coefficient size(α and β)	0.5

3.3 Result Analysis

This article mainly calculates the average delay time of vehicles at the intersection as the evaluation index. The vehicles are equally distributed to different time periods to form the number of queues in Table 2, and it is assumed that the vehicles arrive at the intersection evenly. Under the signal control strategy in this article, the maximum green light time for traffic flow in the straight direction is 50 s, and the maximum green light time for traffic flow in the left turn direction It is 30 s, and the average headway of vehicles passing through the intersection is 2 s. The saturated flow rate is both 2000 pcu/h, and the average delay time of the vehicle under the fixed period signal is about 23.10 s. Under this signal control strategy, through the proper combination and optimization of the sub-phases, the priority phase rotation order can be obtained; by calculating the time each time the vehicle passes through the intersection, the total vehicle delay time is obtained, and the average vehicle delay time is calculated to be about 17.34 s. Numerical results show that the use of the traffic node management strategy based on this paper can reduce the average delay time of vehicles.

Table 2. Number of vehicles queued in different periods

Period	East	West	South	North
8:04–8:05	4	6	6	6
8:05–8:06	3	4	7	5
8:06–8:07	5	3	5	4
8:07–8:08	3	5	5	5
8:08–8:09	5	4	4	8
8:09–8:10	2	8	5	4
8:10–8:11	6	6	8	7
8:11–8:12	5	4	4	5
8:12–8:13	3	6	5	5
8:13–8:14	5	5	6	6
8:14–8:15	4	3	4	8
8:15–8:16	3	4	5	5
8:16–8:17	4	6	5	4
8:17–8:18	3	5	7	5
8:19–8:20	4	6	6	6

Since the simulated road network scene area is large enough, the layout of the main streets in the urban area can be simulated more accurately. The nodes of the designed vehicles are randomly released on the road and start to move in two directions at the same time, which ensures the randomness of the simulation. In summary, all the basic simulation parameters are summarized with high reliability.

4   Conclusion

In the era of mobile Internet, traditional traffic signal control concepts and technologies have gradually been unable to meet the needs of urban traffic development. For this reason, an urban traffic node management scheme based on 5G wireless network architecture is designed. The experimental results show that the method in this paper can reduce the number of vehicles in line and reduce the average delay time of vehicles, thereby effectively improving the utilization efficiency of road traffic resources and alleviating urban traffic problems.

The design method in this paper can effectively realize the management of traffic nodes and reduce the average delay time of vehicles. In future research, the goal should be to improve the efficiency of node management, improve the urban traffic node management plan, and further improve the level of urban traffic management.

Fund Projects. Guangxi High School Young Teachers Improve Scientific Research Ability Project in 2019 "The Research of Nanning Rail Transit Emergency Pre-warning System" (2019KY0935).

Nanning Talents Small Highland Special Fund Support Plan in 2018 "The Research of Nanning Rail Transit Emergency Connection Auxiliary Decision System" (2018019).

References

1. Li, W., Jiang, M., Chen, Y., et al.: Estimating urban traffic states using iterative refinement and Wardrop equilibria. IET Intel. Transport Syst. **12**(8), 875–883 (2018)
2. Anzani, M., Javadi, H.H.S., Modirir, V.: Key-management scheme for wireless sensor networks based on merging blocks of symmetric design. Wireless Netw. **24**(8), 2867–2879 (2018)
3. Liang, Z., Huang, Y., Cao, Z., Liu, T., Wang, Y.: Creativity in trusted data: research on application of blockchain in supply chain. Int. J. Performability Eng. **15**(2), 526–535 (2019)
4. Fu, Y., Wang, S., Wang, C.X., et al.: Artificial intelligence to manage network traffic of 5G wireless networks. IEEE Network **32**(6), 58–64 (2018)
5. Li, Z., Wang, J.: Security storage of sensitive information in cloud computing data center. Int. J. Performability Eng. **15**(3), 1023–1032 (2019)
6. Sundan, A.P., Jha, R.K., Gupta, A.: Energy and spectral efficiency optimization using probabilistic based spectrum slicing (PBSS) in different zones of 5G wireless communication network. Telecommun. Syst. **73**(1), 59–73 (2020)
7. Vidal, P.J., Olivera, A.C.: Management of urban traffic flow based on traffic lights scheduling optimization. IEEE Lat. Am. Trans. **17**(1), 102–110 (2019)
8. Ji, Y., Ding, X., Li, H., et al.: Layout design of conductive heat channel by emulating natural branch systems. J. Bionic Eng. **15**(3), 567–578 (2018)
9. Liu, Y., Lyu, Y., Bttcher, K., et al.: External interface-based autonomous vehicle-to-pedestrian communication in urban traffic: communication needs and design considerations. Int. J. Hum.-Comput. Interact. **36**(5), 1–15 (2020)
10. Liu, Z., Wang, C.: Design of traffic emergency response system based on internet of things and data mining in emergencies. IEEE Access **7**(99), 113950–113962 (2019)
11. Liu, S., Liu, D., Srivastava, G., Połap, D., Woźniak, M.: Overview and methods of correlation filter algorithms in object tracking. Complex Intell. Syst. **7**(4), 1895–1917 (2020). https://doi.org/10.1007/s40747-020-00161-4
12. Liu, S., Lu, M., Li, H., et al.: Prediction of gene expression patterns with generalized linear regression model. Front. Genet. **10**, 120 (2019)
13. Fu, W., Liu, S., Srivastava, G.: Optimization of big data scheduling in social networks. Entropy **21**(9), 902 (2019)

Multi Source Information Matching and Fusion Method in Wireless Network Based on Fuzzy Set

Quan-wei Sheng[✉]

Changsha Medical University, Changsha 410219, China

Abstract. The current wireless network information fusion matching method is based on the calculation of the uncertainty of a large number of data samples, and gets the information fusion results according to the corresponding decision. However, due to the uncertainty of data samples, it is difficult to complete statistics, which leads to low processing accuracy and poor information fusion effect of information fusion matching method. In order to solve the above problems, a multi-source information matching and fusion method based on fuzzy set is studied. After matching with multi-source information, Markov distance is used to allocate the information. According to the probability of supporting information conflict, information fusion rules are formulated, and multi-source fusion is realized by using fuzzy sets. Simulation results show that the proposed information fusion method can reduce the energy consumption of network nodes by 81.4%, and has high fusion accuracy and reliability.

Keywords: Fuzzy set · Wireless network · Multi-source information · Matching fusion

1 Introduction

Multi source information fusion is also called multi-sensor information fusion. It comprehensively analyzes the incomplete environmental information collected in local environment in various forms and ways, and processes the redundant or contradictory information between the collected information in many aspects, so as to achieve the analysis of information complementarity and uncertainty of information, and further to the overall system The lines are expressed in a complete and consistent way [1–3]. The accuracy of planning, decision-making and emergency response can be enhanced by multi-source information fusion, thus reducing the risk brought by decision-making. The multi-source information fusion can improve the robustness and reliability of the system compared with the single source information. The robustness and reliability of a system relying on single information is relatively poor. If the information source fails (sensor failure, transmission delay, signal distortion, etc.), the performance of the whole system will be fatally affected, that is, it can not work normally. With the increasing amount of information obtained by multi-sensor system, it leads to different factors in time domain, space field, credibility, expression mode, information emphasis and practical application, and puts forward new requirements for information processing and

© ICST Institute for Computer Sciences, Social Informatics and Telecommunications Engineering 2022
Published by Springer Nature Switzerland AG 2022. All Rights Reserved
S. Liu and X. Ma (Eds.): ADHIP 2021, LNICST 417, pp. 51–61, 2022.
https://doi.org/10.1007/978-3-030-94554-1_5

work management. At present, no matter in the theoretical method and implementation technology of research, the technology of information fusion needs to be opened up. When multi-source information fusion is combined, the uncertain information provided by multi-sensor system is expressed in probability by Bayes estimation principle. Secondly, the independent decision elements are divided reasonably in a whole sample space. Then, the decision elements can be processed by Bayes conditional probability formula. Finally, the decision elements can be processed by using Bayes conditional probability formula According to the given rules, the system decision is output, and the final decision of the system is generally taken as the maximum posterior probability. The method needs to give the prior probability distribution of target information of each sensor through a large number of experiments, which is difficult to achieve in practice. In addition, when the results of decision-making are increased, the calculation of Bayes estimation will be very complex to affect the real-time performance of the processing system. The robust hypothesis is used to verify whether the test data is consistent and the information matching fusion is realized by using the statistical decision theory. Because of the fusion estimation based on the maximum and minimum decision of Luban, there are some limitations in the fusion of uncertain information. The data fusion based on Markov chain is a linear weighting method by using multiple sensors' collected values. It is necessary to measure the data with common observation values, which is difficult to measure, and affects the accuracy of information matching fusion. In the process of multi-source information collection, because of many reasons, information contains a lot of uncertainty, and fuzzy sets can reflect the uncertainty to the maximum extent in the reasoning process, thus optimizing information matching fusion. Based on the above analysis, this article will study the method of wireless network multi-source information matching and fusion based on fuzzy sets, and verify the feasibility of the method through simulation experiments, and solve the problems in the traditional method.

2 Multi Source Information Matching and Fusion Method in Wireless Network Based on Fuzzy Set

2.1 Reliability Allocation of Multi Source Information in Wireless Networks

In this paper, Markov distance is used as an important basis for reliability allocation, and the corresponding basic reliability distribution function is needed to map the relationship between distance and reliability. The Markov distance between multi-source information in wireless network is defined as follows:

If M is set as n element population (n indexes are investigated in total), the mean vector is $\mu = (\mu_1, \mu_2, \cdots, \mu_n)$ and covariance matrix is $\sum (\sigma_{ij})_{n \times n}$, the calculation formula of Markov distance between sample $X = (X_1, X_2, \ldots, X_n)$ and population M is as follows [4]:

$$d^2(X, M) = \frac{(X - \mu)^2}{\sigma^2} \tag{1}$$

Mahalanobis distance has the following characteristics: (1) Mahalanobis distance between two points is independent of measurement unit; (2) Mahalanobis distance between two points calculated by standardized data and centralized data is the same; (3) mutual interference between different variables can be effectively eliminated [5].

The selection of the basic reliability allocation function based on Mahalanobis distance needs to meet the following two main principles

(1) The final distribution result sum of all reliability is 1, and the single distribution result is greater than or equal to 0 and less than or equal to 1;
(2) The distribution function should be a monotonic decreasing function, that is, with the increase of Mahalanobis distance between the target and the known sample, the reliability of the distribution function gradually decreases;

The common distribution functions satisfying the above two basic principles are inverse trigonometric function and exponential function, as shown in the following formula [6]:

$$y = \frac{1}{2} - \frac{1}{\pi} \arctan\left(\frac{x}{a} - b\right), a, b > 0$$
$$y = a^{bx}, 0 < a < 1, b > 0 \tag{2}$$

In the above formula, a and b are the parameters of the distribution function. Combined with the distribution trend of the basic reliability distribution function of the two functions, the exponential function is selected as the basic reliability distribution function model based on Mahalanobis distance. According to 3σ principle, the specific values of distribution function parameters a and b are determined. According to the monotonicity of the exponential function, when the Mahalanobis distance between the tested target and the sample set is greater than 3 times of the standard deviation of the known sample data set, the initial reliability allocation result is less than ε value, so the mapping accuracy of the basic reliability allocation function can be adjusted by setting the threshold value ε. Therefore, there are the following relationships:

$$\varepsilon = a^{3\sigma} \tag{3}$$

In order to make the exponential basic reliability distribution function proposed in this paper better adapt to the statistical distribution of different sample data, the distribution function is transformed into the following form:

$$y = \left(\varepsilon^{1/(a\sigma_{max})}\right)^{x}, 0 < \varepsilon < 1 \tag{4}$$

In the above formula, parameters a and ε are the adjustment coefficients of the basic reliability allocation function; x is the Mahalanobis distance between the target to be tested and each known sample data set; y is the mapped initial reliability allocation result. According to the process shown in the figure below, the reliability of multi-source data in wireless network is allocated [7] (Fig. 1).

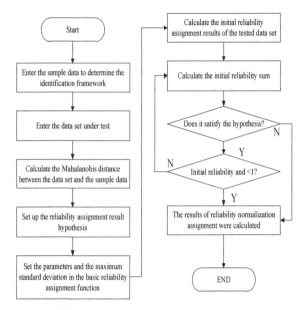

Fig. 1. Reliability allocation process of wireless network information

After the reliability of multi-source information is assigned, the matching relationship between wireless network information and entities is established to facilitate information fusion.

2.2 Multi Source Information and Wireless Network Entity Matching

The process of candidate pair generation based on attribute index can be divided into three parts: inverted index construction, attribute weight calculation and candidate entity pair generation.

Inverted index construction traverses all entity k in multi-source data source, and then traverses attribute p in k. With the attribute name and attribute value as the composite key and the index value as the collection of entity ID, the mapping relationship conforms to the following form [8]:

$$(p.name, p.value) \rightarrow \{ID_1, ID_2, \cdots, ID_n\} \tag{5}$$

In the above formula, $p.name$ is the name of the data source entity property; $p.value$ is the value of the data source entity attribute. The attribute name and the attribute value can reflect the property's discrimination. Define the weight of the attribute as follows:

$$W_{(p.name,p.value)} = \frac{1}{|Svc|} \tag{6}$$

In the above formula, $|Svc|$ represents the number of entities with the same attribute name and corresponding same attribute value; $W_{(p.name,p.value)}$ represents the entity attribute weight. The weight of different attributes can be calculated by inverted index.

For each group of ID set, enumerate any two entity pairs, give a certain weight $W_{(p.name, p.value)}$ or accumulate on the basis of the original weight. Finally, the entity pairs with all weights are traversed. If the accumulated weight of the entity pairs is greater than a certain threshold, they are regarded as candidate entity pairs [9].

The similarity between the two candidate entities is calculated according to the following formula:

$$sim(k_1, k_2) = \omega_1 \times sim(k_1.N, k_2.N)$$
$$+ \omega_2 \times sim(k_1.P, k_2.P) + \omega_3 \times sim(k_1.T, k_2.T) \tag{7}$$

In the above formula, $sim(k_1.N, k_2.N)$ is the similarity of entity name; $sim(k_1.P, k_2.P)$ is the similarity of entity attribute; $sim(k_1.T, k_2.T)$ is the similarity of entity context; ω_1, ω_2 and ω_3 represent the corresponding weights of the three. According to the descending order of the similarity between the candidate entities, the matching relationship between the wireless network multi-source information and the entities is determined. Combined with the matching relationship between entities, the basic synthesis rules of multi-source information fusion are formulated according to the D-S theory.

2.3 Making Rules of Multi-source Information Fusion

In this paper, we use the idea that the probability of supporting information conflict is distributed according to the average support degree of each information, and obtain the following composition rule [10]:

$$\begin{cases} m(A) = \sum\limits_{\substack{ij \\ \cap A_i = A}} \prod\limits_{j=1}^{m} m_j(A_i) + rq(A) \\ \forall A \subset U, A \neq \Phi \\ m(\Phi) = 0 \\ r = \sum\limits_{\cap A_i = \Phi} \prod\limits_{j=1}^{m} m_j(A_i) \\ q(A) = \frac{1}{n} \sum\limits_{i=1}^{n} m_i(A) \end{cases} \tag{8}$$

Where $q(A)$ is the average support degree of proposition A and r is the conflict probability. In order to improve the shortcomings of the above information fusion rules, the correlation coefficient between information is introduced to deal with conflict information, and the correlation coefficient between information can be calculated according to the following formula:

$$d_{12} = \frac{\sum\limits_{A_i \cap B_j = A_k \neq \Phi} m_1(A_i) m_2(B_j)}{\sqrt{\left(\sum m_1^2(A_i)\right)\left(\sum m_2^2(B_j)\right)}} \tag{9}$$

In the above formula, m_1 and m_2 are the basic trust allocation functions of two conflict information E_1 and E_2 under the D-S theory recognition framework, and A_i and B_j are the units corresponding to the basic trust allocation functions m_1 and m_2 respectively. Correlation coefficient d_{12} is used to describe the similarity between two information. The greater the value of d_{12}, the better the certainty. When d_{12} is 1, it indicates that the two information E_1 and E_2 are completely similar; when d_{12} is 0, it indicates that the two information E_1 and E_2 are completely conflicting.

It is defined that the amount of information provided by an information is related to the focus element produced by it, then the calculation formula of information capacity provided by information under D-S theory is as follows:

$$C(E) = \sum_{i=1}^{n(A_i)} \frac{m(A_i)}{|A_i|} \tag{10}$$

In the above formula, $|A_i|$ is the cardinality of the focus element; $n(A_i)$ is the number of focus elements. For the focus element A_i, if $m(A_i)$ is 0, the information capacity is 0, indicating that this information does not contain any useful information. If $|A_i|$ is 1, the information capacity is 1, indicating that this information contains the most useful information. Therefore, before information fusion, the information capacity of each information is calculated, and then normalized to get the rule of multi-source information fusion in wireless network. According to the rules of wireless network multi-source information fusion, the fuzzy set theory is used to realize the fusion of multi-source information in wireless network.

2.4 Realize the Multi-source Information Matching and Fusion in Wireless Network

In order to solve the problem of multi-sensor data fusion, the observation data of the sensor needs normalization (fuzziness) processing (fuzziness), which corresponds to the range of [0, 1]. The theory of fuzzy set provides a normalized tool for people, that is, the fuzziness process of sensor observation information can be realized by constructing appropriate membership function.

The membership degree $u \in [0, 1]$ of fuzzy set is introduced to represent the local decision value of sensor. For the measured value x of each sensor, the membership function $u(x)$ can be used to map it to a number u (membership degree) in the interval [0,1]. The size of u reflects the support degree of the sensor for a certain hypothesis, which can also be called basic probability assignment. Here are some membership functions commonly used in fuzzification, but in practical application, we should build practical membership functions according to specific problems.

For the sensor that can determine the mean value $\bar{x}$ and the deviation δ of the parameter, if the measured value is, the membership function can be used:

$$u(z) = \begin{cases} 1 - \frac{|z-\bar{x}|}{2\delta}, & |z - \bar{x}| < 2\delta \\ 0, & |z - \bar{x}| \geq 2\delta \end{cases} \tag{11}$$

2 For multiple groups of measured values, the maximum value $\max f$ and minimum value $\min f$ can be determined. If the measured value is $f(u_i)$, the membership function can be used as follows:

$$M_f(u_i) = \frac{f(u_i) - \min f}{\max f - \min f} \tag{12}$$

If multiple sensors are used to detect multiple objects, and if the number of sensors in the number of sensors supported for an object can be determined as n, the membership function can be used as follows:

$$A(u_i) = \frac{n}{N} \qquad (13)$$

Where u_i is the i object. Simultaneous interpreting the membership function of fuzzy sets according to different sensors' acquisition of multi-source information. The fuzzy similarity matrix is established to calculate the support and reliability of different information. The reliability is taken as the weight, and the average basic trust function of information is obtained by weighted average. According to the information object fusion rules constructed by D-S theory, the final data fusion result is obtained by n-1 fusion operation on N pieces of information. So far, the research of multi-source information matching and fusion method based on fuzzy set is completed.

3   Simulation Experiment

The multiple information sources of wireless network make it more difficult to effectively manage the information of wireless network, and cause a large number of effective data abuse and loss of sensor. In this paper, a multi-source information matching and fusion method based on fuzzy set is proposed. This section will verify the effectiveness of the method.

3.1   Experiment Content

The simulation experiment was carried out in NS 2 test software. In order to reflect the superiority of the above proposed fusion method based on fuzzy sets, the fusion matching method studied above is compared with Bayes estimation based information fusion method and Markov chain based data fusion method. The experiment mainly compares the energy consumption and accuracy of the network nodes when the information is transmitted after fusion. Through the analysis of the experimental data, the information fusion effect of the method is studied.

3.2   Experimental Process

The information fusion matching method based on fuzzy set is named as method 1, the information fusion method based on Bayes estimation is named as method 2, and the data fusion method based on Markov chain is named as method 3. Three different methods are used to fuse the information in the experimental network. The whole process is monitored by the test software and the experimental data are obtained.

The energy consumption of information transmission between nodes is an important index to evaluate the fusion method. The data of each communication node in the network detected by the network communication monitoring software is analyzed, and the final experimental conclusion is obtained.

3.3 Experimental Result

Three kinds of information fusion methods are applied in the experimental network to fuse the information transmitted in the network. The monitoring software detects the energy consumption of network communication nodes, and obtains the comparison results of node energy consumption as shown in the following table (Table 1).

Table 1. Energy consumption comparison of network communication nodes/mJ

Wireless network node	Method 1	Method 2	Method 3
1	1.71	9.41	11.48
2	1.67	8.73	12.30
3	1.73	9.30	10.51
4	1.81	9.22	12.13
5	1.75	8.85	17.96
6	1.68	8.41	14.52
7	1.64	8.83	17.33
8	1.66	9.64	11.49
9	1.79	9.27	9.84
10	1.63	8.96	11.91
11	1.72	9.63	9.93
12	1.69	9.82	10.11
13	1.83	9.81	10.19
14	1.67	9.34	11.88

According to the analysis of the above table, the energy consumption between different nodes in method 1 is lower than that in method 2 and method 3. The difference of energy consumption of different nodes in the wireless network of application method 1 is small, while the difference of energy consumption of different nodes in the network of application method 3 is large, and the difference of energy consumption of nodes in the wireless network of application method 2 fluctuates in a certain range.

Calculate the average energy consumption of wireless network nodes in the process of this experiment when transmitting the information fused by different methods. The average energy consumption of wireless network nodes in application method 1 is 1.71 MJ, that in application method 2 is 9.23 MJ, and that in application method 3 is 12.26 MJ. Application method 1 can reduce the energy consumption of 81.4% at least.

The accuracy comparison of three wireless network information fusion methods in information fusion is shown in Fig. 2.

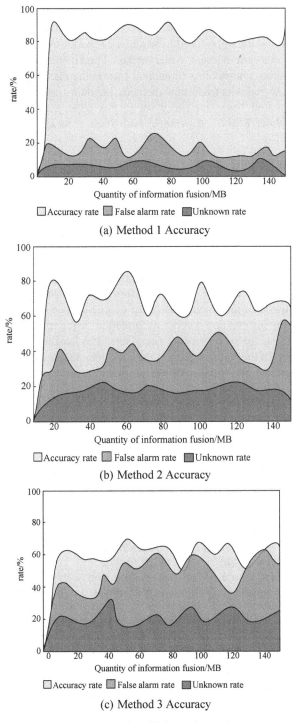

(a) Method 1 Accuracy

(b) Method 2 Accuracy

(c) Method 3 Accuracy

Fig. 2. Comparison results of information fusion accuracy

Analysis of the above figure shows that the correct rate, false alarm rate and unknown rate of method 1 are lower than those of method 2 and method 3. With the increase of the amount of information to be fused in wireless networks, the accuracy of method 1 fluctuates slightly in a certain range, while method 2 and method 3 fluctuate sharply. It shows that the accuracy and stability of method 1 are better than the other two methods.

To sum up, compared with traditional methods, the multi-source information matching and fusion method based on fuzzy set in this paper can reduce node energy consumption by about 81.4% on average, with high accuracy, strong stability and reliability.

4 Concluding Remarks

With the increasing demand for various kinds of information, multi-sensor information fusion technology has become an effective means of data processing and decision support. Because of the redundancy of information between different kinds of sensors, the system has better fault tolerance ability. Aiming at the problems of the current common information fusion methods, this paper studies the multi-source information matching fusion method based on fuzzy set. After the information is distributed by Mahalanobis distance, the multi-source information and entities are matched. According to the probability of supporting information conflict, information fusion rules are formulated, and multi-source fusion is realized by using fuzzy sets. The feasibility is verified by simulation experiment. In the future research, the experimental parameters will be further changed to improve the performance of the information fusion method.

References

1. Shi, Z., Zhou, X., Li, K., et al.: Cyberspace security monitoring technology based on multi-source information fusion. Comput. Eng. Des. **41**(12), 3361–3367 (2020)
2. You, H., Shi, H., Yang, Y., et al.: Intelligent fusion diagnosis method for multi-source information of power grid fault based on improved D-S evidence theory. Guangdong Electric Power **33**(11), 16–25 (2020)
3. Hu, X., Zhang, W., Xu, J.: A method of multi-source data fusion based on absolut grey fusion algorithm. Electron. Warfare Technol. **35**(05), 37–41 (2020)
4. Ke, J., Lu, X.: Intelligent recommendation algorithm for social networks based on fuzzy perception. Inf. Technol. **44**(02), 130–134 (2020)
5. Liu, Y., Hui, H., Lu, Y., et al.: Multi-source information fusion indoor positioning method based on genetic algorithm to optimize neural network. J. Chin. Inertial Technol. **28**(01), 67–73 (2020)
6. Xiao, L., Luan, X.: Predictive control method of matching deviation quality for car door based on data fusion method. Agric. Equip. Veh. Eng. **57**(12), 29–34 (2019)
7. Ge, X.: Multi-channel information fusion method of wireless network based on grid system. J. Jilin Univ. (Inf. Sci. Ed.) **37**(06), 617–622 (2019)
8. Liu, S., He, T., Dai, J.: A survey of CRF algorithm based knowledge extraction of elementary mathematics in Chinese. Mob. Netw. Appl. (2021). https://doi.org/10.1007/s11036-020-01725-x

9. Liu, S., Pan, Z., Cheng, X.: A novel fast fractal image compression method based on distance clustering in high dimensional sphere surface. Fractals **25**(4), 1740004 (2017)
10. Liu, S., Fu, W., He, L., Zhou, J., Ma, M.: Distribution of primary additional errors in fractal encoding method. Multimed. Tools Appl. **76**(4), 5787–5802 (2014). https://doi.org/10.1007/s11042-014-2408-1

Weak Signal Acquisition and Recognition Method for Mobile Communication Based on Information Fusion

Wu-lin Liu[1], Feng Jin[1], Hai-guang He[1], and Yu-xuan Chen[2(✉)]

[1] Information and Communication College, National University of Defense Technology, Xi'an 710106, China
[2] Fuyang Normal University, Fuyang 236300, China

Abstract. In traditional communication signal acquisition, the effect of weak signal recognition is poor. Therefore, a weak signal acquisition and recognition method for mobile communication based on information fusion is designed. The mobile communication signal location model is established; the TDOA algorithm is used to locate the mobile communication signal; the receiver is used to capture the mobile communication signal. M-QAM modulation technology is used to modulate the parameters of mobile communication signal transmission mode, and the feature classification model of mobile communication signal is established. The modulation types of mobile communication signals are identified by using cyclic spectrum, and the simulation is carried out. The modulation types are classified by using mobile communication signal feature classifier and information fusion technology, and the recognition of mobile communication signals is completed. Experimental results show that the recognition rate of this method is 24% higher than that of traditional method 1 and 34% higher than that of traditional method 2. This method is based on information fusion. The information is fused by combining classifiers to classify modulation types, which significantly improves the recognition accuracy.

Keywords: Information fusion · Mobile communication · Weak signal acquisition · Recognition

1 Introduction

Communication signal transmission has a wide range of applications in many fields, but there are also many challenges. As an important part of communication countermeasure, the acquisition and identification of communication signal is more and more difficult. The specific reason is that with the application of communication if, frequency hopping and spread spectrum technology, the signal may be submerged in noise, which makes the signal-to-noise ratio of the received signal at the communication signal acquisition and identification receiver very low, and if it is a weak signal in the communication signal, This situation will get worse [1–3]. With the increase of traffic, the communication system needs to process the weak signal under the condition of big data samples.

S. Liu and X. Ma (Eds.): ADHIP 2021, LNICST 417, pp. 62–73, 2022.
https://doi.org/10.1007/978-3-030-94554-1_6

In addition, in the modern high-tech war, the electromagnetic environment is becoming more and more complex. In addition to the conventional mobile communication signals, some uncommon and new modulation types of mobile communication signals appear from time to time. Usually, the characteristics of these new modulation types of mobile communication signals are complex and difficult to identify, but the threat level is relatively high, here referred to as the weak signal of mobile communication. It is of great significance to capture and recognize them.

Data fusion is a comprehensive information processing technology about how to use multi-source information in collaboration to obtain more objective and essential knowledge of the same thing or objective. In recent years, researchers have put forward a variety of information fusion algorithms from different perspectives. Because of the different starting point and destination of the problem, different methods of information fusion algorithm processing are different. The fusion information of weak signals in mobile communication is from multiple classifiers. Traditional classification methods mostly use feature extraction or neural network classifier. For the case of sufficient samples, neural network classifier has good effect, but the actual samples collected in communication confrontation environment are often limited, and the problem of poor generalization ability of neural network leads to its limited application in radiation source identification.

Relevant scholars have conducted research on this and made some progress. Li Kun et al. proposed a narrow-band power line carrier communication signal recognition algorithm [1], which uses the wavelet transform amplitude variance value and high-order cumulant of the carrier communication modulation signal as the identification feature parameters, and uses an improved support vector machine method to design signal recognition It improves the robustness of power line noise signal removal and avoids problems such as under-learning and over-learning in the signal processing process. This method can improve the accuracy of signal capture, but it takes longer to capture the signal. Li Changba et al. proposed an automatic modulation recognition method for communication signals based on deep learning [2], which extracts communication signal features based on feature extraction, uses self-encoding technology to obtain spectrum in complex electromagnetic environments, and conducts feature set training based on deep learning technology. Automatic recognition of MQAM communication signal modulation mode. The method in this paper has better classification anti-jamming recognition ability, but poor signal capture ability.

For the above methods, there is a poor signal capturing ability. Problems such as low signal recognition accuracy. For this reason, this article introduces the information fusion method to capture the weak signal of mobile communication. The signal feature classifier is established according to the conventional features, the modulation pattern is recognized through multiple classifiers, and the information fusion technology is adopted to use the complementarity and redundancy of the performance of each classifier to improve the accuracy and anti-interference of the capture and recognition results. Effectively enhance the mobile communication weak signal capture and recognition effect.

2 Weak Signal Acquisition and Recognition Method for Mobile Communication Based on Information Fusion

2.1 Mobile Communication Signal Location

The apparatus and method for searching signals in a mobile communication system can process input signals in a parallel manner. A signal searcher device in a mobile communication system may include a first sequential storage device and a second sequential storage device for storing codes and input signals, respectively. The signal searcher may include a plurality of despreading devices capable of despreading input signals containing components in a parallel manner using codes. When the length of coherent accumulation is multiple of the number of despreading devices, the additional buffer will store the accumulated despreading signal with sum component whose number is equal to the number of offsets searched. The system completes the positioning of air targets through the cooperation of different mobile base stations and a system receiver. When the target enters the detection area, the base station radiation signal that irradiates the target will be reflected, and part of the energy of the reflected wave signal will be received by the receiving station. The time delay of target echo signal can be obtained by generalized correlation between direct wave signal and target reflection signal. As shown in Fig. 1, the radar target positioning model based on mobile communication signal at a certain time is presented.

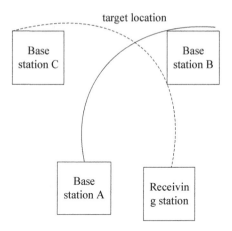

Fig. 1. Passive location model of mobile communication signal

When the echo signal of each base station is delayed, TDOA algorithm can be used to locate the target. At the same time, the coordinates of base station i are defined as $X_i = [x_i, y_i]$ and $i = 1, 2, \cdots, N$ (N is the number of base stations), and the target is located at $x = [x, y]$. Without considering the noise, it is assumed that the arrival time difference between the signal transmitted by base station i and that transmitted by base station j is $\sigma_{i,j}^*$. since the propagation speed of electromagnetic wave in the air is constant, it can be concluded that:

$$r_{i,j}^* = c\sigma_{i,j}^* = R_i - R_j \times \|x - X_i\| - \|x - X_j\|* \qquad (1)$$

In formula (1), $R_i = \|x - X_i\|$ represents the distance from the target to base station i, $R_j = \|x - X_j\|$ represents the distance from the target to base station j, and $*$ represents the actual value without noise. $r_{i,j}^*$ represents the distance difference formed by multiple base station pairs. Since the echo time delay of each base station is known, the distance difference formed by multiple base station pairs can be obtained without considering the noise, thus forming a hyperbolic equation group. By solving the equation group (that is, finding the focus of the hyperbola), the position of the target can be obtained. According to the properties of the equations, the hyperbolic equations constructed by using different base stations as reference base stations to calculate the delay difference are equivalent to the equations constructed by using a specific base station as reference base station. By solving the equations, the target mobile communication signal can be located. After positioning, the receiver is used to capture the mobile communication signal.

2.2 Mobile Communication Signal Acquisition

For most of the mobile communication signals, coherent integration and incoherent integration are used for acquisition, while for the signals which only contain Doppler shift, two-dimensional search is used for acquisition. Generally, the carrier to noise ratio of strong signal in mobile communication is 40 dB·Hz. But when users use mobile communication, mobile communication signal receiving devices are often in weak signal environment, such as indoor, garage, forest and so on. The signal is weakened due to occlusion and multipath effect, which leads to the decrease of receiver sensitivity. The receiver needs to extend the dwell time to improve the processing performance. The main methods include coherent integration and incoherent integration. Coherent integration can improve the sensitivity of receiver acquisition by extending the signal length, but the signal length will be affected by the mobile communication signal message data bits. Incoherent integration is not affected by data information, but it will produce square loss. Therefore, when processing a certain segment of target signal, the receiver can design different integration strategies, and reasonably combine coherent integration and incoherent integration, so as to coordinate the shortcomings between them and obtain the optimal acquisition performance of the receiver. The general capture process diagram is shown in Fig. 2.

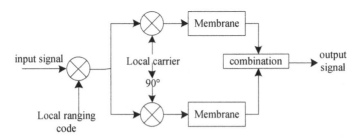

Fig. 2. Coherent/incoherent acquisition process

In some environmental conditions, when the signal quality is weak, extending the length of the signal sequence can improve the signal-to-noise ratio, which is the principle of coherent integration. Assuming that the signal is converted to baseband, only the complex number of signals with Doppler shift is included. Because the received signal has the uncertainty of Doppler frequency shift and code delay, the carrier tracking loop can not track the signal change directly. Because the received signal is a spread spectrum signal formed by the direct sequence spread spectrum modulation and carrier modulation of the communication data, the signal receiving power is very small, so the signal must be de expanded firstly. At the same time, due to the dynamic of the receiver carrier, the Doppler frequency shift is uncertain, and the code acquisition must be searched in the whole code phase and frequency domain at a fixed interval. Therefore, the acquisition of spread spectrum signal must complete the initial synchronization of pseudo code phase and the initial estimation of carrier frequency difference. During the search process, the code phase stepping amount is half code phase unit, and Doppler frequency step amount is a Doppler frequency shift unit, then a code phase search unit and a Doppler search unit constitute a search unit in the two-dimensional search space, as shown in Fig. 3.

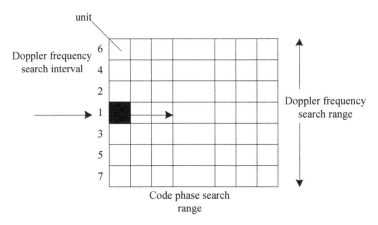

Fig. 3. Two dimensional sequence search

As shown in Fig. 3, through two-dimensional sequence search, when the correlation result is higher than the preset signal detection threshold, the mobile communication signal acquisition is completed.

2.3   Parameter Modulation of Mobile Communication Signal Transmission Mode

After the acquisition of signal is completed, m-QAM modulation technology is used to modulate the transmission mode parameters of mobile communication signal. In modulation, besides the system transmission BER and band utilization rate with general significance, other parameters related to modulation methods should be considered, such as: peak to average ratio of m-QAM modulation signal, Euclidean distance between constellation points and minimum phase offset of signal. For different transmission systems, the requirements for these parameters are different. The peak to average ratio of m-QAM signal: the magnitude of the peak to average ratio of m-QAM signal reflects the anti-linear distortion ability of m-QAM signal, especially the nonlinear distortion caused by the nonlinear power amplifier. The larger the peak to average ratio of m-QAM signal, the worse its anti-linear distortion performance. Minimum Euclidean distance of m-QAM signal: the minimum Euclidean distance is the minimum distance between constellation points on m-QAM signal constellation. This parameter reflects the ability of m-QAM signal to resist Gaussian white noise. The maximum of m-QAM signal can be obtained by optimizing the constellation distribution of m-QAM signal, and thus the m-QAM modulation scheme with better anti-interference performance can be obtained. Minimum phase offset of m-QAM signal: the lowest phase offset is the minimum offset of constellation phase of m-QAM signal. This parameter reflects the phase jitter ability of m-QAM signal and sensitivity to clock recovery accuracy. It can also modulate the transmission formula parameters of mobile communication signal by optimizing m-QAM signal. After the parameter modulation of the transmission mode of mobile communication signal is carried out by m-QAM modulation technology, the classification model of mobile communication signal characteristics is established.

2.4 The Feature Classification Model of Mobile Communication Signal is Established

In the complex environment of communication electronic countermeasure, the intercepted communication signals need to be preprocessed. Based on the determination of the modulation type of communication signals, the steady-state classification features are extracted by using the time domain, frequency domain and time-frequency domain methods, so as to establish the steady-state Feature Classification Library. The communication signal feature classification model is shown in Fig. 4.

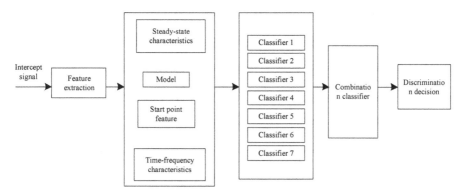

Fig. 4. Communication signal feature classification model

Secondly, according to the signals classified by steady-state features, the instantaneous phase variance method, recursive graph method and phase variance method based on higher-order cumulant are used to detect the starting point of the communication signals classified by steady-state features, so as to realize the further classification of communication signals. Then, in the process of feature analysis and extraction of communication emitter signal, it is usually necessary to convert the measured values reflecting the characteristics of communication emitter signal into feature space. However, this conversion process is usually not linear, which makes the features of communication emitter signal are usually high-dimensional feature vectors, which mainly show time-frequency image features, high-order statistics features and high-order spectrum features, and these feature vectors may overlap in the high-order feature space, resulting in the reduction of classifier efficiency and classification probability. Therefore, by using the time-frequency image feature analysis and extraction method, the high-order features of the intercepted communication signal are further extracted by selecting the time-frequency image features with less dimension and low complexity, so as to form a transient feature classification library, which can improve the classification efficiency of the classifier and the classification probability. Finally, the classifier design method of combining single feature classifier and combined feature classifier is adopted to improve the classification performance of multi feature vectors of the designed classifier, so as to improve the classification and recognition ability of communication emitter signals. After the feature classification model of mobile communication signal is established, the modulation type of mobile communication signal is identified.

2.5 Modulation Type Recognition of Mobile Communication Signal

Modulation type is the main steady-state feature of signal differentiation. Therefore, the same type of signal is selected by modulation type, and prior information for other feature parameters is extracted and classified. Because the cyclic spectrum can show better noise resistance on non-zero cycle frequency and Gaussian white noise only appears on the zero cycle spectrum, the cyclic spectrum can be used to recognize different modulation types of signals [11]. If the mean $j(t)$ of continuous time stochastic process $a(t)$ and the autocorrelation function $H(t, \varepsilon)$ of random process $a(t)$ are periodic and the period is T, then $a(t)$ is called generalized cyclic stationary random process, namely:

$$\begin{cases} j(t + T) = j(t) \\ H(t + T, \varepsilon + T) = H(t, \varepsilon) \end{cases} \tag{2}$$

As shown in formula (2), the process is a random cyclic process, then the autocorrelation function $H(t, \varepsilon)$ can be expanded by Fourier series:

$$H(t, \varepsilon) = \sum_{\beta} H^{\beta}(\varepsilon) \exp(k2\pi\beta t) \tag{3}$$

In formula (3), $\beta = \frac{n}{T}, n = 0, 1, 2, \cdots$ and β represent the cyclic spectrum. $H^{\beta}(\varepsilon)$ is the cyclic autocorrelation function, k is a Fourier parameter [12], where the cyclic autocorrelation function can be expressed as:

$$H^{\beta}(\varepsilon) = \frac{1}{T} \int_{-T/2}^{T/2} H\left(t + \frac{\varepsilon}{2}, t - \frac{\varepsilon}{2}\right) \exp(k2\pi\beta t) dt \tag{4}$$

The results of Fourier transform for $H^{\beta}(\varepsilon)$ are as follows:

$$D^{\beta}(f) = \int\limits_{\infty}^{\infty} H^{\beta}(\varepsilon) \exp(-k2\pi\beta t) d\varepsilon \qquad (5)$$

In formula (5), $D^{\beta}(f)$ represents the cyclic spectral density function of cyclic autocorrelation function [13]. The modulation type of mobile communication signal is identified by cyclic spectral density function. Then, the mobile communication signal features are classified by the classifier.

2.6 Classification and Recognition Using Mobile Communication Signal Feature Classifier

Firstly, the intercepted signal individuals in the communication countermeasure environment are extracted from the multi angle feature analysis. Firstly, through preprocessing, the signal sequence (interval) which is stable and suitable for feature analysis is extracted. For the obtained signal analysis interval, the frequency domain analysis method can be used to extract the frequency domain characteristics of the individual signal. However, because the small characteristics of the communication signal radiation source are more irregular non-stationary, nonlinear and non Gaussian, it is difficult to use the traditional signal processing method for the characteristics of spurious components and parasitic modulation, while small wave analysis and time-frequency distribution are very good tools for non-stationary signal analysis. Therefore, the transient characteristics are extracted by fractal method in time domain, and then a small sample classifier is designed. Aiming at the results of small sample condition and statistical learning theory, single classifier and combined classifier are mainly used to improve the classification ability of signal feature vector under small sample condition. The information of different classifiers is fused to classify the characteristics of mobile communication signals. The recognition of L modulation pattern can be regarded as a pattern recognition problem of L category $\phi_i(i = 1, 2, \cdots, L)$. Suppose that each class has N_i M dimensional samples indicating the class, φ_i is the mean value of ϕ_i, $\sum$ is the covariance matrix of all class samples, the sample set composed of all class ϕ_i samples is U_i, B_i^l is the $l(l \leq N_i)$ sample in U_i, B is the sample with unknown style, and the classifier settings are as follows. If the maximum value of the sample in U_i in the $l(l = 1, 2, \cdots, M)$ coordinate is max il and the minimum value is min il, then there is a "super box" in M dimensional space composed of [min $i1$, max $i1$] $\times$ [min $i2$, max $i2$] $\times \cdots$ [min iM, max iM] $\in R^M$. It defines the possible areas of class ϕ_i samples in the feature space. However, in many cases (for example, the samples are arranged along the volume diagonal of the "super box"), the "super box" obtained by the above method will include large adjacent areas while surrounding the sample set. We use B_i to transform the coordinate system of the space, and find the "super box" in the transformed space to make it "tight" as much as possible. The "super box" classifier will contain L "super boxes" corresponding to each type. When the unknown sample B falls into the "super box" surrounding the class ϕ_i sample, it is considered that it may belong to the ϕ_i class. If no "super box" is included, it will be rejected. It is sent to the nearest neighbor classifier. Suppose that the Mahalanobis distance from B to ϕ_i is:

$$d_i(B) = \sqrt{(B - \varphi_i)^T \sum\nolimits^{-1} (B - \varphi_i)} \tag{6}$$

The mahalanobian distance between Class B and class ϕ_i is calculated by formula (6). The Markov distance classifier uses the minimum distance criterion to judge the unknown sample category. The Markov distance is used to improve the reliability of the recognition results. According to the recognition results of all training samples by using the criterion of Markov distance and minimum distance, a maximum E_i of Markov distance is calculated for each class ϕ_i, so that the samples identified as ϕ_i in the training set and the Markov distance of ϕ_i is less than 95% of the samples in the subset of E_i belong to class ϕ_i. The Markov distance classifier uses E_i to verify the recognition result ϕ_i, and when $d_i(B) \geq E_i$, it refuses to recognize the sample. In conclusion, the feature vector B with length 20 extracted from the signal segment with unknown modulation style is first sent to the "super box" classifier. If the sample is not rejected, it is sent to the nearest neighbor classifier. Nearest neighbor classifier either refuses B, or identifies it as an ϕ_i class. If the latter situation occurs, the result of the "super box" classifier should be referred to. If B is not included in the "super box" of corresponding ϕ_i, B is rejected, otherwise it will be sent to the Markov distance classifier. The Markov distance classifier will calculate the Markov distance from class ϕ_i. if it is greater than or equal to the threshold E_i, B will be rejected, otherwise, the class of B is determined as ϕ_i. After selecting the modulation styles to be identified, collect the communication signals (mobile communication weak signals) belonging to these styles. After preprocessing, a series of signal segments are obtained, each segment contains 4000 sampling points, and then, according to the feature extraction method in the feature library, each signal segment is extracted, and a 20-dimensional feature vector is formed and the corresponding modulation category is given. The vector obtained is composed of the original training sample set. The "super box" classifier is established according to the training sample set. The nearest neighbor classifier and Markov distance classifier can recognize the modulation pattern within the set range, and realize the weak signal recognition of mobile communication based on information fusion.

3   Experiment

The proposed method based on information fusion for mobile communication weak signal acquisition and recognition is compared with the traditional acquisition and recognition method 1 and the traditional acquisition and recognition method 2 to compare the recognition accuracy of different methods in the process of acquisition and recognition.

3.1   Experimental Process

In order to ensure the smooth progress of the experiment, TDOA measurement algorithm is used to test the accuracy of signal location. The results are shown in Fig. 5.

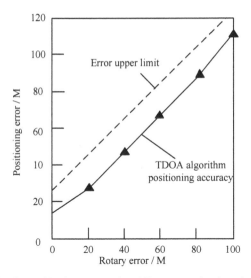

Fig. 5. Positioning result of mobile communication signal

It can be seen from Fig. 5 that the positioning results of mobile communication signals tested by TDOA measurement algorithm are within the allowable error range, which meets the experimental requirements. The experiment is further carried out by using the method of weak signal acquisition and recognition based on information fusion. Firstly, the intercepted radiation source signals are classified and classified. Then, the instantaneous and transient features such as the starting point feature and time-frequency image feature are used respectively. The combined classifier is used to recognize and classify the intercepted radiation source signals. It is assumed that the signals from 10 different sources intercepted are continuous phase, the sampling frequency is 600 MHz and the sampling signal is 10 ms. When the SNR of intercepted signal is 10 dB, the intercepted radiation source signal is identified by using the above characteristics, and the recognition accuracy results are obtained, and compared with the traditional method 1 and traditional method 2.

3.2 Experimental Results and Analysis

The recognition results obtained by using the proposed weak signal acquisition and recognition method of mobile communication based on information fusion, traditional method 1 and traditional method 2 are shown in Fig. 6.

It can be seen from Fig. 6 that the recognition accuracy of traditional method 1 is about 75%, that of traditional method 2 is about 65%, and that of the proposed method is about 99%. Through the analysis, it is found that the proposed method is based on information fusion, through the combination of classifiers to fuse the information and classify the modulation types, the recognition accuracy is significantly improved, which is about 24% higher than the traditional method 1 and 34% higher than the traditional method 2.

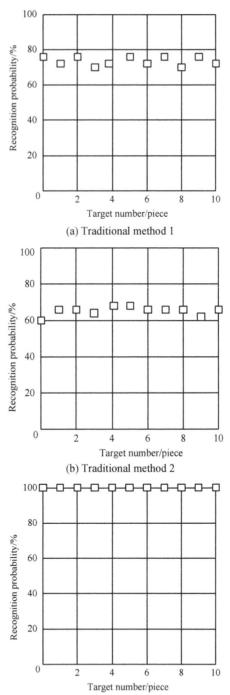

(a) Traditional method 1

(b) Traditional method 2

(c) A weak signal acquisition and recognition method for mobile communication is proposed

Fig. 6. Comparison results of recognition accuracy

4 Concluding Remarks

Aiming at the low recognition accuracy in the acquisition and recognition of weak signal in mobile communication, a method of weak signal acquisition and recognition based on information fusion is designed. Through the design and comparison experiment, the proposed method based on information fusion has higher recognition accuracy. It is hoped that the method can provide reference for the research of weak signal acquisition and recognition in mobile communication.

References

1. Li, K., Zhao, H., Wang, Y.: Narrowband power line carrier communication signal recognition algorithms. Commun. Technol. **52**(05), 1041–1048 (2019)
2. Li, C., Yang, J., Huang, Z., et al.: Automatic modulation recognition of communication signal based on deep learning. Space Electron. Technol. **16**(01), 49–54,74 (2019)
3. Li, J., Zhang, Z.: Improved algorithm for pattern matching signal recognition based on clustering algorithm. J. Shanghai Dianji Univ. **22**(01), 46–49 (2019)
4. Liu, P., Liu, T., Wang, S., et al.: Bearing fault diagnosis method based on information fusion and fast ICA. J. Vib. Shock **39**(03), 250–259 (2020)
5. Liu, Y., Xie, T., Xu, Y., et al.: Condition assessment of thin-walled robot bearing based on muliti-sensor data fusion. J. Mech. Strength **41**(05), 1042–1047 (2019)
6. Qiu, Z., Zhao, Z., Zhan, J.: Communication Signal Recognition Algorithm Based on Feature Extraction, vol. 40, no. 03, pp. 1–6 (2020)
7. Li, R., Hou, K., Leng, P., et al.: Modulation identification of radar and communication signals based on convolutional neural network. Aerosp. Electron. Warfare **36**(02), 7–13 (2020)
8. Xu, M., Hou, J., Wu, P., et al.: Convolutional neural networks based on time-frequency characteristics for modulation classification. Comput. Sci. **47**(02), 175–179 (2020)
9. Yao, Y., Peng, H.: Automation modulation recognition of the communication signals based on deep learning. Appl. Electron. Tech. **45**(02), 12–15 (2019)
10. Xie, B.: Research on automatic modulation recognition system for nonconforming communication signals. Electron. Des. Eng. **27**(03), 150–154 (2019)
11. Liu, S., Liu, G., Zhou, H.: A robust parallel object tracking method for illumination variations. Mob. Netw. Appl. **24**(1), 5–17 (2018). https://doi.org/10.1007/s11036-018-1134-8
12. Liu, S., Fu, W., He, L., Zhou, J., Ma, M.: Distribution of primary additional errors in fractal encoding method. Multimedia Tools Appl. **76**(4), 5787–5802 (2014). https://doi.org/10.1007/s11042-014-2408-1
13. Liu, S., He, T., Dai, J.: A survey of CRF algorithm based knowledge extraction of elementary mathematics in Chinese. Mob. Netw. Appl. (2021). https://doi.org/10.1007/s11036-020-01725-x

Design of Network Public Opinion Information Intelligent Retrieval System Based on Wireless Network Technology

Quan-wei Sheng$^{(\boxtimes)}$

Changsha Medical University, Changsha 410219, China

Abstract. Wireless sensor routing protocol affects the retrieval work of the system. The traditional intelligent retrieval system of network public opinion information is difficult to obtain the event probability due to the selected technology, which leads to the weak ability of the system to retrieve massive data. This paper designs an intelligent retrieval system of network public opinion information based on wireless network technology. In terms of hardware, Lucene search engine architecture and rs323 bus circuit are designed; in terms of software design, network public opinion information similarity calculation model is designed, wireless sensor routing protocol is set based on wireless network technology, and network public opinion information intelligent retrieval logic is established. In the experiment, the amount of public opinion information to be retrieved is 100, 5000 and 100000 respectively. In the same test period, when the amount of network public opinion information to be retrieved is large, the retrieval system based on wireless network technology can obtain the target information, while the amount of target information obtained by traditional system is far less than expected.

Keywords: Wireless network technology · Network public opinion information · Intelligence · Retrieval system

1 Introduction

In the face of the rapid growth of Internet public opinion information, some scholars refer to the content of [1] and design a Python based retrieval system. However, according to the practical application of other systems, it is found that the correspondence between the retrieval algorithms and models of most traditional systems is weak. Therefore, taking the Python based retrieval system as a reference, this paper studies the intelligent retrieval system of network public opinion information based on wireless network technology [1]. The so-called wireless network refers to the network that can realize the interconnection of various communication devices without wiring. Wireless network technology covers a wide range, including not only global voice and data networks that allow users to establish long-distance wireless connections, but also infrared and RF technologies that optimize short-distance wireless connections. Literature [2] Designing a public opinion information search system for Baidu search engine, relying

S. Liu and X. Ma (Eds.): ADHIP 2021, LNICST 417, pp. 74–86, 2022.
https://doi.org/10.1007/978-3-030-94554-1_7

on Baidu search engine, mainly researching crawler, search engine, text information mining and other technologies, should develop a complete solution for the application needs of network public opinion monitoring work. It can greatly improve the speed of searching public opinion information on Baidu website and Baidu Tieba and the accuracy of obtaining information. However, the target information acquisition volume of the system is far from expected.

Based on this, this article starts with wireless network technology and designs an intelligent retrieval system for network public opinion information that is more in line with expectations.

2 Hardware Design of Network Public Opinion Information Intelligent Retrieval System

2.1 Design Lucene Search Engine Architecture

Lucene, as a full-text search engine architecture, needs to be developed twice on the basis of Lucene if it is to be fully applied to intelligent retrieval of network public opinion information. Lucene is known to provide the analysis interface, index engine, storage management and query engine to the retrieval system. According to the above basic modules, Lucene core logic architecture is constructed, as shown in Fig. 1 below.

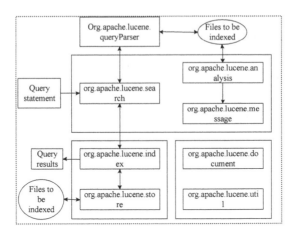

Fig. 1. Lucene search engine architecture

Through the basic components of the logical structure, we can see that there is a calling relationship between the various modules when Lucene is running. The interface module uses the parser to analyze the index text, and writes the analyzed data into the index file through the infrastructure common module. When users query, the query statements are sent to the index core module through the interface module, and the query results are read through the interface module Take the data in the index file, get the retrieval results, and execute the feedback operation [3].

2.2 Design rs323 Bus Circuit

When the system receives network public opinion information, there are synchronous transmission data and asynchronous transmission data. It needs to design rs323 bus circuit to ensure that the system can receive all kinds of information. The rs323 interface has 9 pins or 25 pins. The rs323 with 25 pins is selected in the design. Under the transmission rates of 50, 100, 600, 1200, 2400, 4800, 9600 and 19200 baud respectively, it is tested whether the rs323 allows 2500pf capacitive load. On the basis of expanding the communication distance, it is ensured that the hardware has the ability to suppress noise and other interference information. The bus design is shown in Fig. 2 below Circuit [4].

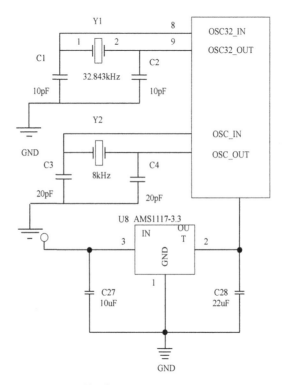

Fig. 2. Bus circuit design

Considering that the system is connected with the computer, MAX3232 chip is used as the transceiver chip of rs323 bus circuit. The MAX3232 chip has two channels, and it has a dedicated low-voltage transmitter output stage. Through two receivers and two drivers, it can realize the data receiving and sending with the data rate of 240kps, and ensure the efficiency of the retrieval system in processing massive network public opinion data. So far, the hardware design of the library and information resources integration system is completed.

3 Software Design of Network Public Opinion Information Intelligent Retrieval System Based on Wireless Network Technology

3.1 Design the Similarity Calculation Model of Public Opinion Information on the Network

When retrieving network public opinion information, we need to perform the retrieval task according to semantic similarity. The common semantic feature of two concepts is the similarity between semantics. This index can be represented by $sim(x, y)$, and x and y represent the conceptual elements between the two information. In the form of calculation, similarity is required to meet the following calculation requirements: make similarity index between, so $sim(x, y) \in [0, 1]$. When the two information are completely similar, $sim(x, y) = 1$; when the two information is completely different, then $sim(x, y) = 0$. The similarity index is symmetrical. Therefore, according to the above calculation requirements, combined with the influence of retrieval semantic distance on similarity index, according to the decreasing function relationship, assuming the degree of change between similarity index and semantic distance is $k^{-dis(x,y)}$, the influence of network public opinion information on similarity index can be described by the following formula:

$$f(x, y) = \frac{C(x) + C(y)}{|C(x) - C(y)| + 1} \tag{1}$$

In the formula, $C(x)$ and $C(y)$ represent the content level of information x and y respectively. When $f(x, y) > 1$, the range of similarity index is not met. Therefore, divide f by 2 times the depth of layer d, so that the influence of hierarchy factors on conceptual similarity is kept within the range, and a new calculation result is obtained:

$$f(x, y) = \frac{C(x) + C(y)}{2d[|C(x) - C(y)| + 1]} \tag{2}$$

According to the above calculation formula, with the increase of concept level, the similarity between them is also increasing. At the same time, with the expansion of the difference between the two concept levels, the similarity between retrieval elements is gradually decreasing [5]. Therefore, the change degree between the two elements is obtained

$$dis(x, y) = C(x) + C(y) - 2F(f(x, y)) \tag{3}$$

In the formula: $dis(x, y)$ represents the semantic distance between two concepts; $F(f(x, y))$ represents the level of the minimum classification of two concepts. By quantifying the influence of semantic distance and hierarchical factors on the similarity index, we get the following results:

$$Z_1(x, y) = \frac{k^{-dis(x,y)} \cdot q_{x,y}}{2d \cdot (dis(x, y) + 1)} \tag{4}$$

In the formula: $q_{x,y}$ is the sum of levels $C(x) + C(y)$. At the same time, the logarithmic function is introduced and the nonlinear function is used to evaluate the semantic similarity:

$$Z_2(x, y) = \log_2\left(1 + \frac{|s(x) \cap s(y)|}{|s(x) \cup s(y)|}\right) \tag{5}$$

In order to avoid infinite values of x and y, 1 is added to the true part of logarithmic function to make its value 0. Through the above calculation process, the semantic similarity calculation model is designed, and parameters w_1 and w_2 are added to adjust the similarity index:

$$sim(x, y) = \begin{cases} w_1 Z_1 + w_2 Z_2 & x \neq y \\ 1 & x \neq y \end{cases} \tag{6}$$

According to the above calculation formula, the design of network public opinion information similarity calculation model is realized [6].

3.2 Wireless Sensor Routing Protocol Based on Wireless Network Technology

When wireless sensor networks transmit data, there are a lot of uncertainties among the cloud security situation elements, so wireless network technology is used to determine the uncertainty of cloud security situation elements. Wireless network technology uses credibility function H to calculate the uncertainty relationship between various elements of cloud security situation:

$$H(U, V) = W(U, V) - Y(U, V) \tag{7}$$

In the formula: U represents the result caused by uncertain reasons; V represents the uncertain reasons between cloud security elements; W and Y represent the growth of trust and distrust respectively. The parameter calculation equations of the two functions are as follows:

$$W(U, V) = \begin{cases} 1, & ifP(U) = 1 \\ \dfrac{\max\left(P\left(\frac{U}{V}\right), P(U) - P(V)\right)}{P(U)} \end{cases} \tag{8}$$

$$Y(U, V) = \begin{cases} 1, & ifP(U) = 0 \\ -\dfrac{\min\left(P\left(\frac{U}{V}\right), P(U) - P(V)\right)}{P(U)} \end{cases} \tag{9}$$

In the formula, $P(U)$ is the prior probability of the result; $P\left(\frac{U}{V}\right)$ is the conditional probability of X when the uncertainty occurs. When n different security events cause a security event, if the event is $n_1, n_2, \ldots, n_n$, then n credibility can be obtained by wireless network technology. The uncertainty relationship between cloud security situation elements can be measured by using the results obtained. The comprehensive trust value is calculated according to the direct trust value, indirect trust value and energy trust value. According to the actual number of packets received and sent, the direct trust value between nodes is predicted:

$$X^{now(1)} = \mu(\omega_1 S_1 + \omega_2 S_2)^{pre} + (1 - \mu)(S_1 + S_2)^{now} \tag{10}$$

In the formula, μ represents the weight of historical trust value; $(1 - \mu)$ represents the weight of current trust value; ω_1 and ω_2 represent different volatile factors; S_1 and S_2 respectively table send and receive packets, accounting for the proportion of total packets. Direct trust value is used to calculate indirect trust value between different nodes. Suppose $A_n = (A_1, A_2, \ldots, A_m)$ is a set of neighbor nodes, where m is the actual number of public nodes, so the calculation result of indirect trust value is as follows:

$$X^{now(2)} = \frac{1}{n} \sum_{a \in A_n}^{n} \left(X_{ia}^{now(1)} \cdot X_{aj}^{now(1)} \right) \tag{11}$$

In the formula, $X_{ia}^{now(1)}$ and $X_{aj}^{now(1)}$ represent the direct trust values between node i and node a, node a and node j respectively. Calculate the energy trust value, the formula is:

$$E_j = \frac{S_1 E_j}{E_0} \tag{12}$$

In the formula, $S_1 E_j$ is the residual energy value; E_0 is the initial energy value. The comprehensive trust value is calculated according to the above three formulas:

$$X = H_1 X^{now(1)} + H_2 X^{now(2)} + H_3 E_j \tag{13}$$

In the formula: H_1, H_2 and H_3 are the prediction weights of wireless network technology. According to the basic definition of compressed sensing theory, if the dimension of measurement matrix is $M \times N$, then the column vector of $N \times 1$ dimension can be represented by b, then:

$$y = \Phi b = \left\{ \begin{array}{cccc} \alpha_{11} & \alpha_{12} & \cdots & \alpha_{1N} \\ \alpha_{21} & \alpha_{22} & \cdots & \alpha_{2N} \\ \vdots & \vdots & \ddots & \vdots \\ \alpha_{M1} & \alpha_{M2} & \cdots & \alpha_{MN} \end{array} \right\} \left\{ \begin{array}{c} b_1 \\ b_2 \\ \vdots \\ b_N \end{array} \right\} \tag{14}$$

$$y_i = \sum_{j=1}^{N} \alpha_{ij} b_j \tag{15}$$

In the formula: Φ represents sparse matrix; α_{MN} represents transformation coefficient in $M \times N$ dimension; α_{ij} represents transformation quantity in row i and column j. By multiplying and adding N nodes b_j, M measurement results are obtained. The row elements of the sparse matrix are used as the transmission path projection of WSN, and the matrix is divided into M sub regions:

$$\Phi = \begin{Bmatrix} \alpha_{11} & 0 & \alpha_{12} & 0 & \cdots & 0 & \alpha_{1\frac{N}{M}} & 0 \\ 0 & \alpha_{21} & 0 & 0 & \cdots \alpha_{2\frac{N}{M}} & 0 & 0 \\ 0 & 0 & 0 & \alpha_{31} & \cdots & 0 & 0 & 0 \\ 0 & 0 & 0 & 0 & \cdots & 0 & 0 & 0 \\ \vdots & \vdots & \vdots & \vdots & \ddots & \vdots & \vdots & \vdots \\ 0 & 0 & 0 & 0 & \cdots & 0 & 0 & \alpha_{M\frac{N}{M}} \end{Bmatrix} = \begin{Bmatrix} \Phi_1 \\ \Phi_2 \\ \Phi_3 \\ \Phi_4 \\ \vdots \\ \Phi_M \end{Bmatrix} \qquad (16)$$

According to the data in the formula, the matrix contains multiple regions and many 0 elements [7, 8]. The non-zero elements in the block matrix are multiplied by the corresponding data, and then the data is transmitted to reduce the energy consumption. According to the above process, ant colony algorithm divides cluster tree wireless sensor network into zones, and executes the routing protocol according to the established search method.

3.3 Establishing Intelligent Retrieval Logic of Network Public Opinion Information

In order to implement the search routing protocol designed above, it is necessary to formulate a verification search logic matching with it. Based on the characteristics of character type data, the verification retrieval logic was developed by stage selection. The occurrence frequency of known characters has certain deviation. For example, the occurrence probability of data in ordinary text may be greater than the occurrence probability of data. Therefore, after the index is generated, the character value is mapped to the corresponding bucket number according to the frequency hiding operation of region division, and then the order of all bucket numbers is straightened out to extract the frequency information characteristics of hidden characters [9]. According to the data obtained in the research process, the frequency order of English letters is given, as shown in Table 1 below.

Table 1. English alphabet frequency sorting table/%

Letter	Frequency	Letter	Frequency	Letter	Frequency
A	8.17	J	0.15	S	6.33
B	1.49	K	0.77	T	9.06
C	2.78	L	4.03	U	2.76
D	4.25	M	2.41	V	0.98
E	12.7	N	6.75	W	2.36
F	2.23	O	7.50	X	0.15

(*continued*)

Table 1. (*continued*)

Letter	Frequency	Letter	Frequency	Letter	Frequency
G	2.02	P	1.93	Y	1.97
H	6.10	Q	0.10	Z	0.07
I	6.97	R	5.99	–	–

In the process of hiding frequency operation, the characters are assigned to each bucket according to the frequency of characters, and then converted into the corresponding bucket number, so that users can infer the corresponding relationship between the bucket number and characters according to the number of buckets and character frequency. Combined with the letter frequency in Table 1, the basic verification process of the verification algorithm is designed. In the first step of verification, input the letter frequency and the number of buckets in Table 1, and calculate the frequency value that each bucket can hold

$$r = \frac{1}{n}(1 \pm 10\%) \tag{17}$$

In the formula: n represents the number of buckets; $(1 \pm 10\%)$ represents the floating range of splitting the same character into different buckets in the subsequent allocation process. The second step is to ensure that there is a corresponding mapping relationship between characters and bucket numbers according to the following restrictions: first, assign characters to each bucket in order until the end of allocation; second, when a character with the highest frequency cannot be placed in the remaining space in the bucket, it is proved that the bucket can end the allocation task and start to allocate the next bucket. In the third step, hash function is used to capture the random value of bucket number. The verification process of the function is controlled by the following formula:

$$hash(*) = s\varphi key mod\, 1(0, 1) \tag{18}$$

In the formula: s represents the hash table size; $hash(*)$ represents the downward rounding of the expression by the multiplication hash function, and the value range of parameter φ is $(0, 1)$; $\varphi key mod\, 1$ represents the fractional part of the numerical product of key multiplied by $(0, 1)$ interval [10]. In order to make the verification mode and encryption mode have a matching relationship, the default parameter $\varphi = 0.618$, through the above calculation and analysis process, the network public opinion information intelligent retrieval logic is established, and the network public opinion information intelligent retrieval system based on wireless network technology is realized.

4 Experimental Study

4.1 Purpose of the Experiment

Taking the retrieval system designed in this paper as the test object, find out the parts of the system that have conflicts, and then adjust for such problems to ensure the use of the system in the actual work. Test whether the hardware of the system can connect stably and whether the redesigned hardware is compatible with the software. Finally, the system is applied to the actual working environment. Under different test conditions, the differences between the designed retrieval system and the traditional system are compared.

4.2 System Stability and Reliability Test

In order to ensure that the designed system has more powerful retrieval function and test the stability of the system, the test results of the system stability are shown in Fig. 3 below.

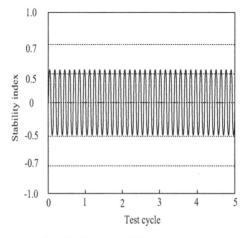

Fig. 3. System stability test results

According to the distribution state of the curve in the figure, the stability index test results of the system fluctuate evenly between −0.5–0.5, and there is no big difference between the stability indexes in each test stage. It can be seen that the designed system meets the stability test requirements. Set up system test data sets, use different data sets to record different scales of network public opinion information, test the retrieval efficiency of the system. The following Table 2 is the setting table for data sets of different sizes. The size of each small field is required to be between 2.5 and 3.0 mb.

Table 2. System test data set

Serial number	Number of files	Total dataset size
1	100	234.35 MB
2	500	1.86 GB
3	1000	3.43 GB
4	5000	17.75 GB
5	10000	38.43 GB

The design of the retrieval system as the experimental group, set up the general retrieval target of the retrieval system. The data in Table 3 below are the comparison results between the expected value and the actual retrieval time of the system.

Table 3. Comparison of retrieval time of Internet public opinion information

Number of files	Expected value	Actual value of the designed system
100	115 ms	115.7 ms
500	630 ms	663.2 ms
1000	1345 ms	1399.8 ms
5000	5750 ms	5884.1 ms
10000	13280 ms	13497.5 ms

In order to further illustrate the above comparison test results, the relationship curve between the actual value and the expected value of the system is drawn, as shown in Fig. 4 below.

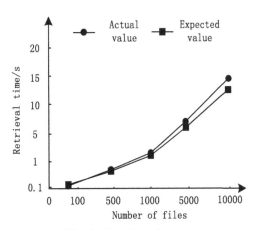

Fig. 4. Search result curve

According to the test results shown in Fig. 4, under the control of wireless network technology, the designed system achieves the retrieval efficiency which is highly similar to the expected value.

4.3 System Retrieval Ability Test

Taking the designed retrieval system as the experimental group, the traditional system 1 and the traditional system 2 designed by the traditional method as the control group A and the control group B respectively, the three groups of systems are applied to the actual network public opinion information retrieval work. The experiment set up three retrieval information quantity, which were 100, 5000 and 100000. Test the ability of the system to retrieve network public opinion information under different volumes, and the results are shown in Fig. 5 below.

According to the test results shown in Fig. 5, when the number of network public opinion information to be retrieved is 100, the three groups of systems have better retrieval effect; when the number of network public opinion information to be retrieved is 5000, the amount of retrieval data obtained by the two traditional systems in the same test period is less than 4000, while the designed system can retrieve 5000 public opinion information; when the number of network public opinion information to be retrieved is 5000, the number of retrieval data obtained by the two traditional systems is less than 4000 When the number of public opinion information is 100000, the retrieval results of the two traditional systems drop sharply again. In the same test period, the public opinion information obtained by the two systems is less than 80000. However, the system designed in this paper obtains 100000 complete network public opinion information at the same time. According to the above test results, the system designed in this paper also has excellent data retrieval effect in the face of huge network public opinion information.

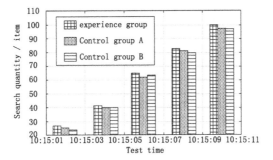

(a) Test results of 100 pieces of public opinion information

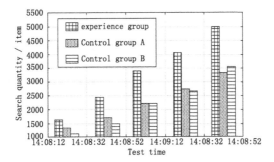

(b) Test results of 5000 pieces of public opinion information

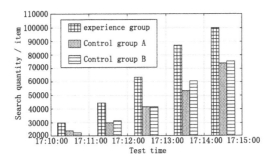

(c) Test results of 100000 pieces of public opinion information

Fig. 5. System retrieval effect test

5   Concluding Remarks

This research is based on the shortcomings of the traditional retrieval system, through the wireless network technology to make up for the shortcomings of the traditional system, and further obtain more satisfactory retrieval results. But combined with the design of the system software, the calculation of software design is more complex, prone to system crash. In the future, we can design a set of intelligent control algorithm to realize the operation of the software with a simpler calculation process, and provide more accurate test results for the intelligent retrieval of network public opinion information.

References

1. Xin, L.: From assessment to establishment: construction and development of cases assess catalog from a utilization perspective. J. Sichuan Univ. (Soc. Sci. Edn.) **02**, 161–173 (2020)
2. Tang, G., Zhao, C., Li, J., et al.: Research on public opinion information search system relying on Baidu search Engine. Comput. Digit. Eng. **047**(011), 2785–2790 (2019)
3. Xu, M., Zhang, X.: A small footprint keyword spotting system for voice control. J. Signal Process. **36**(06), 879–884 (2020)
4. Wu, D., Tao, S., Yan, P., et al.: Content-based instrument image retrieval system. Electron. Meas. Technol. **42**(11), 112–117 (2019)
5. Gong, P., Lyu, C., Gong, Y.: Research on object storage and retrieval system for space application data. Appl. Res. Comput. **36**(03), 833–837 (2019)
6. Li, G., Zhu, T., Liu, B., et al.: Optimization of the multi-media information retrieval system of digital library for big data. Inf. Sci. **37**(02), 115–119 (2019)
7. Yao, J., Cheng, Y.: Design and implementation of video retrieval system based on deep learning. Comput. Meas. Control **27**(06), 231–235 (2019)
8. Liu, S., He, T., Dai, J.: A survey of CRF algorithm based knowledge extraction of elementary mathematics in Chinese. Mob. Netw. Appl. (2021). https://doi.org/10.1007/s11036-020-017 25-x
9. Liu, S., Pan, Z., Cheng, X.: A novel fast fractal image compression method based on distance clustering in high dimensional sphere surface. Fractals **25**(4), 1740004 (2017)
10. Liu, S., Fu, W., He, L., Zhou, J., Ma, M.: Distribution of primary additional errors in fractal encoding method. Multimedia Tools Appl. **76**(4), 5787–5802 (2014). https://doi.org/10.1007/ s11042-014-2408-1

Remote Speed Control Method of Brushless DC Motor for Wheeled Robot Based on Embedded Wireless Communication Technology

Shao-yong Cao and Wei-Jie Tang[✉]

School of Industrial Automation,
Zhuhai College of Bejing Institute of Technology, Zhuhai 519085, China

Abstract. According to the functional requirements of mobile robot, a remote speed control method of Brushless DC motor for wheeled robot based on embedded wireless communication technology is designed. Compared with the traditional microcontroller, using high-performance DSP2812 as the CPU of the robot simplifies the peripheral circuit and improves the high integration and stability of the method. At the same time, embedded wireless communication technology and fuzzy PID control algorithm are used to make up for the defect of single control algorithm and improve the dynamic performance of motor. The experimental results show that the remote speed control method of Brushless DC motor for wheeled robot based on embedded wireless communication technology improves the speed and control accuracy of motor, and improves the motion performance of robot.

Keywords: Embedded · Wireless communication · Wheeled robot · DC motor · Speed control

1 Introduction

The DC motor of wheeled robot has the characteristics of simple structure, strong stability, wide speed range and large output torque. It is widely used in drilling exploration, metallurgical processing, medical equipment, intelligent robot and other fields, and has achieved good results. For DC motor, accurate speed regulation is particularly important. In the early days, the DC motor speed control method of wheeled robot was generally realized by the circuit built by the simulator, which had limited speed range and low precision. After the birth of single chip microcomputer, there are many DC motor speed control methods with single chip microcomputer as the control core [1]. These methods take the speed of the motor as the controlled object, take the speed and current of the motor as the feedback of the method, and use PID and other conventional control methods to realize the accurate speed regulation of the DC motor. However, with the increase of the complexity of industrial production process, the traditional control method has been unable to adapt, so more complex control algorithm must be adopted [2]. Because of the limitation of memory space of MCU, these advanced algorithms can't work. Therefore,

© ICST Institute for Computer Sciences, Social Informatics and Telecommunications Engineering 2022
Published by Springer Nature Switzerland AG 2022. All Rights Reserved
S. Liu and X. Ma (Eds.): ADHIP 2021, LNICST 417, pp. 87–99, 2022.
https://doi.org/10.1007/978-3-030-94554-1_8

the DC motor speed control method based on arm, DSP and embedded wireless communication emerges as the times require, which provides higher control precision and wider speed range. The emergence of embedded wireless communication technology solves this problem and provides a new idea for precise speed regulation of DC motor. The traditional PLL circuit is basically analog-to-digital hybrid circuit [3]. Although the steady-state performance is good, the dynamic performance and anti-interference ability are poor. Based on the above analysis, in order to ensure that the DC motor speed control method can provide better speed control performance, and ensure that the method can run more complex control algorithm, this paper designs a remote speed control method of Brushless DC motor for wheeled robot based on embedded wireless communication technology [4]. Firstly, the brushless DC motor of wheeled robot is debugged, and then the parameters of the brushless DC motor are optimized. By controlling two quantities of current, the speed regulation of the brushless DC motor for wheeled robot is realized.Embedded wireless communication is used to realize high precision all digital phase locked loop technology and PWM motor drive signal. DSP is used to generate high-precision high-speed reference signal to form a set of high-performance DC motor speed control method.

2   Long-Distance Speed Control Method of Brushless DC Motor for Wheeled Robot

2.1   Transfer Function of Long-Distance Speed Regulation of DC Motor

In the speed regulation method of DC motor of wheeled robot, most of the motors are separately excited DC motors, and their equivalent circuits are shown as follows (Fig. 1):

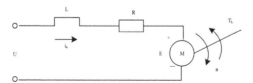

Fig. 1. Equivalent control circuit of DC motor

In most cases, the debugging method of Brushless DC motor for wheeled robot adopts variable voltage to realize the speed regulation of DC motor, so the input of the method is armature voltage m, and the output of the method is motor speed n. It can be seen from the figure that the armature voltage u is equal to the sum of the motor armature back electromotive force e, the resistance R voltage and the inductance l voltage, namely:

$$e + i_a R + L\frac{di_a}{dt} = u \tag{1}$$

Among them, the motor armature back electromotive force e can be calculated by the following formula:

$$e = C_e n/u \tag{2}$$

Where g is the electromotive force constant of the motor. Based on this, the physical equation of motor can be obtained as follows:

$$C_e n + i_a R + Le \frac{di_a}{dt} = \Delta u \tag{3}$$

In addition, under the ideal no-load condition, the mechanical motion equation of the brushless DC motor for wheeled robot is known as:

$$T_e - T_L = \frac{GD^2}{375} \frac{dn}{d\Delta ut} \tag{4}$$

Where $T = Cn$ is electromagnetic torque; TL is motor torque; GD2 is flywheel inertia, then.

$$T_i T_m \frac{d^2 n}{dt^2} + T_m \frac{dn}{dt} + n = \frac{u}{C_e} \tag{5}$$

In which:

$$T_i = \frac{L}{R} \tag{6}$$

$$T_m = \frac{GD^2}{375} \frac{R_d}{C_m C_e} \tag{7}$$

After further optimization, the transfer function of the motor mathematical model can be obtained:

$$W(s) = \frac{X_c}{X_r} = \frac{1/C_e}{T_d T_m s^2 + T_m s + 1} \tag{8}$$

Based on the above algorithm, the brushless DC motor for wheeled robot is calculated at a long distance to ensure the accuracy of control parameters.

2.2 Parameter Optimization of Brushless DC Motor for Wheeled Robot

Several optimization measures adopted in the structural design of the brushless DC motor for wheeled robots are further discussed, and some advantages of the brushless DC motor in operation are mentioned, and the performance of the motor with this structure in control is the focus of our further research [5]. In order to achieve good control of brushless DC motor for disc coreless wheeled robot, it is necessary to make a more in-depth theoretical discussion on its two main characteristics, coreless structure and array structure, which are different from ordinary disc motor, in order to provide a clear and reliable theoretical basis for the control application of this kind of motor, and standardize the basic design data of the motor based on this, as shown in the following Table 1:

Table 1. Basic parameters of brushless DC motor for wheeled robot

Design data	
Power (W)	200
Speed (R / min)	375
Phase number	3 (Y connection)
Input frequency (Hz)	50
Number of stator (module)	1
Polar logarithm	8
Number of coils (3 phases)	48
Turns per phase	640
Wire diameter (mm)	1 × 0.6
Inner diameter of permanent magnet (CM)	10.4
Outer diameter of permanent magnet (CM)	18
Air gap (CM)	0.8
Winding thickness (CM)	0.6
Thickness of permanent magnet (CM)	1.2
Efficiency (%)	80
power factor	0.99
Cooling system	Self cooling
Insulation class	F

For the brushless DC motor for wheeled robot, its permanent magnet rotor generates a constant magnetic field. When three-phase symmetrical current is applied to its three-phase symmetrical winding, a rotating magnetic field will be generated in the air gap [6]. According to the unified theory of motor, the two magnetic fields must remain relatively static to produce stable electromagnetic torque and drive the motor to rotate at synchronous speed [7]. Since the speed (frequency) of the stator magnetomotive force of the self-control synchronous motor is controlled by the rotor speed, the quantity that can be controlled is the amplitude and phase of the stator current. By controlling the two quantities of the current, the purpose of speed regulation of the brushless DC motor for wheeled robot can be achieved [8]. Based on this, the vector of the synthetic magnetomotive force of the stator is calculated

$$\overline{F_j^s} = N_s \overline{i_j^s} = N_s \frac{3}{2} I_m e^{j(\lambda_0 + \theta)} = F^s e^{j(\lambda_0 + \theta)} \tag{9}$$

Where N is the number of turns of stator coil and the amplitude of stator magneto-motive force, I is the maximum value of phase current. The purpose of vector control is to improve the performance of torque control, and the final implementation is still to control the stator current (AC). Because the physical quantities (voltage, current, electromotive force and magnetomotive force) on the stator side are all alternating current, and their space vectors rotate at synchronous speed in space, it is not convenient to adjust, control and calculate them. Therefore, coordinate transformation is needed to transform the physical quantities from static coordinate system to synchronous rotating coordinate system [9]. Use the information of numbers to record the shape of English or Chinese characters. See the table for the pin description of system operation (Table 2).

Table 2. Description of DC motor operation pin

Pin number	Pin name	Level	Pin function description
1	VSS	0V	Power ground
2	VCC	−5V	Power supply positive
3	V0	H/L	Contrast (brightness) adjustment
4	RS(CS)	H/L	Rs = "H" indicates that db7-db0 are display data Rs = "L" indicates that db7-db0 is the display instruction data
5	DB 0	H/L	Tri state data line
6	DB 1	H/L	Tri state data line
7	DB 2	H/L	Tri state data line
8	DB 3	H/L	Tri state data line
9	/RESET	H/L	Reset terminal, low level valid
10	VOUT	–	LCD driving voltage output terminal

Using single-chip microcomputer to drive stepping motor by software, not only can we freely set the speed, rotation angle and rotation times of stepping motor within a certain range through programming method, but also can conveniently and flexibly control the running state of stepping motor to meet the requirements of different users. Standing in the synchronous rotating coordinate system, each space vector of the motor becomes a static vector, and each space vector in the synchronous coordinate system becomes a direct flow. According to several forms of torque formula, the relationship between the torque and each component of the controlled vector can be found, and the component values of the controlled vector required for torque control can be calculated in real time. According to these given quantitative real-time control, the control performance of DC motor can be achieved.

2.3 Realization of Remote Speed Control of DC Motor

In the speed control of Brushless DC motor for wheeled robot, how to achieve high performance control of motor instantaneous torque is a key problem. The basic requirements of motor torque control are: fast response, high precision, small torque ripple, high efficiency and high power factor. The output torque control of Brushless DC motor for wheeled robot can be attributed to AC / DC axial flow control [10]. Different combinations of transverse axial flow will affect the efficiency, power factor and output torque of the control mode. Based on the given torque, how to determine the AC and DC current is actually the problem of stator current vector control. The stator structure of asynchronous motor is similar to it. In the air gap, when the coil draws symmetrical AC current, it will produce the same rotating magnetic field as the induction motor. When the rotor has synchronous speed, the direction of the air gap magnetic field produced by the stator is the same, so the air gap magnetic field produced by the two rotors does not have relative speed, but has spatial potential angle difference. Two relatively static magnetic fields in the air gap interact to produce electromagnetic moment, which drives the rotor to rotate synchronously. When the stator current changes, the stator air gap magnetic field also changes, resulting in electromagnetic torque, which makes the load and speed different. Based on this, we need to further optimize the distribution structure of motor vector speed regulation (Fig. 2).

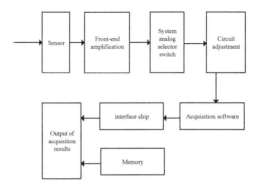

Fig. 2. Optimization of distribution structure of motor vector speed regulation

The adaptive compensation of microcontroller is a kind of multivariable and strongly nonlinear, so it is difficult to establish an accurate mathematical model [11]. This paper introduces the structure and classification of the self-adaptive compensation of single-chip microcomputer, gives the basic equations of the self-adaptive compensation under linear condition and ideal state, and establishes the mathematical model of the self-adaptive compensation after coordinate transformation. The controller module is mainly used to realize the motor speed control algorithm, but its output range does not match with the PWM module, so the designed controller module must be able to control the frequency of PWM signal output by the PWM module. The low frequency noise will be produced after the loop filter processing of the output of the frequency and phase detector. Therefore, the active p-type controller is used for further filtering processing, and its integral link can also effectively reduce the steady-state error. The active p-type controller also includes proportional link and integral link [12]. Using two adders and two multipliers, the pipeline computing process of embedded wireless communication can be easily realized.

The model equation of synchronous motor is not only complex, but also the degree of electric coupling is related to the rotor position, that is to say, the equation is indeterminate. The mathematical model of Brushless DC motor for wheeled robot in dq coordinate system can be obtained by coordinate transformation of stator equation of Brushless DC motor for wheeled robot in a, B and C coordinate system.

Voltage equation:

$$u_d = \frac{d\psi_d}{dt} - \omega\psi_q + R_1 i_d \tag{10}$$

$$u_q = \frac{d\psi_q}{dt} + \omega\psi_d + R_1 i_q \tag{11}$$

$$0 = \frac{d\psi_{2d}}{dt} + R_1 i_{2d} \tag{12}$$

$$0 = \frac{d\psi_{2q}}{dt} + R_1 i_{2q} \tag{13}$$

Flux linkage equation:

$$\psi_d = L_d i_d + L_{md} i_{2d} + L_{md} i_f \tag{14}$$

$$\psi_q = L_q i_q + L_{md} i_{2q} \tag{15}$$

$$\psi_{2d} = L_{2d} i_{2d} + L_{md} i_d + L_{md} i_f \tag{16}$$

$$\psi_{2q} = L_{2q} i_{2q} + L_{md} i_q \tag{17}$$

Electromagnetic torque:

$$T_{em} = p\left(\psi_d i_q - \psi_q i_d\right) \tag{18}$$

Mechanical motion equation:

$$J\frac{d\Omega}{dt} = T_{em} - T_L - R_\Omega \Omega \tag{19}$$

According to the vector control method of synchronous motor [13], the principle of vector control method is designed as shown in the Fig. 3.

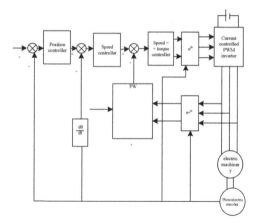

Fig. 3. Vector control principle of DC motor

In the figure, the rotor position sensor detects the position angle B of the brushless DC motor for the disc coreless wheeled robot, the torque controller controls the current, and the controllable PWM inverter supplies three-phase current to the brushless DC motor for the disc coreless wheeled robot. After the method initialization, the main program cyclically reads the control instruction (provided by SCI interrupt service program) and the motor speed data (provided by timer interrupt service program), and submits it to the Fuzzy-PID control algorithm module to calculate the new control quantity. The calculation result is sent to the comparison register, and PWM waveforms with different

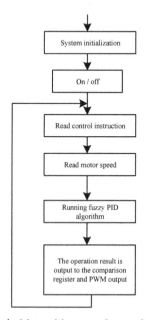

Fig. 4. Motor driver speed control step

duty ratios are output, and the motor speed is controlled by the motor driver. The specific steps are as follows (Fig. 4):

The single-step debugging method and full-speed debugging method are combined organically, so that it can find errors faster and more accurately than the single-step debugging method. Full-speed execution can be matched with set breakpoints to roughly determine the error range. Step-by-step can know the execution of each instruction in the program in detail, and you can easily know whether the instruction is correct by comparing the result of instruction running. On the basis of realizing the control of a single stepping motor, this method can also synchronously control multiple stepping motors. As shown in the following figure, further optimize the control of motor speed (Fig. 5):

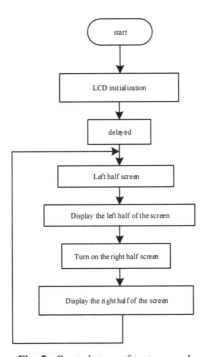

Fig. 5. Control steps of motor speed

Combining single-step debugging with full-speed debugging based on the above steps can make the whole debugging process find errors more quickly and accurately, thus improving the debugging efficiency. Full speed execution can be matched with set breakpoints to roughly judge the error range. By comparing the running results of instructions, we can know the running situation of instructions in the program, so it is easier to judge whether the instructions are correct or not. Taking a single stepping motor as the control object, the stepping motor of the multi-wheeled robot platform is synchronously controlled, so as to better improve the speed control effect of the DC motor of the wheeled robot and ensure the control accuracy.

3 Analysis of Experimental Results

Methods after the design is completed, in order to ensure the effect in energy saving, it needs to be applied to an example for simulation experiment. Only one diesel generator set is used in the experiment, and the parameters of main components are shown in the Table 3.

Table 3. Experimental parameters

Name	Parameter
Maximum engine power	80 KW
Generator efficiency	95%
Rated speed of generator	900 r·min^{-1}
Rated voltage of battery pack	350 V
Rated capacity of battery	100 Ah
Starting voltage of battery pack	50%

Generator set is composed of engine and generator. The fuel consumption of generator set is the fuel consumption of engine. The fuel consumption rate X of diesel engine can be regarded as a nonlinear function of speed s and torque Pi of diesel engine. The fuel consumption rate of the engine is:

$$X = f(P_i, s) \tag{20}$$

Assuming that the speed of the generator is fixed, the fuel consumption rate is a linear function of the engine power. The above speed identification algorithm is applied to the direct torque control method of Brushless DC motor for wheeled robot in the previous section, and the simulation is carried out. In the case of no-load and load, the waveforms of identification speed, actual speed and given speed are shown in the figure.

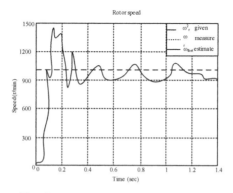

Fig. 6. Test results of traditional methods

The structure and principle of other parts of the simulation are the same as those in the previous section, which will not be repeated here (Figs. 6 and 7).

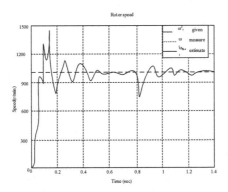

Fig. 7. The results of this method are as follows

It can be seen from the figure that during the speed rising period, the estimated speed is quite different from the actual speed value, especially when it rises to the maximum speed point. When the motor runs in steady state, the estimated speed is almost the same as the actual speed; When the load is running, especially when the speed fluctuates, the estimated speed can track the fluctuation of the actual speed in real time, which shows that the model reference adaptive speed estimation method has high dynamic performance. This is because DSP2812 is used as the CPU of the robot, which simplifies the peripheral circuit and improves the integration and stability of the method. Embedded wireless communication technology and fuzzy PID control algorithm are used to make up for the shortcomings of single control algorithm and improve the dynamic performance of the motor.

In order to further verify the effectiveness of this method, comparative experiments of motor control accuracy are carried out, and the control accuracy results of different methods are shown in Fig. 8.

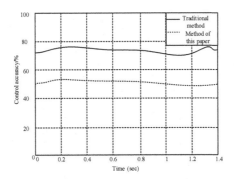

Fig. 8. Control accuracy results of different methods

As can be seen from Fig. 8, the motor control accuracy of the traditional method is about 50%, while the control accuracy of the method in this paper is about 75%, which indicates that the model reference adaptive speed estimation method has high control accuracy, and can ensure that the DC motor can effectively track the control target in the remote speed regulation.

4   Conclusion

Combined with embedded wireless communication technology, the remote speed control method of Brushless DC motor for wheeled robot is optimized. The method consists of two parts: control and drive. PLL and controller are implemented in FPGA. DSP is used to generate high accuracy reference signal. The design process of the method is introduced in detail. The high performance DSP2812 is used as the CPU of the robot, which simplifies the peripheral circuit and improves the integration and stability of the method. Embedded wireless communication technology and fuzzy PID control algorithm are used to make up for the shortcomings of single control algorithm and improve the dynamic performance of the motor. The simulation results show that the motor speed control method designed in this paper has good performance and can track the control target quickly. In the future development, this method will be applied to ensure the accuracy of the test, find out the details and further improve.

References

1. Chen, X., Liu, G., et al.: Sensorless optimal commutation steady speed control method for a nonideal Back-EMF BLDC motor drive system including buck converter. IEEE Trans. Industr. Electron. **67**(7), 6147–6157 (2019)
2. Song, W., Chen, H., Zhang, Q., et al.: On-chip embedded debugging system based on leach algorithm parameter on detection of wireless sensor networks. Math. Probl. Eng. **2020**(93), 1–7 (2020)
3. Zheng, S., Zheng, S.: Intelligent hospital and traditional Chinese medicine treatment of cerebrovascular dementia based on embedded system. Microprocess. Microsyst. **81**(1), 103661 (2021)
4. Qi, Z., Shi, Q., Zhang, H.: Tuning of digital PID controllers using particle swarm optimization algorithm for a CAN-based DC motor subject to stochastic delays. IEEE Trans. Industr. Electron. **67**(7), 5637–5646 (2020)
5. Chen, N., Jiao, X., Tan, R., et al.: A multi-parameter speed control model of traveling wave ultrasonic motor. Int. J. Appl. Electromagn. Mech. **64**(1–4), 457–464 (2020)
6. Das, S., Sarkar, P.P., Chowdhury, S.K.: Frequency tunable printed antenna with 90% size reduction for wireless communication applications. J. Instrum. **14**(11), 11010 (2019)
7. Bag, B., Biswas, P., Biswas, S., et al.: Wide-bandwidth multifrequency circularly polarized monopole antenna for wireless communication applications. Int. J. RF Microw. Comput.-Aided Eng. **29**(3), 21631.1–21631.11 (2019)
8. Alibakhshikenari, M., Khalily, M., Virdee, B.S., et al.: Double-port slotted-antenna with multiple miniaturized radiators for wideband wireless communication systems and portable devices. Prog. Electromagnet. Res. **90**(10), 1–13 (2019)
9. Shahjalal, M., Hasan, M.K., Chowdhury, M.Z., et al.: Smartphone camera-based optical wireless communication system: requirements and implementation challenges. Electronics **8**(8), 913–915 (2019)

10. Guerrero, E., Guzmán, E., et al.: FPGA-based active disturbance rejection velocity control for a parallel DC/DC buck converter-DC motor system. IET Power Electron. **13**(2), 356–367 (2020)
11. Liu, S., Liu, D., Srivastava, G., et al.: Overview and methods of correlation filter algorithms in object tracking. Complex Intell. Syst. **7**(3), 1895–1917 (2020)
12. Liu, S., Lu, M., Li, H., et al.: Prediction of gene expression patterns with generalized linear regression model. Front. Genet. **10**, 120 (2019)
13. Fu, W., Liu, S., Srivastava, G.: Optimization of big data scheduling in social networks. Entropy **21**(9), 902 (2019)

Information Collection Method of Organic Vegetable Diseases and Insect Pests Based on Internet of Things

Zhi-heng Song[✉], Hang Zhang, and Nai-xiang Li

College of Computer and Information Engineering, Tianjin Agricultural University, Tianjin 300384, China

Abstract. The traditional method of collecting pest information of organic vegetables is not wide enough, the intelligent method of collecting pest information of organic vegetables under the Internet of things is studied. Firstly, the data of intelligent terminal of the Internet of things is collected through the Internet of things technology, including spatial location data collection, numerical data collection to carry out all-round data collection and data fusion of organic vegetable pest information to get complete organic vegetable pest information, and then use JSON to code the data of Internet of things intelligent terminal, and then use wireless communication technology and select C/s Finally, through interpolation data processing and spatial clustering data processing, combined with Internet of things technology, the data processing of Internet of things intelligent terminal is carried out. Through the Internet of things technology for data collection, data coding, data transmission and data processing, the intelligent collection of organic vegetable pest information can be realized. Through the experimental comparison, it can be concluded that the intelligent collection method of organic vegetable pest information under the Internet of things is wider than the traditional method, which proves the effectiveness of the intelligent collection method of organic vegetable pest information under the Internet of things.

Keywords: Internet of things · Organic vegetables · Pest information · Intelligent collection

1 Introduction

With the development of the economy and the improvement of people's living standards, people are paying more and more attention to food safety issues [1]. Organic vegetables are naturally grown and non-polluting vegetables. The development of organic vegetables provides an ideal choice for solving current food safety problems [2]. In particular, the cultivation of organic vegetables is focused on natural production [3]. Therefore, the development of organic vegetables has a broad market space [4]. First, the industrialization of organic vegetable cultivation is conducive to environmental protection. Second, the organic vegetable industry can provide quality agricultural products to the society.

S. Liu and X. Ma (Eds.): ADHIP 2021, LNICST 417, pp. 100–109, 2022.
https://doi.org/10.1007/978-3-030-94554-1_9

Third, the organic vegetable industry can obtain good economic benefits. Fourth, the organic vegetable industry can increase the market competitiveness of domestic agricultural products and promote the coordinated development of the economy.Fifth, the organic vegetable industry can increase employment opportunities. Organic vegetables have no chemical residues, have a good taste, and have proven to be more nutritious than ordinary vegetables. Organic food is known as the "sunrise industry" and has a broad market. Organic vegetables require vegetables to be used in the process of planting, such as pesticides, fertilizers, growth regulators, etc., and can not use genetically modified technology, and must be certified by an independent agency. Nowadays, the demand for safe food is increasing, and the development of organic vegetables provides an ideal choice for solving current food safety problems. However, the occurrence of pests and diseases will have a serious impact on the yield and quality of organic vegetables. Therefore, the information collection is the only way to ensure the high yield and quality of organic vegetables. In order to effectively manage, it is necessary to collect long-term information on the biological habits of pests and diseases, summarize the occurrence and succession, optimize and improve pest control techniques, thereby improving the ability of pest control and prevention, and exerting pest control in organic vegetables. The role of stable production and increased production. Under the popularization of the Internet of Things, the application of Internet of Things technology in agricultural production, management and management has become extensive. The combination of Internet of Things technology and agricultural applications has gradually formed the agricultural Internet of Things. The traditional manual collection method is time-consuming, labor-intensive and inefficient, lacking timeliness and accuracy, and it is difficult to meet the requirements of modern agricultural information collection. Therefore, the intelligent information collection methods are studied.

In this paper, with the help of the advantages of the Internet of Things, combined with GPS technology, an intelligent terminal data collection method is designed, which collects the information of areas where the pests and diseases of organic vegetables are seriously damaged, and collects the numerical data of related pests and diseases objects by field investigation. Multi-source data are fused to make the data show spatial correlation. The data of intelligent terminal of Internet of Things is encoded by programming language, and the data is processed by interpolation and spatial clustering algorithm. Intelligent collection of diseases and insect pests information of organic vegetables was completed.

2   Intelligent Collection Method Under the Internet of Things

2.1   IoT Intelligent Terminal Data Acquisition

The occurrence of plant diseases and insect pests is complex, but it usually happens in two ways: spot occurrence and occurrence area [5]. Plant diseases and insect pests of organic vegetables have object attribute and space attribute. Object data include the occurrence grade and time of plant diseases and insect pests of organic vegetables. It is necessary to collect the object attribute and spatial attribute of plant diseases and insect pests of organic vegetables at the same time. In the collection of object attributes, The occurrence grade needs technicians to count the incidence rate, according to the evaluation standard of the occurrence grade of diseases and insect pests, the diseases and insect pests are classified and treated.

Spatial Position Data Acquisition

Spatial location data acquisition is the main content of mobile GIS data acquisition. There are two basic modes of data acquisition: one is to collect geographic information entities in x, y.

The coordinates are entered clockwise, and the second is the representation of geographic entities using points, lines, polygons, and grid adjacencies [6]. GPS survey is a new generation of satellite navigation and positioning system established with the rapid development of modern science and technology. It has the characteristics of globality, all-weather, high precision, automation and high benefit. GPS measurement has been widely used because of its all-weather, real-time and precise 3D navigation, good anti-jamming and confidentiality. With the popularization of smartphone, GPS model can be used to measure and collect space position. Therefore, GPS and Internet of Things technology are used to collect the spatial position of plant diseases and insect pests in organic vegetables, and the collected information has some points and areas.

Organic vegetable diseases and insect pests may occur in spots, and may exist in the form of spots on current satellite images. It is more suitable to collect organic vegetable pests in the form of spots [7]. Point data collection mainly refers to the recording of pest occurrence locations on maps in the form of points and their transmission to Internet of Things servers [8]. The client uses mobile GIS technology to load the off-line map of the research area and display the current collection position by GPS technology, and uses the drawing engine to map the pest occurrence area.

In areas where there are serious pests on organic vegetables, the pests may be displayed in the form of polygons on the satellite images in the form of patches, and the surface collection shall conform to the actual situation of pests. When collecting data in the fields, the agricultural technicians may use their smart phones to locate the pests and directly display the data on the maps. Then the agricultural technicians may use the mobile APP to load the offline map of the research area, display the location of the current collection area through the GPS technology, record the occurrence of pests on the maps in the form of polygons through the drawing engine, and locate the pests through the smart phones. Then the agricultural technicians may collect the data in the fields and directly display the data on the maps. Then the agricultural technicians may use the mobile APP to load the offline map of the research area, display the location of the current collection area through the GPS technology, record the occurrence of pests on the maps through the drawing engine, and then transmit the collected pests through the Internet of Things client to the server center through the network.

Numerical Data Acquisition

When collecting data, it is necessary to collect the object attributes, such as the occurrence grade of the pests and diseases, the time of collection, the users of the strip field, etc. The essence of this kind of problem is numerical data acquisition. Because numerical data acquisition does not involve spatial location, it can be collected by paper record and electronic record. It is necessary to classify the pests and diseases. The criteria of pests and diseases on organic vegetables in different areas are different [9].

Field investigation by agricultural technicians is needed to determine the grade of plant diseases and insect pests of organic vegetables in an area. There are three kinds of investigation contents: aphids rate, hundred insect numbers and foliage rate. Because pests and diseases are rare, it is difficult to investigate the number of 100 pests. The percentage of foliage and the percentage of aphids are the usual investigation methods. In the field survey, the common methods are three-level sampling, subsection, typical and random. To determine the specific unit, we should use random sampling survey, sampling methods commonly used in the field survey are five-point, diagonal, chessboard, parallel and "Z" type and so on. Five-point sampling method is from the field is the center of the intersection point to the middle point of four corners, such as 5 points sampling, this method is the most common method. Diagonal sampling can be divided into single diagonal sampling method and double diagonal sampling method. Single diagonal sampling method is in the field of a diagonal, at a certain distance to select all the required sample points [10, 11]. The double diagonal sampling method is to distribute the sampling points evenly on the two diagonals of the four corners of the field. The checkerboard sampling method is to divide the field under investigation evenly into many plots, and then to distribute the sampling points evenly among certain plots of the field [12]. The parallel sampling method is suitable for the investigation of diseases and pests with uneven distribution. Z-shaped sampling method has more sampling points in the edge of the field but less in the middle. It is suitable to use this method when there are more migratory pests in the field. Among them, the method recommended by agricultural technicians in field survey is "five-point survey". The five-point method of field survey is shown in Fig. 1.

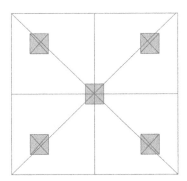

Fig. 1. Five-point survey

Based on the statistics, According to the national criteria for classification of major crop pests (DB65/T 2584–1998), And combined with the actual situation in the study area, Classification of pests and diseases to 5 levels (set M pest rate):1 level, $M \leq 5\%$, Small occurrences; Level 2, $6\% \leq M \leq 25\%$, Moderate mild occurrence; Level 3, $26\% \leq M \leq 50\%$, Moderate occurrence; Level 4, $51 \leq M \leq 75\%$, Moderate bias occurs; Level 5, $M > 75\%$, Recurrence.

Through the collection of numerical data, it is convenient to realize data fusion.

Data Fusion

In order to collect the data of diseases and insect pests, we need to pay attention to the spatial and numerical properties of the data, and the fusion of the two multi-source properties is also very important.

Using IOT technology, through the use of an intermediate nature of the file on the digital image, three-dimensional spatial positioning and other digital properties are combined to complete the external association. The specific operation means is to use extensible markup language to store the digital information or path string of digital image. But in practical application, in the process of association, the continuous read-write operation of data is easy to lead to the increase of data management redundancy and the decrease of security. Moreover, the mode of multi file description can easily lead to the reduction of data integrity in transmission.

The spatial positioning database based on Internet of things technology is composed of spatial data elements of geographic information and geometric measurement. The basic elements are composed of points, lines and planes, the spatial features are composed of basic elements, and the spatial model is composed of spatial features. The spatial models with the same attributes are gathered together to form the spatial geometry layer, and are stored and processed in binary data blocks. In this kind of spatial positioning database, the spatial and numerical properties of spatial data can be stored and integrated. At the same time, it can realize the integration of graph and attribute, and integrate them with the existing enterprise information system, so as to make GIS integrate into IT mainstream.

Spatial database is used to fuse the spatial position of pests. Among them, spatial association can be used to manage this kind of data, and can be directly associated with the database through GIS to read the data information, which can greatly reduce the difficulty of data processing and make the data displayed in ArcGIS Desktop intuitively.

2.2   IOT Intelligent Terminal Data Coding

The positioning device is added to the intelligent terminal. Through the data processing module of C/s, the data can be exchanged offline to complete the collection and transmission of pest information data. Among them, in the data transmission, because this kind of data contains spatial relationship data and numerical data. It is usually encoded using JSON and XML.

The full name of XML is an extended markup language, a structured markup language that allows users to define their own markup language.XML is a lightweight data storage file relative to relational databases with the following characteristics:

The feature of XML file is that the content and the overall structure of the text are completely complementary and interfere. With this feature, the data processing can realize the management of text content and the framework of processing process is completely separate. It has the advantages of interactive operation, uniform format characteristics, and supports diversified coding operations and easy to expand functions.

Based on the Internet of things, the data acquisition and utilization is the way of client-side and server-side interaction. The client is mainly responsible for the data collection. The server performs processing and transmission. But when the client and server interact with each other, the spatial data is required to be transmitted at a high speed. At this time, XML file can not meet the needs of use, choose JSON for data processing, which is a light data exchange mode, and it is easy to transmit data. It can realize the data conversion between different platform systems.

The characteristics of XML file are as follows: its content text and data structure do not interfere with each other. Thus, data processing can effectively separate the management of content text from the management of data processing flow. It has the following advantages: good interactivity, unified format, multiple and diversified coding, scalability and so on. XML files can communicate and interact freely in various operating systems. Basically, all the systems on the market can support XML files. Its own encoding method is easy to record, and it can handle a variety of encoding languages, and it can further expand the scope and format of use according to the documents. Through the architecture of Internet of things, through the server to connect the client for data collection, in which the client for data collection, the server for transmission and processing. XML is not suitable for this kind of data processing because of the speed requirement. At this time, the use of JSON, as a data exchange format with small amount of calculation, has the advantage of fast speed, and can realize the data conversion and transmission between different platforms and systems.

JSON format data can be directly encoded in the server, which is very convenient for the development and subsequent maintenance of the server and client. With this advantage, this method is used for data collection.

2.3   Intelligent Terminal Data Transmission in Internet of Things

At present, C/s transmission processing mode is widely used. Through the ISO structure of the system architecture, fully develop the potential advantages of the hardware of each port, and further reduce the communication energy consumption and time-consuming of the system. Using function distribution, the communication task is divided into several subtasks, and each subtask is divided into different computers. Data collection, expression and processing are realized in the client. The core processing is carried out on the server side. In this paper, GPS positioning device is added here. Data acquisition function module and C/S module are used to collect data. At this time, we can use C/S mode for data transmission, and design a communication module to support.

2.4   Data Processing of Intelligent Terminal in Internet of Things

After collecting the data of diseases and pests, it is necessary to process the data to finally complete the intelligent collection of the information. Its data theory is weighted average, which multiplies the values of the known sample points by the corresponding units, then sums up the total values and divides them by the sum of the units. Assuming that Xi is a known sample point, Wi is a weight corresponding to a known sample point, and Y is a weighted arithmetic mean, the formula for calculating the weighted arithmetic mean is as follows:

$$y = \frac{\sum X_i \times W_i}{\sum (W_i)} \tag{1}$$

The Fourier transform is used to redefine the long-distance composite wave function: it is expressed as a state where many single-frequency waves are superimposed. Among them, the expression of the long-range synthetic wave function is:

$$X(k) = \sum_{n=1}^{N} x(n) \cdot \exp\left(\frac{2\pi}{N}(n-1) \cdot (k-1)\right) \tag{2}$$

In the formula, $X(k)$ represents the value of the long-range synthetic wave function, N represents the number of vegetable pest information, represents the number of vegetable pest information time series, $x(n)$ represents the vegetable pest information time series, and k represents the number of elements in the long-distance pulse wave function.

Fourier transform the $x(n)$ in Eq. (2):

$$x(n) = X(k) / \sum_{k=1}^{N/2} \left[\cos\left(\frac{2\pi \cdot k}{N \cdot dt}\right)\right] \tag{3}$$

In the formula, dt represents the time interval between two points of the long-range synthesized wave.

Suppose i represents the array element, $A(x, y)$ represents a collection point in the long-range synthesized wave, and $F_1(0, F_1)$ represents the focal point of the long-range wave information. Among them, the delay from $A(x, y)$ to $F_1(0, F_1)$ is Δt_{i1}, then the phase difference expression between the two is:

$$\Delta\varphi_{i1} = \frac{2\pi \cdot k}{N \cdot dt} \cdot x(n) \tag{4}$$

In the formula, $\Delta\varphi_{i1}$ represents the phase difference between $A(x, y)$ and $F_1(0, F_1)$.

Using formula (4) to obtain the long-distance information of the k harmonic transmission of $x(n)$ can be expressed as:

$$d_1(k, x) = \left(\frac{\cos \Delta\varphi_{i1} + \sin \Delta\varphi_{i1}}{2N}\right) \tag{5}$$

Remote information can be received in the same way:

$$d_2(k, x) = \left(\frac{\cos \Delta\varphi_{i1} + \sin \Delta\varphi_{i1}}{S}\right) \tag{6}$$

In formula (6), the expression of the long-distance area S is:

$$S = \sum_{i=N}^{N} A(i) \tag{7}$$

In the formula, $A(i)$ represents the value of the amplitude weighting function. If $A(i) = 1$, the receiving distance is not weighted; otherwise, the receiving distance performs amplitude weighting.

The formula for calculating the weight y_i in the inverse distance weighting method is:

$$yi = \left(\frac{d_2(k, x)}{SP_{di}} \right) \tag{8}$$

P_{di} is the distance between the interpolation point and the known sample point. From the formulas (8), we can see that the formula for calculating the weighted average of inverse distance Y is as follows:

$$Y = \frac{\sum X_i \times \frac{1}{P_{di}}}{\sum \frac{1}{P_{di}}} \tag{9}$$

The method of inverse distance interpolation can calculate the occurrence level of pests and diseases sampling strip field by inverse distance weighted average on the basis of organic vegetable sampling strip field. Because organic vegetable sampling strip field is representative in the occurrence grade, and the occurrence is "dot" at first, and gradually develops into "sheet" shape over time, in a certain range of buffers should show the trend of inverse distance change, so this method is adopted. The inverse distance interpolation method is used to process interpolation data.

The K-means clustering and the minimum span tree need to be selected during the grouping process. Through the interpolated data processing and spatial clustering data processing combined with the Internet of Things technology, the IOT intelligent terminal data processing can be realized. Through the Internet of Things technology for data collection, data encoding, data transmission and data processing, intelligent information collection of organic vegetable pests and diseases can be realized.

3   Simulation Experiment

In order to ensure the effectiveness of the intelligent collection method under the Internet of Things, a simulation experiment was designed. Horizontal collection refers to the average collection in the time of occurrence, and the average occurrence level is calculated. The average grade is taken as the level at the time. Longitudinal collection refers to the average level at the time of the year in the 10-year time span, and the average value is used as the level of organic vegetable pests and diseases at that time. The experiment was repeated 50 times, and the results were averaged to improve the accuracy and credibility of the experiment. The results of the horizontal collection experiment are shown in Fig. 2, and the results of the longitudinal collection experiment are shown in Fig. 3.

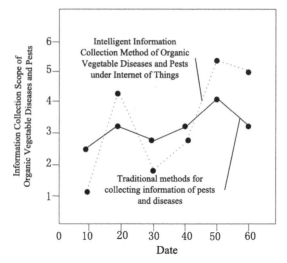

Fig. 2. Lateral acquisition experiment results

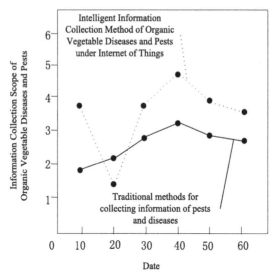

Fig. 3. Longitudinal collection of experimental results

Through experiments, it can be seen that whether it is the horizontal collection of organic vegetable pest information or the vertical collection of organic vegetable pest information, the intelligent collection method under the Internet of Things is more extensive than the information collection using traditional pest information collection methods. This is because the method designed in this paper firstly collects the spatial information through GPS technology, then collects the specific indicators of diseases and insect pests by field investigation, uses the Internet of Things to fuse the two data with multi-source data, so that the data of diseases and insect pests can be displayed in spatial correlation, and uses interpolation and spatial clustering algorithm to process the data, thus reducing redundant data and enhancing the data processing ability.

4 Conclusion

The intelligent collection method of organic vegetable pests and diseases under the Internet of Things enables the agricultural technicians to collect, integrate and analyze the information, which is simpler, more convenient, more scientific and effective, and enables the agricultural technicians to conveniently view and Understand the development trend of pests and diseases. When organic vegetable pests and diseases are likely to occur, it is possible to grasp the trend information and collect prevention and control measures at the first time, which has important application value for the production of organic vegetables and pest control.

Fund Projects. Application and Promotion Project of Tianjin Agricultural Science and Technology Achievements, Integrated application of core information technology for early warning, diagnosis and prevention of vegetable diseases in greenhouse, Project number 201704070.

References

1. Changzhen, Z., Jiahao, C., Deqin, X., et al.: Research on vegetable pest warning system based on multidimensional big data. Insects **9**(2), 66 (2018)
2. Navarro-Miró, D., Iocola, I., Persiani, A., et al.: Energy flows in European organic vegetable systems: effects of the introduction and management of agroecological service crops. Energy **188**, 116096 (2019)
3. Fang, S., Gu, L., Zhang, L.: Research on automatic spraying machine based on intelligent monitoring of agricultural pests and diseases on internet of things. Agric. Mech. Res. **39**(38), 224–227 (2017)
4. Napoli, R., Farina, R., Testani, E., et al.: Potential carbon sequestration in a Mediterranean organic vegetable cropping system. A model approach for evaluating the effects of compost and Agro-ecological Service Crops (ASCs). Agric. Syst. **162**, 239–248 (2018)
5. Chen, G., Kolb, L., Cavigelli, M.A., et al.: Can conservation tillage reduce N_2O emissions on cropland transitioning to organic vegetable production? Sci. Total Environ. **618**(15), 927–940 (2018)
6. Redekar, N., Trammell, C., et al.: Solarization effects on the soil microbiome at an organic vegetable farm in the Pacific Northwest (USA). Phytopathology **108**(10), 160 (2018)
7. Zhao, L., Ren, Z., Wang, J.: Traceability system of organic vegetables based on internet of things. Jiangsu Agric. Sci. **44**(52), 427–430 (2016)
8. Jiang, J., Han, G., Liu, L., et al.: Outlier detection approaches based on machine learning in the internet-of-things. IEEE Wirel. Commun. **27**(3), 53–59 (2020)
9. Edge, S.: More idiocy in the internet of things. New Sci. **238**(3175), 53–54 (2018)
10. Liu, S., Liu, D., Srivastava, G., et al.: Overview and methods of correlation filter algorithms in object tracking. Complex Intell. Syst. **7**, 1895–1917 (2020)
11. Liu, S., Lu, M., Li, H., et al.: Prediction of gene expression patterns with generalized linear regression model. Front. Genet. **10**, 120 (2019)
12. Liu, S., Bai, W., Zeng, N., et al.: A fast fractal based compression for MRI images. IEEE Access **7**, 62412–62420 (2019)

Research on Coding Method of Microscopic Video Signal Based on Machine Learning

Hai-xiao Gong[1,2(✉)] and Jie He[1,2]

[1] School of Data Science and Software Engineering,
Wuzhou University, Wuzhou 543002, China
[2] Guangxi Colleges and Universities Key Laboratory of Image Processing and Intelligent
Information System, Wuzhou University, Wuzhou 543002, China

Abstract. At present, the commonly used microscopic video signal coding methods have poor processing ability and low coding accuracy. Therefore, this paper proposes a new micro video coding method based on machine learning technology. Firstly, the video coding processing architecture is established by intra prediction, inter prediction, transformation, quantization, entropy coding and loop filtering, and then the coding processing is realized by image segmentation, intra prediction and inter prediction, and the depth decision method is used for depth analysis. The experimental results show that this method can effectively improve the processing ability of microscopic video signal coding, and at the same time enhance the coding accuracy.

Keywords: Machine learning · Microscopic video signal · Video coding processing · Depth decision

1 Introduction

With the continuous updating of mobile terminal technology, smart phones with stronger performance and more functions have become the choice of many consumers. If the two smart phone operating systems represented by Android and IOS make smart terminals more valuable, then the emergence of various types of video application software has substantially changed the way video signals are obtained [1, 2]. Video chat, remote meetings, and various live broadcast software have changed the way people work and live, and real-time video applications have attracted more and more attention. It can be determined that the current low-resolution video will slowly disappear in people's lives. This type of video mainly includes two formats of 360P (480 × 360) and 480P (640 × 480). And 720P (1280 × 720) HD video and 1080P (1920 × 1080) Full HD video will become the mainstream video formats in the future. At the same time, international manufacturers, mainly Samsung, have also introduced displays that support 4K (3840 × 2160). With the continuous deepening of research and development and the reduction of material costs, high-definition displays supporting 2K (2560 × 1440) and 4K (3840 × 2160) will be greatly popularized [3, 4].

The improvement of intra coding depth algorithm and inter coding depth algorithm has become an effective means to reduce the complexity of hevc coding signal. At present, one of the main research areas in this field is to simplify the process of data depth coding [5].

Machine learning is closely related to computational statistics, so it is also called statistical learning method and is an important branch in the field of artificial intelligence. Machine learning hopes that through continuous learning and adjustment of computers, it will eventually enable it to realize human learning capabilities. The essence of machine learning is spatial search and the generalization of functions. The main step is to use the features extracted from the data as input, abstract and construct a certain data model, obtain the information in the data, and use the obtained model to solve the unknown data analysis and prediction. According to the predicted results, the appropriate machine learning model is continuously selected until the generalization ability satisfactory to the unknown data can be obtained.

In HEVC, the improvement of the coding depth algorithm is considered to be the key technology of current HEVC to reduce the complexity of the coded signal to achieve practicality. As a recent research hotspot in academia, machine learning has gained a large range with its powerful learning and prediction capabilities Applications. Therefore, in order to improve the processing ability and coding accuracy of the micro-video signal coding processing method, this paper proposes a new micro-video signal coding processing method based on machine learning technology. First by frame prediction, interframe prediction, transform, quantization and entropy coding and processing module based video coding such as loop filtering architecture, then through the image block segmentation, frame prediction, interframe prediction coding for processing, and carries on the deep analysis of reference depth decision method, thus effectively improve the ability of handling the microscopic video signal encoding, and strengthened the coding accuracy.

2 Microscopic Video Signal Encoding Processing Architecture Based on Machine Learning

The hybrid coding framework based on machine learning is adopted. The framework mainly includes intra prediction, inter prediction, transformation, quantization, entropy coding and loop filtering. The detailed coding framework is shown in Fig. 1.

In Fig. 1, the original video sequence on the left is input into the encoder and then divided in the order of image group, frame, slice and coding unit CU. These different levels of division levels can make full use of the information of the coded unit and reduce the coded data. the amount. The encoder will have a set of initial coding parameters before specific coding, which are assigned to GOP level, frame level and coding block unit level [6–8]. However, in the actual encoding process, the encoding may not be performed as expected. When each layer is encoded, the parameters will be recalculated according to the encoding quality of the reference image and the complexity of the current image [9–11].

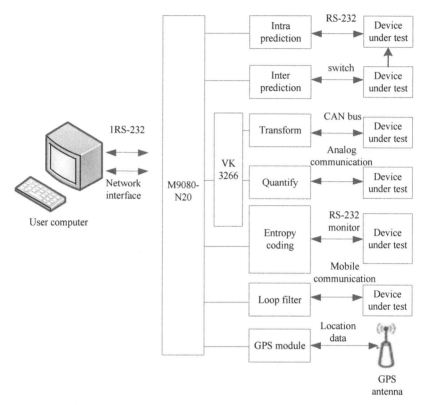

Fig. 1. Architecture of micro video signal coding and processing based on machine learning

It can only be used for reference in the first frame. At the specific coding unit level, each frame is first divided into the largest coding unit with 64 × 64 pixels. Specifically to the coding process of each LCU, the coding unit can continue to be decomposed into prediction unit and transformation unit. The encoder will traverse all the mode selection combinations and decide whether to continue the division by calculating the cost of each combination mode. This process is repeated until all encoding modes are traversed, and the optimal encoding mode is obtained. The encoding mode is stored, and the information compression of the current encoding block is completed through the process of transformation, quantization and direct encoding. After the encoder completes the encoding of the current encoding block or the slice where the encoding block is located, the obtained encoding parameter information and the preset information of the previous encoding block are uniformly packaged, and the necessary network propagation information is added for network transmission. After receiving the corresponding code stream, the receiver carries out the video and decoding process according to the reverse flow of the coding process and the pre-defined syntax rules.

Although Fig. 1 still uses a hybrid coding framework, the main coding modules are the same as those in the previous coding framework. But hybrid coding based on machine learning introduces new coding techniques in almost every module. The purpose of this design is to better handle high-resolution video signals. As a new segmentation mode, the flexible partition structure based on quadtree can represent video content more flexibly and effectively, enabling the encoder to select a more appropriate encoding mode according to the local characteristics of the image. 35 intra-prediction modes including 33-degree prediction, DC prediction mode and Planar prediction mode can better express the local features of high-definition video, match the complex and diverse textures in the video, obtain better prediction results, and achieve effective removal of spatial redundancy purpose. The asymmetric prediction unit division and motion merging technology makes good use of spatial correlation to reduce the motion parameter redundancy between adjacent blocks. The adaptive motion vector prediction technique also uses the correlation between spatial domain and temporal domain to remove the motion parameter redundancy [12, 13]. The sample adaptive compensation technique adds the corresponding offset value to the pixel according to the statistical characteristics of the pixel after the block filtering, which further improves the texture performance characteristics of the image and can improve the subjective quality and compression efficiency of the image.

3 Microscopic Video Signal Coding Processing Based on Machine Learning

3.1 Image Segmentation

In this paper, a quadtree based coding partition structure is introduced into the coding standard, which mainly includes coding unit Cu, prediction unit Pu and transformation unit tu. the purpose is to make the encoder better capture the characteristics of high-resolution video signal. In the coding process, the input video sequence is first divided into a series of coding tree units. Although the concept of coding tree is similar to that of macroblock in the previous video coding standard H.264, there is a big difference in the allowed pixel size between them.

The coding unit CU is the most basic unit of coding, and the CTU is the basic unit for dividing the CU. According to the quadtree division strategy and the video content adaptive division mechanism, the CTU can be evenly divided into four identical CUs, and each CU can still be divided in a recursive manner until the most suitable division for the image content is selected structure. CTU is finally divided into multiple CUs, and the size of the CU may be the same as or different from the CTU. The structure of a possible coding unit CU is shown in Fig. 2.

PU is the prediction unit, which is the basic unit to transmit the prediction information such as image index, prediction vector and prediction direction. Its root node is CU. There are two modes of PU division, one is symmetrical and the other is asymmetrical. There are four symmetrical and four asymmetrical modes. The specific segmentation shape of PU is shown in Fig. 3.

The use of an asymmetric rectangular PU partition structure can further match the prediction unit with the object boundary in the image. The PU division is related to the prediction mode. In the intra prediction mode, $2N \times 2N$ and $N \times N$ PU divisions can be used. At this time, the shape of the PU is all square; in the inter prediction mode, the

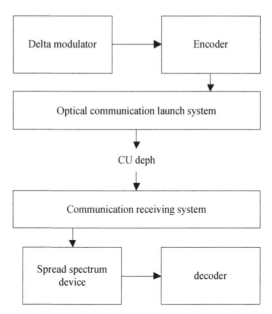

Fig. 2. Structure diagram of coding unit CU

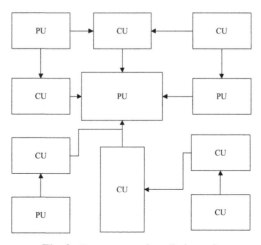

Fig. 3. Pu structure of prediction unit

above 8 types of PU can be used; But when the skip prediction mode is adopted, only the 2N × 2N PU division method can be adopted.

TU is a transformation unit, but its root node is CU. TU is the basic unit used in the process of transformation and quantization, which is used to represent the residual information of the current block after transformation. The shape of TU is square, and the size of TU supported in hevc is 4 × 4–32 × 32, so each CU may contain one or more TU. In the CU block using inter mode coding, TU can exist across the Pu boundary, while in the CU block using intra mode coding, TU cannot exist across the PU boundary, that is to say, TU is limited to a single PU. The former is suitable for all PU sizes, while the latter is only suitable for the case of intra PU size of 4×4.

The root nodes of PU and TU are both CU, but their partition methods are not limited. This structure greatly increases the flexibility of the encoder. With the concepts of coding tree unit CTU, coding tree block CTB and Quadtree Partition Method, the compression rate of high-resolution video can be further improved.

3.2 Intra Prediction

The principle of intra prediction is to use the correlation in the image space to predict the current pixel value using the reconstructed pixels in the adjacent position, and use the difference between the predicted value and the current pixel value as the basis for subsequent processing to reduce the amount of data transmission. In intra-frame prediction, the design of prediction modes with multiple angles will help remove pixel redundancy more effectively. The intra-frame prediction has been further optimized. It has increased the brightness image intra-frame prediction modes from 9 to 35, including Planar mode prediction, DC prediction, and angle prediction in 33 different directions.

The 35 prediction modes in the microscopic video signal coding process based on machine learning are defined on the basis of the PU, but the specific intra prediction process of the encoder is based on the TU as the basic prediction unit. Microscopic video signal coding based on machine learning makes full use of the directional characteristics of the image in the intra-frame prediction design, and the selection process of its prediction mode is also more complicated. The main process is to first build the MPM list of the current frame; after the list is built, use the Hadamard transform instead of the integer DCT transform to roughly select the 35 intra-frame modes; finally, perform normal coding on the roughly selected 8 modes and the MPM prediction mode. The prediction mode with the least cost is selected as the optimal coding mode, and the current block is compressed and coded.

3.3 Inter Prediction

Video signal sequence has strong temporal and spatial correlation, which is also the theoretical basis of video compression coding. The content similarity of the coding blocks in the same position between the current frame and the previous and subsequent frames is high. Inter frame prediction is an important technology to remove the temporal redundancy of video sequence in the coding standard.

In addition to traditional techniques such as motion estimation and reference frame management, more complex interframe motion vectors and prediction mechanisms, including Merge technology and AMVP technology, are introduced in the process of interframe prediction. The similarity between Merge technology and AMVP technology is that both use spatial correlation and temporal correlation to reduce the redundancy of motion parameters. The whole process mainly consists of two steps:

Step 1: select the motion parameters of the adjacent Pu according to a certain order, and establish the candidate MV list;

Step 2: according to the parameters in the candidate MV list, select the best performance as the motion parameters of the current PU.

Although merge technology and amvp technology are basically the same in the selection of motion parameters. This difference is mainly manifested in the following two aspects:

(a) In merge mode, there is no difference in MV, which is directly predicted by the adjacent PU. In amvp mode, there is a difference value of MV, and the encoder only encodes the difference value;

(b) There are also differences between merge mode and amvp mode in the way of establishing candidate MV list and the length of the list.

4   Coding Depth Decision of Microscopic Video Signal Based on Machine Learning

In order to cope with the high-efficiency coding and decoding of high-definition video and to improve coding efficiency in essence, this article has made a relatively large improvement to the coding technology in H.264/AVC. Including quadtree coding block division, intra and inter coding depth selection algorithms. The main process of this link is to traverse all the programming modes of the existing coding block, and perform the coding and rate-distortion optimization process for each coding mode, and then determine whether to continue the division. The optimal coding mode in the current mode is selected for coding without further division. The coding mode is shown in Fig. 4.

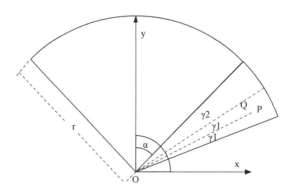

Fig. 4. Video signal coding mode

The essence of video compression is to seek the balance between the video distortion D and the number of bits r of information. The optimal solution is selected by using RDO. This section briefly introduces the video distortion degree and rate distortion optimization technology.

4.1 Video Signal Distortion

Subjective quality assessment and objective quality assessment are two common mechanisms for measuring video signal distortion. Because the evaluation standard of subjective quality assessment is not easy to be quantified, it is difficult to carry out and its accuracy cannot be guaranteed. Therefore, scholars mostly use objective quality assessment methods. There are many ways to evaluate the objective quality of the video signal, the purpose of which is to evaluate the distortion of the current video signal from multiple objective angles. The main measurement methods are square error, mean square error, absolute error sum and peak signal-to-noise ratio, the specific formula is as follows:

$$SSD = \sum_{x=0}^{W-1}\sum_{y=0}^{H-1} |\hat{f}(x, y) - f(x, y)|^2 \tag{1}$$

$$MSE = \frac{1}{WH} \sum_{x=0}^{W-1}\sum_{y=0}^{H-1} |\hat{f}(x, y) - f(x, y)|^2 \tag{2}$$

$$SAD = \sum_{x=0}^{W-1}\sum_{y=0}^{H-1} |\hat{f}(x, y) - f(x, y)| \tag{3}$$

$$PSNR = 10 \log_{10} \frac{(255)^2 WH}{\sum_{x=0}^{W-1}\sum_{y=0}^{H-1} |\hat{f}(x, y) - f(x, y)|^2} \tag{4}$$

In the above formula, $f(x, y)$ and $\hat{f}(x, y)$ respectively represent the original value and the reconstructed value at (x, y), w is the width of the image in the current video, and H is the height of the image.

4.2 Rate-Distortion Curve

The relationship between bit rate and distortion is a convex function, which also determines the relationship between them, as shown in Fig. 5. For a given rate distortion curve, when the distortion D is given, the required minimum rate R can be obtained, which is only a theoretical value. In practical application scenarios, due to various resource constraints, the optimal theoretical value can not be achieved.

In the actual encoding process, a certain encoding mode will be selected, and the distortion D and the rate R in the corresponding mode will be obtained respectively, which are shown in Fig. 6 as a series of operating points. The outer envelope curve is obtained by connecting the operating points, which is the solid line in Fig. 6. The curve also points out the optimal performance that the encoder can achieve in the actual encoding process. In the actual encoding process, given the code rate R, find the operating point closest to the rate-distortion curve by traversing all the encoding mode costs. This is also the main process of RDO.

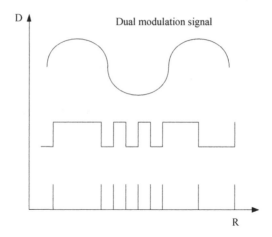

Fig. 5. R-D rate-distortion function curve diagram (R-D curve)

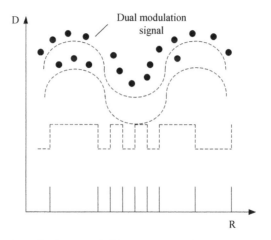

Fig. 6. Schematic diagram of video signal encoding rate-distortion curve

K-means algorithm is used for coding. As the simplest clustering algorithm in machine learning, K-means clustering algorithm is famous for its efficient classification. The main idea of the algorithm is to cluster n point $x_1, x_2, x_3, \ldots, x_n$ into $k(k < n)$ cluster $S = \{S_1, S_2, \ldots S_k\}$, so that each point belongs to the nearest cluster set. The minimum clustering principle is shown in Eq. (5):

$$\arg \min_S \sum_{i=1}^{k} \sum_{x_j \in S_i} |x_j - \mu_i|^2 \tag{5}$$

Among them, μ_i is the center point of each cluster, and x_j represents any point belonging to the cluster set S_i.

The k-means clustering algorithm can be regarded as a process of continuous iteration, adjustment and optimization. The main steps of the algorithm are as follows:

Step 1: Select k points from all the input n sample points as the initial cluster center points;

Step 2: According to the selected k cluster centers, traverse all points except the k cluster center points, find the cluster center closest to each point, and assign it to the corresponding cluster;

Step 3: When a cluster is completed, that is, all points in the sample have corresponding cluster centers, recalculate the mean of each cluster to adjust the cluster center point, and traverse all the points again to find the nearest cluster Center point, add this point to the corresponding cluster;

Step 4: Repeat Step 3 until the cluster centers of the previous two iterations no longer change.

According to the definition and steps of clustering algorithm, assuming that the final clustering process is completed t times, the time complexity of the whole process is $O(n \times k \times t)$, which requires a lot of calculation. If the online training method is used for k-means clustering, it will cost a lot of time, so this paper chooses the off-line training method for k-means clustering, according to the training center point to judge which cluster the input point belongs to in the actual coding process.

In this paper, the texture complexity of the image is represented by the difference between the pixels in the coding region.

The definition of complexity is shown in Eq. (6):

$$C_{3\times3} = \frac{1}{3 \times 3} \sum_{t=1,j=1}^{i=3,j=3} \left(P_{i,j} - \frac{1}{3 \times 3} \sum_{i=1,j=1}^{i=3,j=3} P_{i,j} \right)^2 \qquad (6)$$

Among them, $P_{i,j}$ represents the specific pixel value whose current position is (i, j), and $C_{3\times3}$ represents the texture complexity of the entire block.

Due to the richness and variety of image content, it is impossible to completely determine whether the current CU continues to be divided based on the complexity of the current pixel block. At this time, the complexity of the four sub-blocks needs to be further calculated. For example, for a block with a size of 64 × 64, 64 $C_{8\times8}$, 16 $C_{16\times16}$, and 4 $C_{32\times32}$ need to be calculated in sequence. In the whole process, only 64 8 × 8 area averages need to be calculated, and the combined calculation can be obtained. The average value of other pixels is calculated, and the amount of calculation in this part can be ignored.

If the complexity of the current coding block is small, the complexity of the corresponding four sub blocks is also small, which means that the pixel distribution of the coding block is balanced, and the video content is similar, so it is not necessary to continue to divide. On the contrary, if the complexity difference between the current block and the sub block is large, especially when the complexity difference between the four sub blocks is large, the current coding block should continue to be divided.

5 Experimental Analysis

Podk technology and symmetric encryption algorithm are commonly used in traditional video signal encoding and processing methods. In the process of encrypting the video information, PODK technology mainly uses the accurate data monitoring device to encode the information in real time, but this method has weak data processing capacity; while the symmetric encryption algorithm uses the secret key to establish the communication protocol for the security data of the online communication network, so as to ensure the security of the online confidential information, so as to improve the security Realize coding. However, this algorithm is weak in the perception of offline network information, which can not guarantee the overall perception of network secrets. In order to verify the effectiveness of the method designed in this paper, experiments were designed to compare the application performance of the three methods, so as to verify the feasibility of the method proposed in this paper.

In the process of encoding the confidential information in the video, it is necessary to set the security sensing module in every link of the whole communication system. Security module also contains information isolation, authentication and encryption functions. The way of video information transmission in the network is selective. Therefore, in this experiment, a variety of sensing methods are used to perceive the transmitted data in different channels, so as to judge its security.

The data perception layer in the experiment includes confidential information source, confidential information collection system and confidential security guarantee system. Figure 7 shows the topology of the operating environment of this experiment.

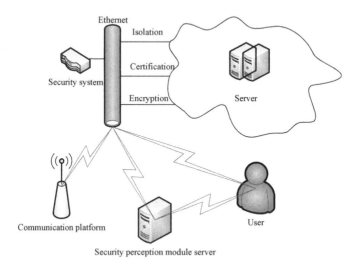

Fig. 7. Topology diagram of experimental operating environment

In the process of the experiment, the video signal coding first needs to transmit the video information to the space network in the system. The transmission process of the video signal in the network is different from that in the processing space. The transmission of the video signal in the network space depends on the security protocol, while the data transmission in the processing space is a secure form of information security isolation Lose. The video signal first enters the acquisition system of related equipment from the perception module of the system, and finally transmits to the data terminal, and arrives at the perception system of video signal through the security processing and security transmission protocol of the data terminal. When the data passes through the isolation module, it encodes the video signal, transmits the encoded information to the intelligent terminal, and then uploads it to the cyberspace. The special server and data processor in the cyberspace encode the information security. After the terminal interface in cyberspace works stably, the staff can observe the working state of the perceptual information data in the system in real time, and the terminal data in different states changes dynamically.

In the experiment, after the environment of the perception system is stable, the video signal can be received and the recognition side coding can be performed. This article first compares the video signal processing capabilities of three different systems, and the results are shown in Fig. 8.

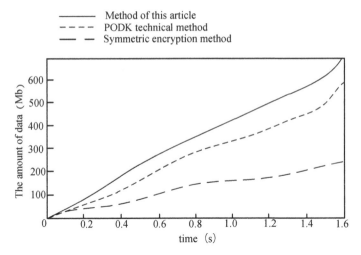

Fig. 8. Comparison of video signal processing capabilities of three different methods

According to the comparison results in Fig. 8, the method in this paper has the strongest processing ability for video signal. This is because this system uses rs245 standard serial port to receive more video signals, and can carry on the preliminary processing to the video information, slow down the later data processing work, and uses machine learning to perceive and process the data to be processed to ensure the security and efficiency of data processing in the form of encryption. However, the two traditional methods mainly use the common function operation mode, which has certain limitations and can not effectively process the real-time video signal.

On this basis, this paper compares the video signal coding accuracy of the three methods, and the results are shown in Fig. 9.

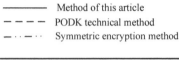

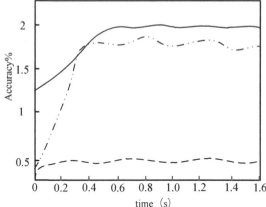

Fig. 9. Comparison of encoding information accuracy under three methods

It can be seen from Fig. 9 that the coding information perception ability of the method in this paper is the strongest, and can perceive coding data with an accuracy of 0.5% in a short time, while the perception ability of the traditional method is about 2% on average. This is because the method in this article uses a self-designed SD flash memory card, which not only has the function of data caching, but also accurately extracts and recognizes data, which increases the work intensity of the coding information recognition module. The method in this article also has an independent coding module that can Independently complete the perceptual operation of encoded information, which greatly improves the perceptual ability of encoded information. Traditional methods mainly perceive the online coding information, but ignore the offline confidential information perception, which leads to a decrease in the average coding ability of the system.

6 Conclusion and Prospect

In this paper, a new micro video signal coding method is designed by using machine learning method. The key technologies and algorithms are designed, and the basic control work is completed by anonymous data controller. The experimental results show that the method designed in this paper has stronger processing ability and higher coding accuracy.

However, due to the limitation of research time and other environmental conditions, the method in this paper still has a series of problems to be improved. In the following research, the method in this paper will be further improved from the perspective of improving the conversion speed between network protocols.

Fund Projects. National Natural Science Foundation of China: The Key Technologies about Fast Coding and Quality Controlling of Fractal Image Compression (61961036).

Basic Ability Improvement Project for Young and Middle-aged Teachers in Guangxi: Research on 3D Terrain Rendering for Large Scene Oblique Photography (2020KY17019).

Natural Science Foundation of Guangxi:Research on the Key Technologies about Decoder for Reliable Transmission of HEVC for Microscopic Video (2020JJA170007).

References

1. Wang, J., Di, Y., Rui, X.: Research and application of machine learning method based on swarm intelligence optimization. J. Comput. Meth. Sci. Eng. **19**(2), 1–9 (2019)
2. Antolinez, F.V., Rabouw, F.T., Rossinelli, A.A., et al.: Observation of electron shakeup in CdSe/CdS Core/Shell nanoplatelets. Nano Lett. **19**(12), 8495–8502 (2019)
3. Zhang, C., Zhou, Y., Guo, J., Wang, G., Wang, X.: Research on classification method of high-dimensional class-imbalanced datasets based on SVM. Int. J. Mach. Learn. Cybern. **10**(7), 1765–1778 (2018). https://doi.org/10.1007/s13042-018-0853-2
4. Zeng, W., Xu, H., Li, H., et al.: Research on methodology of correlation analysis of sci-tech literature based on deep learning technology in the big data. J. Database Manag. **29**(3), 67–88 (2018)
5. Jinnouchi, R., Lahnsteiner, J., Karsai, F., et al.: Phase transitions of hybrid perovskites simulated by machine-learning force fields trained on-the-fly with Bayesian inference. Phys. Rev. Lett. **122**(22), 225701 (2019)
6. Chen, Y., Tao, J., Wang, J., et al.: The novel sensor network structure for classification processing based on the machine learning method of the ACGAN. Sensors **19**(14), 3145 (2019)
7. Qin, F., Xu, D., Zhang, D., et al.: Robotic skill learning for precision assembly with microscopic vision and force feedback. IEEE/ASME Trans. Mechatron. **24**(99), 1117–1128 (2019)
8. Liu, S., Liu, D., Srivastava, G., Połap, D., Woźniak, M.: Overview and methods of correlation filter algorithms in object tracking. Complex Intell. Syst. **7**(4), 1895–1917 (2020). https://doi.org/10.1007/s40747-020-00161-4
9. Liu, S., Lu, M., Li, H., et al.: Prediction of gene expression patterns with generalized linear regression model. Front. Genet. **10**, 120 (2019)
10. Liu, S., Bai, W., Zeng, N., et al.: A fast fractal based compression for MRI images. IEEE Access **7**, 62412–62420 (2019)
11. Wu, Q., Zhang, C., Zhang, M., et al.: A modified comprehensive learning particle swarm optimizer and its application in cylindricity error evaluation problem. Int. J. Perform. Eng. **15**(3), 2553 (2019)
12. Zhou, B., Li, H.: Research on video compression coding based on embedded system. Wirel. Internet Technol. **16**(18), 39–41
13. Xue, R.B., Ma, M.Y., Jin, S., et al.: Surveillance video coding algorithm based on object. Mod. Comput. **27**(3), 41–45 (2019)

Design of General HLA Simulation Federate Based on GPU

Guo-hua Zhu[1], Min Cao[1], Hai-zhou Wang[1], and Li-feng Wang[2(✉)]

[1] School of Artificial Intelligence, Jianghan University, Wuhan 430056, China
zhuguohua1215@tom.com
[2] BaiSe University, BaiSe 533000, China
wanglifeng1101@tom.com

Abstract. In view of the operation efficiency of federate in HLA based simulation scheme, this paper analyzes the method of calling GPU computing resources in HLA federate, and designs a general simulation federate structure based on GPU. By comparing the results of the simulation scheme based on CPU and GPU, it is proved that under the premise of ensuring the stability of simulation federation operation, the simulation Federate based on GPU can greatly speed up the running speed of the Federation members and improve the simulation development and operation efficiency.

Keywords: Simulation · HLA · Federate · GPU · Social information · Wireless network

1 Introduction

HLA (high level architecture) builds simulation system according to the idea and method of object-oriented. It is the technology of dividing simulation members and building simulation federation on the basis of object-oriented analysis and design [1]. HLA mainly considers how to integrate federations on the basis of federates, that is, how to design the interaction among federates to achieve the purpose of simulation. It does not consider how to build federations from objects, but how to build federations under the assumption of existing members, which is an important reason why it is called "advanced architecture". The basic idea of HLA is to use object-oriented method to design, develop and implement the object model of simulation system, so as to achieve high-level interoperability and reuse of simulation federation [2]. In HLA, interoperability is defined as one member can provide services to other members and receive services from other members. HLA itself can not fully realize interoperability, but it defines the architecture and mechanism to realize interoperability among Federation members. In addition to facilitating the interoperability among members, HLA also provides a flexible simulation framework for Federation members. In HLA, the distributed simulation system used to achieve a specific simulation purpose is called Federation, which consists of several interactive simulation object models [3]. HLA defines the basic principles and methods

S. Liu and X. Ma (Eds.): ADHIP 2021, LNICST 417, pp. 124–132, 2022.
https://doi.org/10.1007/978-3-030-94554-1_11

of Federation and federate member construction, description and interaction. It should be noted that the Federation can join a larger Federation as a member.

With the development of computer technology, the practical problems solved by computer simulation technology are becoming more and more complex, and the running speed of simulation systems has been unable to meet the needs of practical problems. GPU (graphics processing unit) has a unique advantage in the field of computer simulation acceleration due to its natural super computing power [4]. Compared with GPU computing, CPU needs to support parallel and serial operations at the same time. It needs a strong versatility to deal with different data types. At the same time, it also needs to support complex universal logic judgment, which will introduce a large number of branch jump and interrupt processing. All these make the internal structure of CPU extremely complex, and the proportion of computing units is reduced. GPU is faced with highly unified, interdependent large-scale data and pure computing environment that does not need to be interrupted. Therefore, GPU chips are much simpler than CPU chips, and GPU designers use more crystal tubes as execution units, which also leads to the super computing power of GPU. For some lightweight computing intensive HLA Simulation members, GPU based application research has made some progress in recent years. At present, scholars in related fields have done some research on the design of general HLA Simulation federate, and achieved some research results. In Ref. [5], this paper introduces the development process of the lbts-sims Federation, and gives experience and lessons in the process of federation development. The data generated by legacy simulation is merged into logical groups to achieve low latency, run-time efficiency and data distribution optimization within Federation. Lbts-sims currently includes four federates: a scene control federate (a legacy ground truth server) and three new local HLA federates. The system can be used as a guiding ideology for HLA developers to integrate legacy simulation with HLA Federation. However, developers customize their own products according to business requirements. There are some problems, such as no unified standard, poor cross platform performance, weak descriptive ability, and inconsistent hardware support for various versions. Based on the above background, this paper designs a general HLA Simulation federate system based on GPU. By introducing the basic implementation of Federation in HLA architecture, a general federation architecture is constructed, and a General Federation programming model based on GPU is designed, so that users can improve the general method of HLA Federation calling GPU processor. This system can speed up the operation of simulation members and improve the efficiency of simulation development and operation.

2   Establishing a General Simulation Member Solution

HLA (high level of Architecture) is a complete set of distributed simulation technical specifications. Its core goal is to solve the simulation support environment of traditional distributed simulation technology. RTI provides general and independent simulation support services. On the issue of resource reusability, this system still can not meet the current needs. Therefore, it is very necessary to research and apply the universal member model framework based on HLA. It can integrate a large number of federate simulation members, design a common member model shared by all simulation members, reduce the degree of resource redundancy waste, and improve the model compatibility attribute of simulation system, as well as the expansibility of simulation. At present, there are universal federates that implement C++ interface, but this paper implements universal federates of java interface.

Regular federation members are only suitable for specific simulation models. The advantage of the general member is that when the simulation program changes, the general member does not modify the simulation model description file (such as XML file) according to the new simulation program, so as to obtain a new simulation model, and solves the problem of programmers when the simulation program changes. The simulation members made major revisions and rewrites, and described the algorithm of the simulation entity through the DLL file, which was used to describe the dynamic behavior of a certain entity in the federation, such as the control algorithm of the PID controller, etc. [6]. The General Federation member is responsible for loading DLL algorithm and reading the member model file XML, and joining it as a member in the Federation, updating the attribute value, receiving the reflection attribute value, receiving the interactive command from the control end, so as to control the status of simulation members (start, pause, stop, etc.) through the control end.

Therefore, according to the processing flow of federate member interaction data, its composition is mainly divided into: member initialization, data receiving layer, data processing layer, data distribution layer, exit federate execution. The processing flow of Java based general federate member interaction data is as Fig. 1.

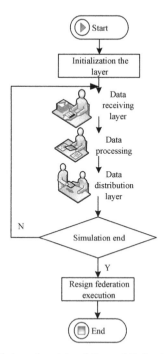

Fig. 1. Processing flow of interaction data of General Federation members based on Java

Initialization the layer, whose functional requirements: prepare the Federation members to participate in simulation promotion, namely, universal Federation member initialization. Its initialization content: Determine basic information such as federated member ID (Member IDentity, identity ID), federated member name (MemberName), determine and load corresponding algorithm components, such as DLL (Dynamic Link Library). The control end has completed the creation of the federation execution (createFederationExecution), waiting for the control command of the control end, the federation members join the corresponding federation execution (joinFederationExecution), declare the relationship between publish and subscribe, and establish the logical port relationship for information and data transmission and reception. Determine the time advancing strategy, whether the federation members are time restricted (Time Regulation), and whether it is time controlled (Time Constrained).

Data receiving layer (DRL), its functional requirements: Receive and store interactive data, provide data support for the data processing layer of federation members [7]. Its main content: according to the ordering relationship declaration in the initialization module, establish the corresponding data relationship storage table, temporarily store the temporary data and obsolete data, such as the latest data will be retained until the third time after receiving the data, and then be replaced.

Data processing layer (DPL), its functional requirements: In accordance with the time to promote service management under the premise of completing the corresponding interactive data processing, in order to get the processing results, and then by the data distribution layer processing. The main contents are as follows: Select the correct received data as the entry parameter of the federate; use the unique algorithm component of the federate to process the received data, get the processing results, and submit the processing results to the data distribution layer.

Data distribution layer (DDL), its functional requirements: according to the published relationship statement to accurately complete the data transmission [8]. Its main contents are: Obtaining the return data of the processing layer; confirming the time advance permission; if it is allowed to advance, it can continue to advance the simulation; confirming the data transmission channel; completing the data transmission.

The functional requirements of design federation execution (RFE): the federate ends the simulation and exits the Federation Execution [9]. Its main content: Receive the command of the control and exiting the federation execution, or the Federation members exit the federation execution by themselves, and the Federation members release their own simulation resources.

By building a general federate, the function of each interface can be realized, and the HLA interface framework and federate model are successfully separated, which makes the development of simulation federate easier and makes the model compatibility and reusability of simulation system greatly enhanced, and provides a foundation for the algorithm implementation of general calling GPU resources.

3   Design of Universal HLA Simulation Federate

The main design goal of universal federate is to enable HLA Simulation members to call GPU computing resources in a unified and simple way. The first is the production of DLL. HLA general simulation members are based on Java. From the point of view of programming language, CUDA and OpenCL both support C/C++ natively. It's troublesome to access other languages. To call GPU computing resources through Java code, you can access CUDA or OpenCL through JNI, or use various GPU Programming libraries based on Java version of JNI, such as jcuda [10].

Our idea is to use C/C++ programming to call the GPU resources, and then make the program into a DLL for Java program to call. By calling the DLL file, we can call the GPU computing resources.

Developing a special simulation member based on the general simulation member program. In the simulation member, we first need to declare the publishing and ordering relationship of object class and interaction class in each simulation member. Data forwarding is to forward the data content of the party who produces the data (Publishing member) to the party who is interested in these messages (subscribing member). The key to solving the problem of data distribution is to make the corresponding relationship between the publishing party and the subscribing party. Declaration management determines the direction of message forwarding, that is, to determine the message source (Publishing member), also to determine the ordering relationship between the message object (ordering member) and the publishing relationship of the simulation member itself, so that each member can accurately receive the information they need [11]. After the simulation operation, with the promotion of simulation, each simulation member constantly updates their object class attributes, and the simulation members constantly generate the data to be processed. During the simulation promotion, the simulation members pass the data to the DLL in the form of input parameters by calling the DLL file. After the data is processed by GPU, the data will be returned to the simulation member. The simulation member controls the whole simulation promotion according to the final received data.

Each simulation federate program is divided into two modules. One module is a general simulation member, which is used to arrange the HLA simulation logic processing part of the system model to run on the computer, in order to complete the HLA simulation federation operation [12]. The other is DLL file, which aims to complete the calculation function of simulation members by calling the GPU computing resources, and speed up the running speed of simulation members through GPU. Among them, the relationship between simulation member module and DLL module is as Fig. 2.

DLL module realizes the calling algorithm of GPU computing resources. The internal operations of DLL file to call GPU computing resources are as follows: firstly, the CPU allocates a block of memory in its internal cache, creates a data table on this memory, and saves the data sent by simulation members to be processed in this data table. As like as two peas, GPU also has a data table that is exactly the same in the internal cache. GPU sets the algorithm according to the algorithm set in advance in CPU, calculates the data table in the GPU internal cache, and finally sends the processed data to CPU, then returns it to the simulation member by CPU [13]. In terms of processing details, first create a view dataview on the internal cache of the CPU through

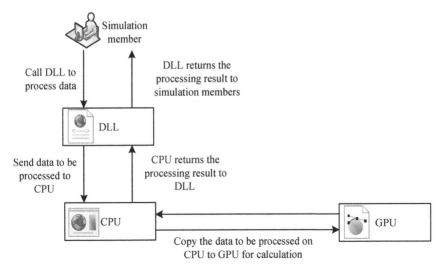

Fig. 2. Relationship between GPU based simulation member module and DLL module

the array_view<float,1>dataView(n,&arr[0]) method. Then run the code on the GPU through the function parallel_for_each(), and the GPU completes the complex calculation processing. Finally, the results obtained after running on the GPU are copied from the GPU to the CPU through the dataView.synchronize() function, so that the program can call the GPU computing resources.

4   Simulation Test

A simulation scheme is designed based on HLA simulation system and p-rti simulation server software. There are three simulation models: aircraft, tank and control center. Aircraft, tank and control center each form a simulation member, and record their coordinate information at each simulation time. The starting coordinates of aircraft and tank are (0, 0) and (100, 100), respectively, and the position information is sent to the control center. After receiving the position coordinates sent by other members, the control center calculates and processes the position coordinate data of aircraft and tank by calling DLL. If the distance between the aircraft and the tank is greater than the set distance, the control center sends a message to let the aircraft and the tank advance, and the aircraft and the tank update their position coordinates according to the specific motion formula. In this way, the simulation will continue until the plane and the tank reach a predetermined distance, and the control center will send the firing instruction to the tank to fire the plane, and then the simulation propulsion will end. The simulation member processing flow based on GPU is as Fig. 3.

The scheme will be implemented in two ways. The first way is that all simulation members' computation processing is undertaken by CPU. The second way is CPU + GPU, in which the HLA logic processing of simulation members is undertaken by CPU and the computation part is undertaken by GPU. In the process of simulation member

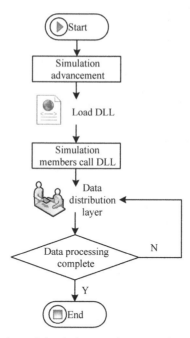

Fig. 3. Flow chart of simulation member processing based on GPU

promotion, the object class continuously sends data to the members who have ordered its object class objects, and receives the data sent by the object class they ordered. In this process, the simulation members need to load the DLL, and realize data processing by calling the DLL many times until the data processing is completed. The simulation steps are set at 1000 steps, 2000 steps and 3000 steps respectively. The simulation running time based on CPU and GPU is compared and analyzed, and the running results are as Fig. 4.

After the HLA Simulation members call the GPU computing resources successfully, they run the simulation cases several times, and all the simulation schemes can be promoted normally, which proves that the HLA Simulation based on GPU is feasible and reliable. Observe the average time of the simulation members when the number of propulsion steps is 1000, 2000 and 3000. It can be seen from the data in Fig. 4 that the GPU takes only about 1/3 of the CPU time when the simulation members have the same number of propulsion steps, which indicates that the HLA Simulation members run faster than the HLA Simulation members running on the CPU after calling the GPU computing resources, which proves that the GPU can optimize the running speed of HLA simulation.

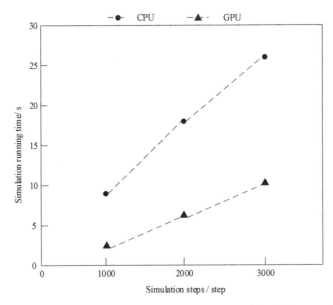

Fig. 4. Comparison analysis results of CPU and GPU based simulation running time

5   Conclusion

This paper aims to optimize the efficiency of HLA Simulation members by using the computing power of GPU. GPU is a special graphics core processor, because of its programming trouble and lack of data tools, it has not been given due attention in the development of HLA based simulation system for a long time. Making full use of GPU's massive data throughput and powerful floating-point computing ability will greatly improve the program performance. This paper designs a federation HLA simulation scheme based on GPU, and constructs a general simulation federation structure based on GPU by analyzing the computing resources of HLA Federation GPU. By allocating the computing part of HLA Simulation members to GPU and using its computing resources to realize floating-point computing processing of simulation data, the time consumption of simulation calculation can be greatly reduced, the running speed of joint members can be accelerated, and the efficiency of simulation development and operation can be improved. The disadvantage of this method is that the design and programming complexity of HLA simulation will be increased, and the development cost will be increased. Therefore, in the following research, we need to optimize the complexity of simulation design and programming, so as to reduce the development costs.

Fund Projects.   Hubei Natural Science Foundation guidance project (2019cfc887).

References

1. Sna, B., Fm, A., Am, C., et al.: Reliable identification of the first transition velocity in various bubble columns based on accurate sophisticated methods. Chem. Eng. Res. Des. **165**, 409–425 (2021)

2. Humood, K., Mohammad, B., et al.: On-chip tunable Memristor-based flash-ADC converter for artificial intelligence applications. IET Circ. Devices Syst. **14**, 107–114 (2020)

3. Geneugelijk, K., Spierings, E.: PIRCHE-II: an algorithm to predict indirectly recognizable HLA epitopes in solid organ transplantation. Immunogenetics **72**(8), 119–129 (2019)

4. Zheng, Z., Jza, B., Dong, Y.A., et al.: CFD simulation of fluidized magnetic roasting coupled with random nucleation model. Chem. Eng. Sci. **229**, 116148 (2020)

5. Savaan, H., Duman, L., Din, M., et al.: Migrating a legacy simulation to HLA: lessons learned integrating with new native HLA simulations. In: IEEE Fall SIW 2003. IEEE (2019)

6. Pd, A., Gza, B., Yc, A., et al.: Reasonable permutation of M2e enhances the effect of universal influenza nanovaccine. Int. J. Biol. Macromol. **173**, 244–250 (2021)

7. Herman, A.P., Gan, J., Yu, A.: GPU-based DEM simulation for scale-up of bladed mixers. Powder Technol. **382**, 300–317 (2020)

8. Tsai, M., Tian, Z., Qin, N., et al.: A new open-source GPU-based microscopic Monte Carlo simulation tool for the calculations of DNA damages caused by ionizing radiation — part I: core algorithm and validation. Med. Phys. **47**(4), 1958–1970 (2020)

9. Warren, C., Giannopoulos, A., Gray, A., et al.: A CUDA-based GPU engine for gprMax: open source FDTD electromagnetic simulation software. Comput. Phys. Commun. **237**, 208–218 (2019)

10. Ma, B., Gaens, M., Caldeira, L., et al.: Scatter correction based on GPU-accelerated full Monte Carlo simulation for brain PET/MRI. IEEE Trans. Med. Imaging **39**(1), 140–151 (2019)

11. Liu, S., Liu, D., Srivastava, G., Połap, D., Woźniak, M.: Overview and methods of correlation filter algorithms in object tracking. Complex Intell. Syst. **7**(4), 1895–1917 (2020). https://doi.org/10.1007/s40747-020-00161-4

12. Liu, S., Lu, M., Li, H., et al.: Prediction of gene expression patterns with generalized linear regression model. Front. Genet. **10**, 120 (2019)

13. Liu, S., Li, Z., Zhang, Y., et al.: Introduction of key problems in long-distance learning and training. Mob. Netw. Appl. **24**(1), 1–4 (2019)

A Searchable Encryption Method Supporting Result Grouping and Sorting

HaiYang Peng[1,2,3](✉), GuiShan Dong[2,3], Hao Yao[1,2,3], Yue Zhao[1,2,3], and YuXiang Chen[1,2,3]

[1] Science and Technology on Communication Security Laboratory, Chengdu 610041, China
physea@mail.ustc.edu.cn

[2] No. 30 Research Institute of China Electronics Technology Group Corporation, Chengdu 610041, China

[3] China Electronics Technology Cyber Security Co., Ltd, Chengdu 610041, China

Abstract. Aiming at the problem of low search efficiency caused by general searchable encryption schemes that do not sort search results during the search process, we propose a method to support grouping and sorting search results. This method uses the adjacent index structure as the basis, uploads the keywords corresponding to the file to the cloud server in order, and records the upload sequence of each keyword in the file, and stores the upload sequence in the index structure. The index is stored in the form of keyword-keyword sequence number-file address pointer. When the data user performs a keyword search, the keywords on the index are first matched, and then the results are grouped in the form of keyword serial number-file, and the search results are grouped and sorted according to the keyword serial number and returned to the user. The scheme proves the safety of the scheme through the method of bilinear difficulty analysis. At the same time, the scheme can support operations such as index update and deletion, and is easy to maintain.

Keywords: Searchable encryption · Packet sorting · Bilinear pairing

1 Introduction

With the rapid development of cloud computing and cloud storage technology, more and more enterprise and organizational users choose to store on cloud servers that can be flexibly purchased, allowing users to remotely access data stored in the cloud through the client. However, the data stored in the cloud in the clear state always has a large security risk, which means that the value of the data owned by the data owner is also outsourced to the cloud service provider. At the same time, the security of the cloud server is also limited by the security of the cloud service provider. Maintenance capabilities. Servers directly connected to the Internet will be subject to various attacks from the Internet. The series of data privacy breaches reported in recent years have caused great losses to users and greatly stimulated users' need for security. Stored in cloud storage.

© ICST Institute for Computer Sciences, Social Informatics and Telecommunications Engineering 2022
Published by Springer Nature Switzerland AG 2022. All Rights Reserved
S. Liu and X. Ma (Eds.): ADHIP 2021, LNICST 417, pp. 133–143, 2022.
https://doi.org/10.1007/978-3-030-94554-1_12

The cloud server is always semi-honest and will be interested in the data stored on the server. In order to avoid the curious cloud server from reading the data stored on it, the data can be stored in the cloud server in a secret storage method. Effectively prevent the cloud server from viewing the data stored on it, because only the user with the key can decrypt and view the stored data. At the same time, in order to enable data to be searched in a secret environment, searchable encryption technology [1] was proposed. Searchable encryption technology is a technology that can search for keywords in a secret environment and return the corresponding files to the user. However, early Searchable encryption technology only supports single-user, single-keyword search, only supports one interaction, and does not sort the search results. Under this background, the user efficiency is not high, and the searchable encryption technology faces engineering applications and promotion. To solve this problem, we propose a searchable encryption scheme that supports grouping and sorting. The keywords corresponding to the file are uploaded to the cloud server in order, and the upload order of each keyword in the file is recorded. Stored in the index structure, the index is stored in the form of keyword-keyword sequence number-file address pointer. When the user searches, the keywords are first matched, and then the order of the keywords in the matching file is determined. The files with the same keyword sequence number should be grouped and sorted to realize the function of grouping and sorting.

The main advantages of the solution in this article:

1. Design a multi-keyword searchable encryption scheme to realize multi-user search, and add the information of the keyword sequence while establishing the keyword index.
2. Sort the searchable encrypted search results, which can sort the results according to the degree of relevance, and further improve the search efficiency.
3. The index is added, deleted and other technologies, so that the searchable encryption technology has better practicability.
4. Through the bilinear hypothesis for difficult problems, the safety of the search process is proved.
5. Through the comparison of simulation calculations, searchable encryption using the adjacent index structure has certain advantages in terms of storage and computational overhead.

2   Related Research

Searchable encryption technology was first proposed by Song et al. [1]. This scheme opened the research process of searchable encryption, but this scheme could not achieve provable security. Under the condition of strong provable security, in 2006, Curtmola et al. [2] proposed the current searchable symmetric encryption with the best performance, but the length of the retrieval trapdoor and the maximum number of documents in this scheme are linear; literature [3–6] realized multi-keyword column query and multi-dimensional range query. Lu et al. [7] improved the search efficiency of range retrieval by constructing an index structure. In recent years, in recent years, dynamic SSE research has received extensive attention. In 2012, Kamara et al. [8] first proposed a dynamic searchable symmetric encryption (DSSE, dynamic searchable symmetric encryption) solution that supports dynamic updates, and formalized the security of DSSE. DSSE not only supports ciphertext retrieval, but also supports dynamic update operations of searchable ciphertext, such as ciphertext addition, ciphertext deletion, and keyword addition and deletion.

Wang et al. [4] solved the problem of sorting and searching encrypted data for the first time, returning matching files through search and sorting, which greatly enhanced the usability of the system. Fu et al. [9] first studied and solved the problem of personalized multi-keyword sorting search for encrypted data. Zhang et al. [10] proposed a multi-keyword sorting search scheme in the multi-owner model. Wang et al. [11] proposed a verifiable fuzzy keyword search scheme based on symbol tree, which has the verifiability of search results. In 2013, Yu et al. [12] proposed a dynamic sorting SE scheme, which stores the tf value and idf value of each file in the inverted index. When updated, only the idf value is used, and the updated key will be recalculated. Therefore, instead of updating all file vectors, only the auxiliary vector is updated. In 2015, Orencik and Savas [13] proposed a multi-keyword ranking, but it could not provide an accurate ranking. In the same year, Zhang et al. [10] proposed a searchable encryption technology to protect keyword ranking information. This solution achieved the ranking of search results while also protecting the ranking information on the cloud server. In 2017, Miao et al. [14]. The searchable encryption technology is realized through the method based on the attribute, and the sorting of the document level is realized. In 2020, Peng Haiyang and others [15] used segmented index to record the control and management of the search range in the index structure in a contiguous manner, but did not sort the search results. In 2020, Guan et al. [16] designed a The cross-language multi-keyword sorting search with semantic expansion can also speed up the sorting process by designing top-k data. The retrieval protocol is based on a heap binary tree structure.

3 Mathematical Foundation

3.1 Bilinear Pairs

Suppose that G_1 and G_2 are both multiplicative cyclic groups of prime number ρ, and g is the generator of G_1. We call the mapping: $G_1 \times G_2 \rightarrow G_2$ is a bilinear mapping. If the mapping $\mathcal{C}$ satisfies the following 3 properties.

(1) Bilinear. For any $a, b \in Z\rho, g \in G$, there is $(g^a, g^b) = (g, g)^{ab}$. For any $g_1, g_2, g \in G$, there is $(g_1, g_2, g) = (g_1, g)(g_2, g)$. For any $g_1, g_2, g \in G$, there is $(g, g_1, g_2) = (g, g_1)(g, g_2)$.
(2) Non-degeneration. $(g, g) \neq 1$.
(3) Computable. For any $P, Q \in G$, it can be calculated that (P, Q) belongs to G_2.

3.2 Introduction to Adjacency Index

In order to sort the search results in searchable encryption, the order-preserving encryption method can usually be used to sort the order of the search results. The order-preserving encryption is not flexible enough to deal with complex scenarios; it can also appear in the file by keyword. When performing the search process, the results of the documents are sorted according to the frequency of keywords. However, this solution first needs to calculate the frequency of keyword occurrences. The frequency of keyword occurrences does not occur in some scenarios. It does not represent the degree of relevance, and some commonly used words will also occupy a large frequency in the text, and the keyword frequency requires full-text scanning, which brings a certain degree of complexity to the calculation.

This article uses adjacency index. In view of the characteristics of the adjacency index that can store other description information on the index, the secret keywords are linked by a linked list, and the documents with the keywords in the same position are grouped into a group and uploaded according to the sequence number of the keywords. Perform grouping and sorting. In this way, the search results can be sorted relatively quickly, and a group of files with a higher degree of relevance can be found quickly, which improves search efficiency. At the same time, this solution is also applicable to multi-keyword search scenarios.

3.3 Construction, Update, and Deletion of Adjacency Index

The data owner (Do) first identifies the document and establishes a set, which can be expressed as: $\mathbf{D} = \{d_1, d_2,..., d_n\}$, and extracts keywords from the file. The set of keywords can be expressed as $\mathbf{W} = \{w_1, w_2, w_3,..., w_n\}$, form the corresponding index, where the file d_x contains multiple keywords and arranged in order $\{w_{x1}, w_{x2},...... w_{xn}\}$, w_{x1} means that the keyword w_x is in this The first keyword uploaded in the file is used to infer and upload all the keywords contained in the file..

As shown in Fig. 1 Adjacency index structure diagram, the keywords of all documents are combined and encrypted to form an index. The left side of w_1 links to the addresses of all documents containing the keyword w_1, and w_{11} is adjacent to all documents that contain the keyword w_1 and the keyword w_1 is the first upload among all documents The set $\{d_{111}, d_{112},..., d_{11n}\}$ of all file set identifiers, as shown in Fig. 1, for example; w_{12} represents the file containing the keyword w_1 and w_1 is the second uploaded keyword, that is, the keyword In the process of document uploading, w_1 is the second uploaded keyword. The set of identifiers for all files is $\{d_{111}, d_{112},...,d_{11n}\}$, and so on, the meaning contained in the adjacent index structure in the figure can be obtained.

Extract all keywords and use contiguous indexing, and add a keyword sequence identifier (w_{xy}, where w_x represents the keyword, and y represents the upload order of the keyword in the file, that is, keyword Serial number), each keyword sequence identifier links the file ID collection that contains the keyword and the keyword is in the same keyword sequence number in the file, which constitutes an important part of this article. The adjacent index is composed of when performing a search, the keywords are first matched, and then the corresponding files are grouped according to the relevant sequence number of the keyword to obtain the corresponding files. After the files are

Enc (W_1)	w_{11}	id_{111} — id_{112} — id_{113} — ...
	w_{12}	id_{121} — id_{122} — id_{123} — ...
	...	... — ... — ... — ...
	w_{1n}	id_{1n1} — id_{1n2} — id_{1n3} — ...
Enc (W_2)	w_{21}	id_{211} — id_{212} — id_{213} — ...
	w_{22}	id_{221} — id_{222} — id_{223} — ...
	...	... — ... — ... — ...
	w_{2n}	id_{2n1} — id_{2n2} — id_{2n3} — ...
Enc $(\cdots)$	$w\cdots$	... — ... — ... — ...
	$w\cdots$	... — ... — ... — ...
	...	... — ... — ... — ...
	$w\cdots$	... — ... — ... — ...
Enc (W_n)	w_{n1}	id_{n11} — id_{n12} — id_{n13} — ...
	w_{n2}	id_{n21} — id_{n22} — id_{n23} — ...
	...	... — ... — ... — ...
	w_{nn}	id_{nn1} — id_{nn3} — id_{nnn} — ...

Fig. 1. Adjacency index structure diagram

obtained and they are small to large according to the keyword sequence number, all files are in the form of keyword sequence number-document Perform grouping and sorting.

Update, first verify the legal identity of the user, collect the document keywords and sort the keywords, find the position of the keyword in the index according to the keyword information, and determine the sequence number of the keyword in the file, and add it to Correspond to the file collection under the serial number (usually added to the end of the linked list), As shown in Fig. 2 Adjacency index update graph, and map the corresponding file address with the keyword serial number (as shown in the Fig. 2). By orderly updating the file ids on the index, it is possible to ensure real-time and effective ordering of the updated files.

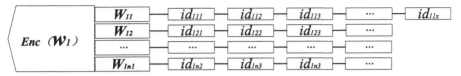

Update, directly insert the end of the linked list
corresponding to the keyword

Fig. 2. Adjacency index update graph

To delete the index, first verify the legal identity of the user, then find the keyword through the index, As shown in Fig. 3 Adjacency index deletion graph, find the position of the file in the index according to the keyword sequence number in the file, and point the pointer of the file to the next file. If you want to delete The file of is at the end of the index, just delete it directly.

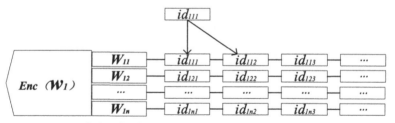

Delete the file. If the file to be deleted is id_{11}, point the address
of the original id_{11} to id_{12}.

Fig. 3. Adjacency index deletion graph

4 System Model

4.1 Introduction to System Basic Model

The solution model is composed of data owners, cloud servers, group users, and internal organizations. Through the cooperation of these parts, the search results can be sorted by groups, according to the order of keywords in the document, and according to the order of keywords in the document. The order in a single document is collected and then sorted.

1. Data owner: In this system, it is a user group with multiple sub-users. The data owner can also be a data user. The data user uploads the keywords contained in the file in an orderly manner and uploads in order of relevance. It generates an index on the cloud server, and then the encrypted file is uploaded to the cloud server.
2. Cloud server: After the user logs in to the server with the user name and password, the cloud server stores the encrypted files uploaded by the user, stores the encrypted index, performs searchable and encrypted search process, and returns the search results to the user who submitted the search request.
3. User group: The user group is the user of data. The data user remotely logs in to the server through his own private key, keywords, user name and password assigned by the internal organization and submits a search request to the server, and the server returns the search results to the user.
4. Internal organization: The internal organization is credible, generates public parameters and assigns user names and passwords to users, and maintains the authority for cloud server and user authentication (Fig. 4).

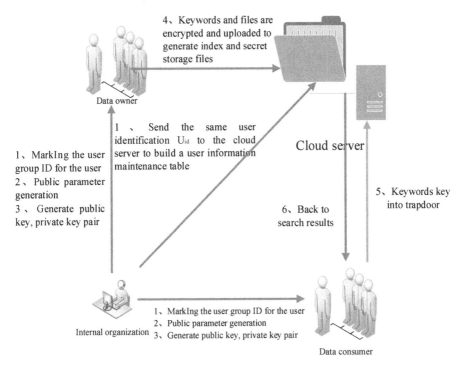

Fig. 4. Searchable encryption scheme composition diagram

4.2 Introduction to the Formal Process

The design purpose of this article is to design a searchable encryption scheme that can group and sort search results according to the location information of keywords, by sorting the file keywords before uploading, and then recording the order information of the keywords in the file to the index In the structure, when the user requests a search, the keyword is generated trapdoor, and the cloud server executes the search process, and matches the file corresponding to the keyword, and then sorts the keywords by the order of the keywords in the document, and finally realizes the search The order of the results.

1) *ParaGen* $(1^n) \rightarrow$ **P**, Mpk, Msk. The algorithm is executed by an internal organization. A security parameter n is input, and the public parameter P of the system is output, the master public key Mpk, and the master private key Msk.

2) *Accredit(U)* $\rightarrow U_{id}$, The algorithm is run by an internal organization, assigns different identity information to each legal user, and internally organizes and maintains user information used by the system.

3) *KeyGen*(**P**, Mpk, Msk, U_{id}) $\rightarrow pk_i, sk_i$. The algorithm inputs the public parameters **P**, the master key Mpk, the master private key Msk, the unique identifier Uid assigned by the internal organization, and outputs the legal user's public key and private key pair (pk_i, sk_i).

4) *Build Index*(**P**, **W**, pk_i, U_{id}) $\rightarrow$ Index (**W**) index establishment, enter the public parameter **P** generated by the system, the public key pk_i of the data owner, and

the keyword information **W** of the file, where **W** contains multiple keys Keyword information group with a certain sequence, unified identity U_{id}, first check whether the unified identity U_{id} of the data owner is in the user information table, if not, refuse service upload; if yes, the keyword is encrypted to construct an adjacency index, The output index Index (**W**).

5) $Enc(pk_i, M) \rightarrow E(c)$ Enter the uploader's public key and file for encryption (it can also be a symmetric key), and enter the cipher text $m = E(c)$.

6) $Trapdoor(U_{id}, w_x, sk_i) \rightarrow T(w_x)$, enter the keyword w_x requested by the user and the user's private key sk_i to generate a keyword trapdoor $T(w_x)$.

7) $Test(T(sk_i), Index, U_{id})$. Enter the trapdoor and the index Index (**W**), the unique identifier U_{id}, and perform the following process: Query the user in the user information table on the server. If the user exists in the information table, the index ciphertext match will be performed to find the keyword Existing ciphertext, and sort the search results by the sequence structure of the keywords, obtain the accessible file addresses, and find the corresponding ciphered files through the keyword and file mapping relationship.

5 Security Proof

5.1 Bilinear Difficult Problem

A variant of the bilinear Diffie-Hellman problem, Given two multiplications of order of large prime q The cyclic group is a generator of, and the bilinear mapping problem is: for a given where calculation If the adversary of any polynomial time solves the problem with a negligible advantage, we consider the problem to be difficult, which is formalized as:

$$\Pr\left[A\left(g, g^a, g^b, g^c, g^{1/a}\right) = e(g, g)^{abc}\right] \geq \varepsilon \tag{1}$$

5.2 Security Certification Process

The security target attacker selects any user U_{id} within the range and the trapdoor information $T(w)$ corresponding to the keyword w that the user can obtain. The main idea is to prove that the attacker cannot obtain effective information about the keyword corresponding to the trapdoor through the known legal trapdoor $T(w)$. In the simplest way, it can be considered that the attacker cannot obtain the effective information from the two legal trapdoors. In the trap door, it is effective to judge whether the trap door comes from the same keyword [2].

System initialization: The system first runs the algorithm to generate the public parameters required by the system, including the master key sk_0 and the master private key pk_0, assigns the corresponding user public and private key pair sk_{ui}, pk_{ui} to the data user, and generates the public and private key pair.

Inquiry and challenge phase 1:

First inquiry: The attacker randomly selects a keyword w to generate a trapdoor to initiate an inquiry. The cloud server calculates the trapdoor $T(w)$ of the keyword w through the trapdoor generation algorithm, and returns the calculated result to the attacker.

Inquiry and Challenge Phase 2:

The attacker again initiates an inquiry and challenge to the ciphertext and index, but the keywords that initiate the inquiry and challenge cannot be the keywords that appeared in the inquiry and challenge phase 1, and the calculated trapdoor is returned to the attacker.

Inquiry phase 3:

The attacker randomly selects one of the two keywords and initiates a query. The query keyword cannot be a keyword that has already been queried. The simple expression is: $w_b, b \in \{0, 1\}$, Use the attacker's key to encrypt and send it to the cloud server.

Output: The attacker outputs guesses $b' \in \{0, 1\}$, If the attacker's guess $b' \in b$ is correct, the attacker is at an advantage during the entire game challenge, and the probability of winning the game is:

$$\varepsilon = \left| \Pr(b' = b) - \frac{1}{2} \right| \tag{2}$$

The hypothesis based on the bilinear difficult problem is: if the advantage of any polynomial time adversary winning in the game is that ε is a negligible function, then the search trapdoor of this scheme meets the trapdoor under the adaptive keyword attack. Distinguishability. That is to say, for the given two trapdoors, the attacker cannot guess the trapdoor corresponding to the keyword from the known information with a non-negligible advantage, and it is also safe to resist semantic attacks.

6   Plan Analysis and Comparison

This paper selects the searchable encryption field in the past few years and the similar literature [10, 14] for comparison, in the case of achieving similar results, from the comparison of computing overhead and storage overhead, analyze the comparison between this paper and theirs. Advantage.

For more convenience, use t_{eo} and t_{e1} to represent the computational cost of performing an exponential operation in G_0 and G_1, respectively, use t_{Ho} and t_{H1} to represent the computational cost of performing a hash operation that maps an arbitrary string to G_0 and G_1. In addition, $|G_0|$ and $|G_1|$ respectively represent the length of the elements of the group G_0 and G_1, and $|Z|$ represents the length of the ciphertext of the symmetric encryption algorithm. Through the unified definition of parameters, this article can intuitively compare the various solutions.

The comparison results are shown in Table 1. Through analysis, it is easier to see that the trapdoor size of Zhang's [10] scheme is $(1 + u)|G_0|$, while Miao's scheme [14] and this scheme are $(2y + 4)|G_0|$. In Zhang's [10] scheme, the storage cost it grows as the number of keywords in the query increases, and this solution increases with the number of keywords. For example, when a large number of keywords are queried and only a small amount of grouping and sorting is involved, the storage overhead of this solution has certain advantages in terms of storage. In addition, the ciphertext size in the Miao [14] scheme is $|F||G_0| + (|F| + jN + N)|G_1|$. The ciphertext size of this scheme

Table 1. Comparing computational overhead and storage overhead

	Miao's plan	Zhang's plan	This plan
Index storage	$(2x + 2 + n)\lvert G_0\rvert$	$2n\lvert G_0\rvert$	$(2x + n)\lvert G_0\rvert$
Trapdoor storage	$(2y + 4)\lvert G_0\rvert$	$(1 + u)\lvert G_0\rvert$	$(2y + 4)\lvert G_0\rvert$
Ciphertext storage	$\lvert F\rvert\lvert G_0\rvert + (\lvert F\rvert + jN + N)\lvert G_1\rvert$	$2n\lvert G_0\rvert$	$k\lvert G_0\rvert + (k + N)\lvert G_1\rvert$
Keygen computing	$(2y + 3)t_{e_0} + yt_{H_1}$	nt_{H_0}	$(2y + 3)t_{e_0} + yt_{H_1}$
Enc computing	$(2\lvert F\rvert + 2x + n)t_{e_0} + xt_{H_1}$	$2nt_{e_0}$	$(2\lvert F\rvert + n)t_{e_0} + xt_{H_1}$
Trapdoor computing	$(2y + 4)t_{e_0}$	$(1 + u)t_{e_0}$	$(2y + 2)t_{e_0}$
Decryption computing	$\lvert F\rvert t_{e_0}$	–	$\lvert F\rvert t_{e_0}$

Note: x: the number of attributes in the system; Y: the number of data user attributes; |F|: the number of undifferentiated files; N: the number of transmission nodes: k: the number of top-k files; j: the number of child nodes of the transmission node; n: the number of keywords in the keyword dictionary; U: the number of keywords queried by the user

is $k\lvert G_0\rvert + (k + N)\lvert G_1\rvert$. In a massive data system, the value of k is obviously much smaller than $\lvert F\rvert$. In this way, compared with the Miao scheme [14], this scheme also has advantages in terms of ciphertext size.

In the table, for the calculation cost of the KeyGen, Enc, Enc Tpdr algorithm, this scheme requires $(2y + 3)t_{eo} + yt_{H1}$, $(2\lvert F\rvert + n)t_{eo} + xt_{H1}$, $(2y + 2)t_{eo}$, From these data comparisons. It is hard to see that compared with the computational cost of the Miao scheme [14], in a large-scale data sharing system, n is the number of files, much larger than other values, the algorithm in this scheme has higher computational efficiency, which means that this scheme is more effective and practical.

7 Summary and Outlook

Aiming at the search scenario of a large number of users and files, this paper proposes a searchable encryption scheme based on keyword location grouping and sorting. Through the adjacent index structure, the order information of the keywords used and the file are mapped and associated with each other, so that the search results are grouped and sorted In the scheme, the index structure of this scheme uses adjacency, which is more efficient and easy to maintain. In terms of safety, this paper adopts the assumption of difficult problems based on bilinear pairs, and the safety has been proved. Through literature surveys, although the solution proposed in this article has certain advantages in some scenarios, it has certain advantages in computing overhead and storage overhead through comparison, but it has not been further optimized in a single group. This article is more critical The scheme of word ordering in this article is also applicable, but it is not explained and introduced in detail. These application scenarios are also urgently needed schemes in reality. In the future, we will do further research on these problems.

Acknowledgements. This work was supported by Sichuan Science and Technology Program (2020YFG0298), Sichuan Science and Technology Program (2021JDRC0077), and Key Laboratory Fund (6142103010711).

References

1. Song, D.X., Wagner, D., Perrig, A.: Practical techniques for searches on encrypted data. In: Proceeding 2000 IEEE Symposium on Security and Privacy, S&P 2000, pp. 44–55. IEEE (2000)
2. Curtmola, R., Garay, J., Kamara, S., et al.: Searchable symmetric encryption: improved definitions and efficient constructions. J. Comput. Secur. **19**(5), 895–934 (2011)
3. Nepolean, D., Karthik, I., Preethi, M., Goyal, R., Vanethi, M.K.: Privacy preserving ranked keyword search over encrypted cloud data. In: Martínez Pérez, G., Thampi, S.M., Ko, R., Shu, L. (eds.) SNDS 2014. CCIS, vol. 420, pp. 396–403. Springer, Heidelberg (2014). https://doi.org/10.1007/978-3-642-54525-2_35
4. Wang, C., Cao, N., Li, J., et al.: Secure ranked keyword search over encrypted cloud data. In: 2010 IEEE 30th International Conference on Distributed Computing Systems, pp. 253–262. IEEE (2010)
5. Wang, C., Cao, N., Ren, K., et al.: Enabling secure and efficient ranked keyword search over outsourced cloud data. IEEE Trans. Parallel Distrib. Syst. **23**(8), 1467–1479 (2011)
6. Cao, N., Wang, C., Li, M., et al.: Privacy-preserving multi-keyword ranked search over encrypted cloud data. IEEE Trans. Parallel Distrib. Syst. **25**(1), 222–233 (2013)
7. Lu, Y.: Privacy-preserving logarithmic-time search on encrypted data in cloud. In: NDSS (2012)
8. Kamara, S., Papamanthou, C., Roeder, T.: Dynamic searchable symmetric encryption. In: Proceedings of the 2012 ACM Conference on Computer and Communications Security, pp. 965–976 (2012)
9. Fu, Z., Ren, K., Shu, J., et al.: Enabling personalized search over encrypted outsourced data with efficiency improvement. IEEE Trans. Parallel Distrib. Syst. **27**(9), 2546–2559 (2015)
10. Zhang, W., Lin, Y., Xiao, S., et al.: Privacy preserving ranked multi-keyword search for multiple data owners in cloud computing. IEEE Trans. Comput. **65**(5), 1566–1577 (2015)
11. Wang, J., Chen, X., Ma, H., et al.: A verifiable fuzzy keyword search scheme over encrypted data. J. Internet Serv. Inf. Secur. **2**(1/2), 49–58 (2012)
12. Yu, J., Lu, P., Zhu, Y., et al.: Toward secure multi keyword top-k retrieval over encrypted cloud data. IEEE Trans. Dependable Secure Comput. **10**(4), 239–250 (2013)
13. Orencik, C., Alewiwi, M., Savas, E.: Secure sketch search for document similarity. In: 2015 IEEE Trustcom/BigDataSE/ISPA, no. 1, pp. 1102–1107. IEEE (2015)
14. Miao, Y., Ma, J., Liu, X., et al.: Attribute-based keyword search over hierarchical data in cloud computing. IEEE Trans. Serv. Comput. **13**(6), 985–998 (2017)
15. Peng, H., Dong, G., Yao, H., Zhao, Y., Chen, Y.: Searchable encryption scheme with limited search scope for group users. In: 2020 International Conference on Networking and Network Applications (NaNA), Haikou City, China, pp. 430–435 (2020). https://doi.org/10.1109/NaNA51271.2020.00079
16. Guan, Z., Liu, X., Wu, L., et al.: Cross-lingual multi-keyword rank search with semantic extension over encrypted data. Inf. Sci. **514**, 523–540 (2020)

Edge Computing Based Real-Time Streaming Data Mining Method for Wireless Sensor Networks

Zhong-xing Huang[1], Xiao-li Ren[2(✉)], Zai-ling Zhou[1], He Zhu[1], and Zhi-li Lin[1]

[1] Guangzhou Metro Design and Research Institute Co., Ltd., Guangzhou 510000, China
[2] Zhongye Design Co., Ltd., Guangzhou 510000, China

Abstract. Traditional data mining techniques are difficult to be directly applied to wireless sensor networks because of the multidimensional and multilayered characteristics of wireless sensor networks. Based on the theory of edge computing, the framework of distributed data mining workflow in wireless sensor networks is optimized, and the flow of distributed data mining in wireless sensor networks is demonstrated. Finally, the design requirements of data mining methods are realized.

Keywords: Edge computing · Wireless sensor network · Data mining

1 Introduction

Due to the large scale and random deployment of WSN, the communication environment, limited energy supply and high failure rate are often impaired, which makes WSN knowledge mining face many severe challenges. A lot of dynamic data will be generated in sensor network applications. In order to ensure the operation effect of the network, the edge computation is used to analyze the network data and extract the knowledge, and the real-time stream data is mined. Therefore, the development of wireless sensor network data mining technology, is essentially to promote the development of real-time intelligent wireless sensor networks. In the literature [6], some scholars put forward a method of integrated feature clustering, which uses matrix representation and convolutional neural network to extract and fuse features, and uses multi-source data structure combined with missing data interpolation method to achieve high accuracy of data mining. In the literature [7], aiming at the data processing of soil, two data mining algorithms, multiple adaptive regression and gene expression programming, are proposed to construct a data model, train the data model and collect features, thus realizing the deep mining of complex data. Traditional data mining can not be directly applied to wireless sensor networks because of its centralization, heavy computation and emphasis on transaction data processing. Therefore, this paper proposes a real-time stream data mining method based on edge computing for wireless sensor networks. By analyzing the data characteristics of wireless sensor networks, the frequent itemsets of data sets are obtained, the data

S. Liu and X. Ma (Eds.): ADHIP 2021, LNICST 417, pp. 144–155, 2022.
https://doi.org/10.1007/978-3-030-94554-1_13

management architecture is constructed by using sensor nodes, the data is compressed to reduce the traffic, and the data mining process is optimized by using the edge computing theory.

2 Real-Time Streaming Data Mining Method for Wireless Sensor Networks

2.1 Composition of Wireless Sensor Network Real-Time Streaming Data Structure

Real-time stream data feature mining in wireless sensor networks is to extract application-oriented, acceptable and accurate data models and patterns from the continuous and fast data streams in sensor networks. In the process of data mining, the data can not be stored and must be processed in time. Data mining algorithm must be effective and fast processing of high-speed data. Traditional data mining algorithms are good at processing and analyzing static datasets, but not suitable for processing large, high dimensional and distributed data generated by wireless sensor networks. Based on this, we first classify the features of real-time stream data in wireless sensor networks as follows (Table 1):

Table 1. Characteristics of real-time streaming data in wireless sensor networks

Name	Traditional data	Wireless sensor network data
Processing architecture	Focus	Distribution
Data type	Static state	Dynamic
Memory usage	Unlimited	Restricted
Processing time	Unlimited	Restricted
Computing power	High	Low
Energy	Unlimited	Limited
Data stream	Static	Successive
Data length	Limited	Infinite
Response time	Non real time	Real time
Update rate	Low	high
Number of scans	Many times	Single time

Data mining is based on the features of different classes of data. The task of real-time stream frequent patterns mining in wireless sensor networks is mainly carried out under the condition of limited computing and storage resources. According to the mining results, we mine the maximal frequent itemsets, closed frequent itemsets, complete frequent itemsets and Twk frequent itemsets for real-time stream data in wireless sensor networks. Furthermore, the random mining algorithm in frequency range based on relative error count can divide the WSN data into probability-based approximation algorithm and deterministic error interval approximation algorithm. The networking style of wireless sensor networks is shown in the following figure (Fig. 1):

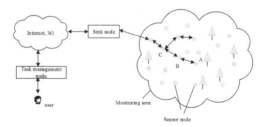

Fig. 1. Composition of wireless sensor networks

Based on the above structure, the sensor node of wireless sensor network is a micro-embedded system. In different application background, the composition of the sensor node is different (Fig. 2).

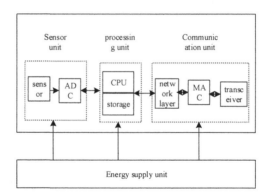

Fig. 2. Sensor node data feature processing module

The sensor unit is responsible for data acquisition, and the processing unit is responsible for data processing and controlling the whole node. In the frequent pattern mining of stream data, we can use the timeliness of stream data and the drift of stream center to combine the two models of landmark window and time attenuation. The technology of frequent pattern mining is mainly based on a dynamic system to form the overall pattern support number, and then calculate the frequency of patterns in the landmark window according to the time attenuation model. The algorithm has high mining precision, low memory cost, and can meet the requirements of high speed stream data processing, and can adapt to different number of transactions, different services and different average length of potential frequent pattern stream data mining.

2.2 Stream Data Mining Association Rule Algorithm Optimization

Stream data mining is to mine the arriving data stream according to a certain sequence, which is different from the mining of static data association rules in that the stream data is high-speed, continuous and without boundary. The unique characteristics of stream data bring a series of problems to the data mining and analysis, so as to identify the effective patterns in the mining cycle more quickly and efficiently. In order to improve the computational efficiency, 1 frequent itemset is proposed to approximate the difference between two datasets. The following formulas are given.

$$\text{error}(D, S) = \frac{|L_1(S) - L_1(D)| + |L_1(D) - L_1(S)|}{|L_1(S)| + |L_1(D)|} \tag{1}$$

In the above algorithm, L data feature set, S data difference, D data interference coefficient. The algorithm for further calculating the degree of difference caused by missing frequent itemsets is:

$$\begin{aligned} S = {}&(L(S) - L(D)| + |L(D) - L(S)D \\ &+ (MMS) - M(D)| + |M(D) - M(S)| \\ &- (0L(D) \cap M(S)|) + (0M(D) \cap L(S)D) \end{aligned} \tag{2}$$

Further optimize the process for calculating the differences in available data mining as follows:

$$\text{error}(D, S) = \frac{\sum\limits_{i=1}^{k} |L_i(S) - L_i(D)| + |L_i(D) - L_i(S)|}{\sum\limits_{i=1}^{k} |L_i(S)| + |L_i(D)|} \tag{3}$$

Each node can collect M kinds of attribute data in N times. This node with multiple sensing elements is called a multimode node. The data monitored by Atr, the i attribute of the sensor node, is a N-long time sequence, s = (sx), where s represents the data collected by the i attribute at j time. The raw data on the sensor is thus abstracted into a matrix:

$$A^0 = \begin{bmatrix} s_0 \\ s_1 \\ \vdots \\ s_{M-1} \end{bmatrix} = \begin{bmatrix} s_{0,0} & s_{0,1} & \cdots & s_{0,N-1} \\ s_{1,0} & s_{1,1} & \cdots & s_{1,N-1} \\ \vdots & \vdots & \ddots & \vdots \\ s_{M-1,0} & s_{M-1,1} & \cdots & s_{M-1,N-1} \end{bmatrix} \begin{bmatrix} t_0, t_1, \cdots t_{N-1} \end{bmatrix} \tag{4}$$

Based on the above algorithm, the mining algorithm is re-run to extract association rules, so as to improve the mining efficiency and reduce the mining cost as far as possible under the limited system resources, and effectively scan the original data set once, and then make an incremental update with the saved results of the previous scan in the next periodic scan, and obtain the frequent itemsets near the support of the adjacent original data set to participate in the estimation of the variation degree of the two data sets, so as to determine whether it is necessary to run the mining algorithm to extract data patterns.

2.3   Implementation of Streaming Data Mining in Wireless Sensor Networks

Wireless sensor network is a kind of data network focusing on the transmission of a large amount of monitoring information. Users need data information rather than hardware devices or sensors themselves. Because of the characteristics of sensor nodes, the research concept of wireless sensor network is quite different from that of traditional information network devices only focusing on data transmission in basic design. Therefore, effective integrated management and operation of the data in the network become the core research technology for optimizing and improving the performance of wireless sensor networks. Wireless sensor networks usually focus on the monitoring data of interest, and take sensor nodes collecting data as the original stream or source of sampling data. Because all nodes build the whole network as the data transmission space or a large amount of data storage base, it can be said that the whole wireless sensor network is the bulk collation and integrated management of the sampled data. The main task and function of wireless sensor nodes are responsible for the sensing, internal storage, interrogation and data mining of the monitoring data, and separate the logical sensing map of the collected data in the monitoring environment from the physical reality of the network, and give the logical transmission structure of the query to the users. Of course, the effective management and processing of the network data from the beginning to the end of the integrated processing of the network through the whole network, the following steps are needed to improve the efficiency of the overall network management of the sensor network. The following steps are needed to consider the overall data (Fig. 3).

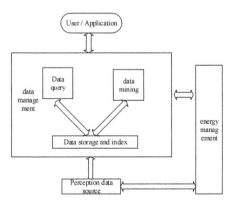

Fig. 3. Data management architecture

In the process of data management, we should consider the transmission path, redundant data and query optimization. Generally, the omni-directional management of data should include obtaining effective information, storing effective data, querying a large amount of stored information, mining deep data and the whole system management technology. The data management of wireless sensor networks is mainly located in the network layer and application layer in the longitudinal network architecture. The main task of the network layer is to provide sampling data and process the original data, and send the final data results to the application layer. Data mining technology is the key part of all data processing. The intuitive definition is to extract or mine the useful "knowl-edge" from the redundant and repetitive sampled data, and use intelligent method to extract data model. Because the energy of sensor nodes and the whole sensor network is limited, we need to process a large number of data in the process of data transfer. Compression of the data at the sensor nodes and transmission of the compressed results can reduce the communication volume of the sensor network and prolong the lifetime of the network (Fig. 4).

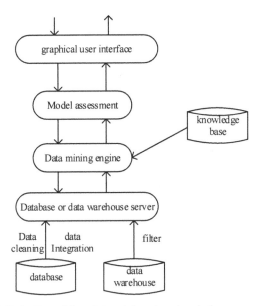

Fig. 4. Optimization of flow data mining steps in wireless sensor networks

We can define perceptual data as a relational database because of the temporal or spatial correlation of perceptual data collected in wireless sensor networks. Then we can mine the data in this relational database, and data mining is to mine the meaningful data contained in many information. Because there is a certain degree of correlation between the data, such as sampled data in a similar or identical point in time, the correlation can be eliminated to the greatest extent by some transformations; but after the adoption of some transformations, the loss of the original data and the error between the predicted data and the original data may sometimes occur, and such compression is called lossy compression. At present, there are many methods to compress the data, but the essence of which is reversible lossless compression and irreversible lossy compression. Data compression is at the cost of certain quality loss, and the quality loss is within the range of error allowed by the condition (Fig. 5).

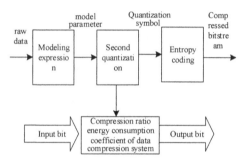

Fig. 5. Data compression processing step optimization

For the correlation of observation data, the approximate data can be obtained by constructing a suitable mathematical model of time series, so that the data amount is less than that of the original time series. Set the sampling data as:

$$\xi = ((t_1, d_1), (t_2, d_2), L, (t_u, d_w)) \tag{5}$$

Secondly, the recursive curve of data feature fitting is given. The function takes t and d in the sampling data sequence as independent variable and dependent variable respectively.

$$d = \alpha + \beta t + \xi, \xi \sim \left(0, \delta^2\right) \tag{6}$$

The least square method is used to fit the above data features linearly, and the data stream features are estimated as follows:

$$\begin{cases} \alpha' = 1/n \sum_{i=1}^{n} d_i - \left(1/n \sum_{i=1}^{n} t_i\right)\beta' \\ \beta' = \left[\sum_{i=1}^{n} t_i d_i - \frac{1}{n}\left(\sum_{i=1}^{n} t_i\right)\left(\sum_{i=1}^{n} d_i\right)\right] / \left[\sum_{i=1}^{n} t_i^2 - \frac{1}{n}\left(\sum_{i=1}^{n} t_i\right)^2\right] \end{cases} \tag{7}$$

Further, the characteristic regression equations of data are obtained.

$$\hat{d} = \alpha' + \beta' t \tag{8}$$

In order to solve the problem of energy consumption in data transmission, an effective data stream management framework is proposed, which aims to process data from different types of systems, aggregate any different and abnormal data streams, and further mine the anomaly feature data of data nodes. According to the initial cluster population value M collected from the data model, the individual characteristics of the optimal population are judged. If the number of cluster features collected is j, the dynamic adjustment range numerical algorithm for abnormal data is:

$$\zeta = \hat{d}\lambda \bigcap^{j-1}{}^{1/}(U - M) \tag{9}$$

Based on the above algorithm, the difference node numeric operators of anomaly data are globally optimized and clustered. In order to speed up the convergence rate of data mining, it is necessary to carry out iterative processing of the above algorithms. Because the automatic data node anomaly mining method is relatively complex and a complete process, the process of data mining is usually relatively cumbersome, time-consuming and prone to bias. Therefore, the process of data mining of anomaly characteristics is optimized as follows (Fig. 6):

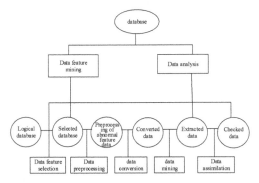

Fig. 6. Abnormal feature data mining procedure

In order to reduce the complexity of data mining to the greatest extent, the steps of anomaly feature data mining are further improved as follows:

1. Randomly collect the characteristic values of data nodes of the security resource pool.
2. Mining the internal and external characteristic data information of the collected data.
3. Further analyzing and transforming the characteristic data, establishing corresponding analysis models, and establishing corresponding SDN data centers.
4. Combine the feature data and SDN data center detection values to achieve automatic mining, obtain and output the mined data feature values.
5. The error Min value is given, then the data meeting the Min value is approximately compressed, and finally the compressed time series is decompressed to obtain the approximate prediction data, thus prolonging the network life cycle and realizing the data feature mining goal.

3   Analysis of Experimental Results

Using Matlab 7.0 as the simulation platform for simulation analysis. In order to verify the practicability of real-time stream data mining in wireless sensor networks based on edge computing, the following comparative experiments are designed. Two computers with the same configuration were used as the experimental object, in which the experimental group was loaded with deep mining method and the control group was loaded with SDN marking principle. A network device with high running stability is selected as the monitoring subject, and the changes of the influence parameters of the experimental group and the control group are recorded respectively.

In order to ensure the rationality of the experimental results, the experimental environment and parameters are set.

The hardware running environment for the experiment was Pentium- Core R980 @ 2.3 GHz dual core CPU, 64 GB memory hard disk.

Experimental software environment: Windows - XP VC+ system, C++ language.

The data of wireless sensor network is collected randomly, and the object is 4200. In order to ensure the efficiency of the experiment, it is always in the range of 18.24–20.04. In this case, the exception data is categorized and analyzed as follows (Table 2):

Table 2. Characteristics of streaming data in wireless sensor networks

Attribute category	Name
SOL	Back, land, neptune, smurf
STN	ftp_write, guess_passwd, imap, multi
R2L	Lpsweep, namp, portsweehop, phf
ROTP	overflow, loadmodule, perl, spy

In the above experimental environment, the experimental parameters are further standardized. Randomly selected feature mining data shall be set up, and the local density characteristic values, sparse characteristic values, support intrusion data flow and other relevant standard parameters of abnormal data shall be regulated, as shown in the following table (Table 3):

Table 3. Data stream standard parameters

Hierarchy	Characteristic value of local density	Support	Pattern (%)	Numerical simulation of sparsity characteristics
A	0.6	5	{d}	18
B	0.9	1	{e, f}	12
C	0.4	2	{a, c}	35
A	0.7	8	{l, e, k}	19
B	0.5	4	{d, l}	25
C	0.8	3	{d, g, t}	34

In the above experimental parameters and environment compared with the traditional method and the actual effect of this method, and record the detection results for subsequent analysis and research. By human intervention, the method of large data mining is changed, and the detailed picture of the change of experimental parameters is drawn according to the form of the specified parameters in the control host. Take 60 min as monitoring time, record the change of the total amount of data mining in big data node after using the method of experiment group and control group respectively (Fig. 7).

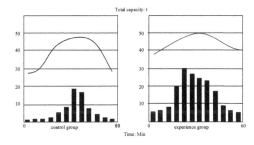

Fig. 7. Data mining load aggregate comparison chart

In the above figure, the actual value level of the column segment is used for data mining, while the curve segment only reflects the basic trend of the physical quantity. The total amount of data mining in the experimental group and the control group increased first and then decreased, but the extreme value of the experimental group was close to 30T, far higher than the extreme value of 19T. Therefore, the data mining effect of this method is better, and more data information can be mined in the same time. In summary, with the application of deep mining method, the total amount of node organization data mining does appear to be on the rise. Furthermore, it reflects the changes of the transmission rate of big data matching paths in the experimental group and the control group during the monitoring time of 60 min (Table 4).

Table 4. Experimental group data mining rates

Monitoring time /(min)	Transmission speed /(T/s)	Average value /(T/s)	Changing trend
5	9.6	10.7	
10	9.9		Rise
15	10.3		Rise
20	10.7		Rise
25	11.2		Rise
30	11.8		Rise
35	11.8		Stable
40	11.5		Decline
45	11.0		Decline
50	10.6		Decline
55	10.2		Decline
60	9.8		Decline

Analysis table shows that with the increase of monitoring time, the data mining rate of the experimental group keeps increasing, stable and decreasing trend, the maximum level reaches 11.8 T/s, and can keep steady state for 5 min (Table 5).

Table 5. Data mining rate of control group

Monitoring time/(min)	Transmission speed/(T/s)	Average value/(T/s)	Changing trend
5	4.2	4.9	
10	5.6		Rise
15	4.2		Decline
20	5.7		Rise
25	4.3		Decline
30	5.5		Rise
35	4.4		Decline
40	5.6		Rise
45	4.1		Decline
50	5.4		Rise
55	4.3		Decline
60	5.2		Rise

The global maximum value is only 5.7 T/s, which is much lower than the extreme value level of the experimental group, and this value can not maintain a stable state for a long time. This shows that under the application of this method, the data mining rate is higher, and the high efficiency can be maintained for a certain period of time, which has better performance than the traditional method. In summary, the edge computation-based wireless sensor network real-time stream data mining method has high practical value, data mining effect is significantly better than the traditional method.

4   Closing Remarks

In wireless sensor networks, sensor nodes will produce a large number of data, which have the characteristics of fast arrival, real-time update and so on, is a typical stream data. The application of stream data mining technology in many fields is more and more extensive, which makes communication and computer function more powerful and can provide better services for users. This paper analyzes and compares the main data mining technologies in wireless sensor networks, and proposes a workflow framework for streaming data mining in wireless sensor networks. The processing flow of WSN stream data mining is explained clearly, and the global pattern can be obtained by the deeper mining of local pattern. With the maturity of wireless sensor network technology, sensor sensing data will be increasingly rich.

References

1. Gombé, B.O., et al.: A SAW wireless sensor network platform for industrial predictive maintenance. J. Intell. Manuf. **30**(4), 1617–1628 (2017). https://doi.org/10.1007/s10845-017-1344-0
2. Lorenz, P., Schott, R., et al.: New path centrality based on operator calculus approach for wireless sensor network deployment. IEEE Trans. Emerg. Top. Comput. **7**(1), 162–173 (2019)
3. Babu, R.G., Karthika, P., Manikandan, G.: Polynomial equation based localization and recognition intelligent vehicles axis using wireless sensor in MANET. Procedia Comput. Sci. **167**(10), 1281–1290 (2020)
4. Gupta, G.P., Jha, S.: Biogeography-based optimization scheme for solving the coverage and connected node placement problem for wireless sensor networks. Wirel. Netw. **25**(6), 3167–3177 (2019)
5. Dugaev, D., Peng, Z., Luo, Y., et al.: Reinforcement-learning based dynamic transmission range adjustment in medium access control for underwater wireless sensor networks. Electronics **9**(10), 1727 (2020)
6. Wang, H., Tan, X., Huang, Z., et al.: Mining incomplete clinical data for the early assessment of Kawasaki disease based on feature clustering and convolutional neural networks. Artif. Intell. Med. **105**(4), 101859 (2020)
7. Jeihouni, M., Alavipanah, S.K., Toomanian, A., et al.: Digital mapping of soil moisture retention properties using solely satellite-based data and data mining techniques. J. Hydrol. **585**(5), 124786–124788 (2020)
8. Maind, A., Raut, S.: Mining conditions specific hub genes from RNA-Seq gene-expression data via biclustering and their application to drug discovery. IET Syst. Biol. **13**(4), 194–203 (2019)
9. Jimenez-Carvelo, A.M., Gonzalez-Casado, A., Gracia Bagur-Gonzalez, M., et al.: Alternative data mining/machine learning methods for the analytical evaluation of food quality and authenticity – a review. Food Res. Int. **122**(AUG), 25–39 (2019)
10. Ma, X., Ji, Y., et al.: Multidimensional visualization of Bikeshare travel patterns using a visual data mining technique: data cubes. J. Beijing Inst. Technol. **28**(2), 79–91 (2019)
11. Liu, S., Liu, D., Srivastava, G., Połap, D., Woźniak, M.: Overview and methods of correlation filter algorithms in object tracking. Complex Intell. Syst. **7**(4), 1895–1917 (2020). https://doi.org/10.1007/s40747-020-00161-4
12. Liu, S., Lu, M., Li, H., et al.: Prediction of gene expression patterns with generalized linear regression model. Front. Genet. **10**, 120 (2019)
13. Fu, W., Liu, S., Srivastava, G.: Optimization of big data scheduling in social networks. Entropy **21**(9), 902 (2019)

Operation Controllable Index Optimization of Virtual Power Plant with Electric Vehicle Based on 5g Technology and Cloud Computing Platform

Ke-jun Qian[1], Shui-fu Gu[1], and Yong-biao Yang[2(✉)]

[1] Suzhou Power Supply Company of State Grid Jiangsu Electric Power Co., Ltd, Suzhou 215000, China
[2] Southeast University, Nanjing 210000, China

Abstract. For the controllable index optimization of virtual power plant with electric vehicles, the traditional optimization method has low environmental benefits due to the reduction of electricity sales and the use of more clean energy. Therefore, this paper proposes the controllable index optimization of virtual power plant with electric vehicles based on 5g technology and cloud computing platform. The cloud computing platform is used to calculate the deviation degree of controllable indexes of power plant operation and decompose the indexes, so as to allocate reasonable resources for the calculation of unit units. Taking the calculation results as the input parameters, an optimization model with power generation cost and environmental benefits as the objective is established. The membership function is used to solve the optimal solution of the model, and the controllable indexes are optimized. The experimental results show that: Based on 5g technology and cloud computing platform, the controllable index optimization method of virtual power plant with electric vehicles has low power generation cost and high environmental benefits, which can meet the operation needs of virtual power plant with electric vehicles.

Keywords: Cloud computing platform · Electric vehicle · Virtual power plant · Index optimization

1 Introduction

Virtual power plant adopts advanced communication mode and software system to aggregate distributed generation, energy storage system, controllable load, electric vehicle and a variety of distributed energy, so that it has sufficient capacity and stable output [1]. Like generation side distributed generation, demand side resources represented by controllable load play an increasingly important role in power system and become an important part of virtual power station [2]. The emergence of virtual power plant, using advanced intelligent measurement, data processing, network communication and other means, not only breaks through the traditional power management mode of power system, but also

© ICST Institute for Computer Sciences, Social Informatics and Telecommunications Engineering 2022
Published by Springer Nature Switzerland AG 2022. All Rights Reserved
S. Liu and X. Ma (Eds.): ADHIP 2021, LNICST 417, pp. 156–168, 2022.
https://doi.org/10.1007/978-3-030-94554-1_14

effectively integrates the flexible load resources of power grid demand side, which is of great significance for the management of power system source and load side [3]. Virtual power plant is a new resource with the identity of "producer" and "consumer", which provides technical support for the development of smart grid [4].

At present, the theoretical research and project implementation of virtual power plants all over the world are mainly from Europe and the United States [5]. The implementation form of the two virtual power plants projects is very different. European countries have implemented some virtual power plant projects, and the implementation situation is also different. At present, some demonstration projects of virtual power plants have been carried out in China. Virtual power plant can manage the distributed renewable energy generating units in the distribution system, solve the problems of safe grid connection and stable operation, and give full play to its advantages in clean energy substitution and environmental friendliness; it can also make use of advanced communication means and demand response technology to save cost and reduce peak load of EV and other controllable loads Valley and other aspects play a positive role [6]. The research on virtual power plant has taken a solid and key step to realize the bright vision of smart grid and energy Internet.

However, electric EV virtual power plants are not stable traditional power plants, large, distributed, random and intermittent, operation indicators are easy to deviate, resulting in low energy efficiency and waste of power resources, and the research on this problem is not comprehensive, this paper uses 5g technology and cloud computing platform to solve the optimal solution of the model with membership function to optimize the operation index of electric vehicle virtual power plant, increase automobile electricity and save power resources.

Cloud computing as a current research hotspot, mainstream information technology companies such as Amazon, IBM, Google have proposed their own cloud computing infrastructure [7]. Cloud computing refers to the supercomputing mode connected through the Internet, including distributed computing, parallel computing and grid computing [8]. Cloud computing is a new computing mode, which integrates distributed operating system, distributed database, grid computing and other technologies. It can make full use of hardware resources and software resources [9]. Therefore, in order to reasonably allocate resources and build an optimization model aiming at power generation cost and environmental benefits, based on the above research background, this paper introduces 5g technology and cloud computing platform to optimize the operation controllability index of electric vehicle virtual power station, and uses cloud computing platform to calculate the deviation degree of controllable index of power plant operation At the same time, the membership function is used to solve the optimal solution of the model, optimize the controllable index, and solve the problems of low environmental benefits and poor environmental protection existing in the traditional index optimization method.

2 Optimization of Controllable Operation Index of Electric Vehicle Virtual Power Station

2.1 Operation Index Decomposition of Virtual Power Plant

The development of 5g also comes from the growing demand for mobile data. With the development of the mobile Internet, more and more devices are connected to the mobile network, and new services and applications emerge endlessly. With the acceleration of the marketization of the power industry, information technology plays an increasingly important role in the process control and operation management of the power industry, and power informatization has become an important factor in promoting the development of the power industry [10]. In order to understand the situation of power production and operation in time, the main problem for power enterprises is how to respond to the new economic situation. At present, small index management method has been applied in many power plants. The calibration value of small index and its completion directly reflect the technical level and management level of operators. In order to ensure the most economical operation mode of the whole plant, various works have been carried out in various power plants to find the most reasonable small operation index quota under various operation conditions as the basis for operation of operators [11–13].

After careful study, it is determined to use the results of the cloud computing platform to calculate the expected value and actual value of each index and parameter under the current working condition, and then get the completion rate of each index, get the deviation degree of each index under the current working condition, and establish the scheme with the load rate and cycle efficiency completion rate under the current working condition as the assessment index, and the small index as the monitoring index. Departments control the completion of major indicators, operators and special engineers focus on analyzing the influencing factors of assessment indicators, and use the deviation degree of each small indicator to determine the small indicators to be analyzed, so as to guide the operation [14].

The framework of cloud computing is that each unit is composed of a single master job scheduling node and a slave task allocation node in each node cluster under the jurisdiction of the unit. The master node is responsible for scheduling all the tasks that constitute a job. The data resources of these tasks are distributed in the user image slices on the storage resources of different slave nodes, and the master node monitors their execution. The slave node is responsible for executing the tasks assigned by the master node. After receiving the assignment from the master node, the slave node starts to find the appropriate computing node for its subordinate storage nodes. First, the node detects its own computing resource surplus. If the remaining computing resources are enough to meet the amount of jobs submitted by users, it will give priority to allocating its own computing resources; if the resources have been exhausted or are not enough to meet the minimum amount of computing resources, it will report to the master node to search for other suitable computing resources in the cloud environment, and then calculate.

According to the actual situation of the virtual power plant with electric vehicles, a hierarchical index system is established, and each index is decomposed as follows.

Since the virtual power plant operates in the unit system, the first level indicators are described as: coal consumption and power supply, which are the company's performance indicators, the calculation formulas of the total length efficiency and unit power plant efficiency of the virtual power plant are as follows:

$$Q_0 = \frac{W_0}{\sum\limits_{i=1}^{N} \frac{W_i}{Q_i}} \tag{1}$$

$$Q_i = Q_g \times Q_J \times Q_D \tag{2}$$

In the formula, Q_0 is the efficiency W_0 is the whole power plant, W_i is the power generation of the whole plant, Q_i is the unit power generation, N is the efficiency of the unit power plant, Q_g is the number of power plants in the unit, Q_J is the boiler efficiency, Q_D is the steam turbine efficiency and the pipe efficiency. The second index is based on boiler efficiency, steam turbine efficiency and pipe efficiency. Boiler efficiency: refers to the utilization rate (degree) of the fuel consumed by the boiler to produce steam by burning fuel coal (oil). There are about 30 indexes affecting boiler efficiency. But the main indexes that can be adjusted in daily operation include: main steam pressure, main steam temperature, boiler evaporation, reheat steam pressure, reheat steam temperature, inlet air temperature of forced draft fan, oxygen content, air leakage coefficient and exhaust gas temperature.

Steam turbine efficiency: it refers to the utilization degree of steam by steam turbine generator set. It indicates the operation economy of steam turbine. The calculation formula of steam turbine efficiency is as follows:

$$Q_J = \frac{3600}{Z_J} \times 100\% \tag{3}$$

The heat rate of steam turbine is expressed in the formula Z_J. There are about 25 indexes affecting the operation economy of steam turbine, but the main indexes that can be adjusted in daily operation include main steam pressure, main steam temperature, regulating stage pressure, reheat steam pressure drop, reheat steam temperature, feed water temperature, circulating water temperature, circulating water temperature rise, condenser end difference, exhaust temperature, condenser vacuum degree.

Pipeline efficiency: including heat loss of high temperature pipeline, leakage loss of system steam and water, and various heat losses not included in efficiency of generator and furnace. There are five indexes affecting the efficiency of pipeline. Heat loss of blowdown and other heat losses not included in boiler and turbine efficiency. There are five indicators that affect the pipeline efficiency. The make-up water rate is taken as the operation department, and the average load rate and cycle efficiency of each value are taken as the team assessment index, and the steam turbine and boiler efficiency index is taken as the assessment index of the steam turbine and boiler specialty, and the load rate index and auxiliary power index are decomposed into the specialty as the performance index. The cloud computing platform is used to analyze the small index deviation and find out the reasons for the evaluation index deviation. The reasons for the deviation of index assessment are shown in the Table 1.

Based on the decomposed indexes and environmental benefits, a multi index optimization model is designed to optimize the controllable indexes of the virtual power plant with electric vehicles.

Table 1. Operation energy consumption index system of electric vehicle virtual power plant

Professional assessment index	Coal consumption impact	Analysis of monitoring indicators	Coal consumption impact
Unit load rate	0.14% affected 1 g/kWh	Power shortage point	1 MPa affects 1.4 g/kwh
	0.3% affects 1 g/kWh	Dispatching power completion rate	1C affects 0.11 g/kWh
Vehicle efficiency	The auxiliary power consumption rate increased by 1%, and the energy consumption increased by 3.0 g/kWh Coal consumption impact 0.14% affected 1g/kWh 0.3% affects 1g/kWh	Main steam pressure	1C affects 0.1 g/kWh
		Main steam temperature	1C affects 1 g/kWh
		Regulating stage pressure	1C affects 1 g/kWh
		Reheat steam temperature	1C affects 1 g/kWh
		Circulating water temperature	1C affects 1 g/kWh
		Circulating water temperature rise	1% affects 3 g/kWh
		Condenser end difference	1C affects 0.12 g/kWh
		Undercooling	Coal consumption impact
		Condenser vacuum	1 MPa affects 1.4 g/kwh
		Feed water temperature	1C affects 0.11 g/kWh
Boiler efficiency	0.14% affected 1g/kWh 0.3% affects 1g/kWh The auxiliary power consumption rate increased by 1%, and the energy consumption increased by 3.0 g/kWh Coal consumption impact	Main steam pressure	
		Main steam temperature	
		Boiler evaporation	
		Reheat steam pressure	
		Reheat to steam temperature	
		Inlet air temperature of forced draft fan	10 C affects 0.12 g/kWh

(*continued*)

Table 1. (*continued*)

Professional assessment index	Coal consumption impact	Analysis of monitoring indicators	Coal consumption impact
		Oxygen quantity	1% affects 1 g/kWh
		Air leakage coefficient	
		Exhaust gas temperature	10 °C affects 1.4 g/kwh
		Combustibles in fly ash	
Pipeline efficiency	The auxiliary power consumption rate increased by 1%, and the energy consumption increased by 3.0 g/kWh	Reheat steam pressure loss	
		Main steam temperature drop	
		Reheat steam temperature drop	
		Feed water temperature drop	
		Water supply rate	1% affects 3 g/kwh
		Heat loss	

2.2 Establish Multi Index Optimization Model

In a dispatch cycle, the objective function of minimizing the generation cost of conventional units is as follows:

$$\min S_0 = \sum_{t=1}^{T} \sum_{m=1}^{M} \left(s_{m,t} v_{m,t} + p_{m,t} + b_{m,t} \right) \tag{4}$$

The formula T represents the number of research periods, the number of unit units M, m is the unit number, $s_{m,t}$ is the total fuel expenditure cost m of conventional units in the period t, $v_{m,t}$ is the operation status of units, $p_{m,t}$ and $b_{m,t}$ the start-up and stop costs of units.

The pollution gases emitted from coal combustion seriously affects the environmental quality and is not conducive to the sustainable development of the environment and resources. Generally speaking, the main pollutants emitted by conventional generating units include CO_2, SO_2, NO_2, etc.

$$\min E_0 = \sum_{v=1}^{R} \sum_{t=1}^{T} \sum_{m=1}^{M} \left(v_{m,t} \left(\begin{array}{c} \alpha_{m,v} + \beta_{m,v} K_{m,v} \\ + \eta_{m,v} \left(K_{m,v} \right)^2 \end{array} \right) \right) \tag{5}$$

In the formula, R the pollutant classification number is expressed, E_0 the total pollutant emission amount is expressed, and the pollutant emission coefficient $\alpha_{m,v}$ of the unit is expressed respectively. The relationship $\beta_{m,v}$ between the coefficient $\eta_{m,v}$ and the output $K_{m,v}$ is quadratic. The more the output, the greater the pollutant emissions.

When dealing with multi index optimization problems, the complex problems can be simplified by single index optimization.

In order to solve the above problems, the method of fuzzy membership degree is introduced. The membership function of the objective function was established to fuzzify it, and the appropriate membership function is selected to solve the problem.

The basis of the problem is to reduce the cost of power generation and pollution emissions as much as possible under the premise of meeting the constraints, so as to ensure optimal environmental benefits. In view of the above situation, for different objective functions, the ascending half straight line shape and the falling half straight line shape are used as the membership function. As can be seen from Fig. 1, the closer the membership degree is to 1, the better the optimization standard is, and the higher the satisfaction degree of the optimization strategy is. The membership functions of the two objective functions are the conventional power generation cost and pollution emissions. The reduced semi linear membership function is used to maximize the consumption of new energy.

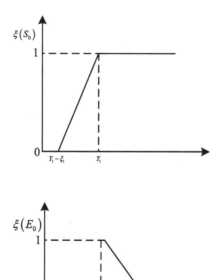

Fig. 1. Membership function corresponding to each objective function

The Fig. 1 shows the target value of power generation cost, ξ_1 represents the expansion value of generation cost with acceptable satisfaction, Y_2 represents the target value of pollutant emission, and ξ_2 represents the expansion value of pollution emission with acceptable satisfaction.

Assuming that the final objective function takes the minimum value of the two membership functions, this value is used to represent the overall satisfaction of the multi index optimization.

$$\xi = \min\{\xi(S_0), \xi(E_0)\} \tag{6}$$

At this point, the original multi index optimization problem is transformed into a nonlinear optimization problem to maximize the satisfaction of single index.

$$\max \xi, \ s.t. \begin{cases} -S_0 + \delta_1 \xi \leq -Y_1 + \delta_1 \\ E_0 - \delta_2 \xi \leq Y_2 + \delta_2 \\ 0 \leq \xi \leq 1 \end{cases} \tag{7}$$

The model is solved according to the solution flow shown in Fig. 2.

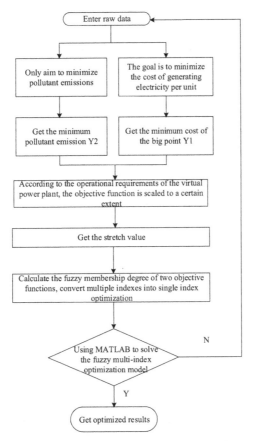

Fig. 2. Flow chart of solving multi index optimization model

Under the environment of MATLAB, the commercial software CPLEX is used to optimize the model, and the multi index model after the optimization of fuzzy membership function is solved. It is concluded that the unit output and pollutant emission under the maximum satisfaction degree are the optimal schemes of the model. So far, the design of controllable index optimization method of virtual power plant with electric vehicle based on cloud computing platform has been completed.

3   Experimental Analysis

3.1   Experimental Parameter Setting

Because the design of this paper is based on environmental efficiency optimization as the goal, electric vehicle type, charging load, motor voltage and power, members, power consumption per kilometer, EV type are related to the utilization efficiency of electric energy and pollutant discharge volume, so these projects as experimental parameters can fully reflect the optimization effect of the design method on environmental benefits, and can guarantee the realization of the experimental purpose. The differences in the experimental parameters will not affect the experimental results, but also make them more comprehensive and objective. According to the above contents, a virtual power plant with electric vehicles is selected as the research object, and the EV types and parameters are shown as Table 2.

Table 2. EV types and related parameters

EV type number of sets	Official vehicle	Business purpose vehicle	Private car		
			To work in an office	Family	Market
Charging load	600	400	2000	7000	1000
Motor voltage and power	12.18	12.18	7.06	3.52	23.96
Number of members	48 V/1000 W	60 V/1200 W	48 V/4000 W	48 V/4000 W	60 V/2800 W
Power consumption per kilometer	4	4	4	4	4
EV type	0.30 kWh/km	0.30 kWh/km	0.15 kWh/km	0.15 kWh/km	0.15 kWh/km

The environmental benefit standards formulated in the experiment are shown in Table3.

Table 3. Environmental benefit standard of polluted gas

Index	Penalty coefficient 10000 yuan/kg	Environmental value/(10000 yuan/kg)
CO_2	0.0075	0.01725
SO_2	0.750	4.50
NO_x	1.50	3.00

Based on the above experimental conditions, the traditional index optimization method based on neural network and the index optimization method based on cluster analysis are used to verify the actual level of different optimization methods with the goal of environmental benefit optimization.

3.2 Experimental Results and Analysis

The experimental results of different optimization methods based on the above experimental conditions are shown in the Fig. 3.

The solid line in the figure represents the cost of power generation, and the dotted line represents the emission of pollutants. Compared with the results in the observation chart, it can be seen from the results that the power generation cost of the index optimization method based on the cloud computing platform is lower than other methods, the highest cost of power generation did not exceed 500 Wh, and the pollutant emission is less than other methods, the emissions are below 20 g. From this feature, it shows that the index optimization method based on the cloud computing platform mainly considers the amount of pollution gas emitted by power generation, uses cleaner unit units to generate electricity, encourages internal load to actively participate in demand response, realizes peak shaving and valley filling, improves the utilization rate of wind power at night, reduces the cost and realizes environmental protection.

Based on the above results, the environmental benefits of different index optimization methods are obtained through calculation, and the differences between different optimization methods are intuitively displayed. The calculation results are shown in Table 4.

It can be seen from the data in the table that the index optimization method based on cloud computing platform has higher index benefit, index benefit all are above 9.45. The traditional index optimization method has lower environmental benefit. This is because the traditional optimization method uses more clean energy due to the reduction of electricity sales, which makes the overall income decline. Combined with the experimental results in the previous section, it can be seen that the design of cloud computing platform based operation controllable index optimization method with electric vehicle virtual power plant has high environmental benefits and more environmental protection.

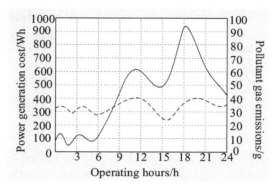

(a) Experimental results of index optimization method based on Neural Network

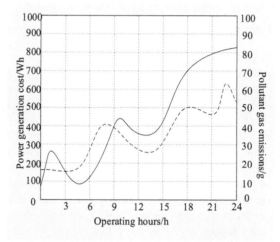

(b) Experimental results of index optimization method based on cluster analysis

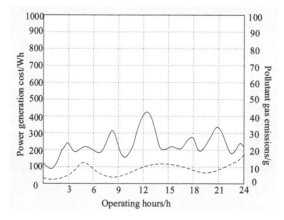

(c) Experimental results of index optimization method based on cloud computing platform

Fig. 3. Changes of power generation cost and pollution emission during power plant operation

Table 4. Calculation results of environmental benefits of different index optimization methods

	Environmental benefits of index optimization method based on neural network	Environmental benefits of index optimization method based on cluster analysis	Environmental benefits of index optimization method based on cloud computing platform
Unit 1	2.63	3.21	9.45
Unit 2	2.94	3.36	9.67
Unit 3	2.29	3.74	9.96
Unit 4	2.45	3.96	9.58
Unit 5	2.51	3.29	9.49

4 Conclusion and Outlook

Firstly, this paper investigates the background of virtual power plant planning and controllable index optimization, and clarifies the purpose and significance of the research. The concept and basic structure of virtual power plant are defined, and the implementation status and academic research results of virtual power plant projects at home and abroad are summarized. Then, starting from the optimization of controllable index of virtual power plant with electric vehicles, aiming at the problems existing in the previous optimization methods, the cloud computing platform is applied in the index optimization. Through comparative experiments, the high performance and high level of the controllable index optimization method of virtual power plant with electric vehicles based on the cloud computing platform is verified, which solves the traditional optimization method. The existing problems in this paper contribute to the operation of virtual power plant including electric vehicles. However, this paper starts with the optimization of controllable indicators of electric vehicle virtual power plants. The main goal of the design plan is to optimize environmental benefits without considering the economic benefit optimization of virtual power plants. Future research could serve as the direction to optimize the economic efficiency of electric vehicle virtual power plants and improve the market share of electric vehicles.

Fund Projects. The Science and Technology Project of State Grid Jiangsu Electric Power Co., Ltd. "Research on Key Technologies of Intelligent Management of Electric (Steam) Vehicles and Operation Mode of Virtual Power Plants".

References

1. Wei, Z.N., Chen, Y., Huang, W.J., et al.: Optimal allocation model for multi-energy capacity of virtual power plant considering conditional value-at-risk. Autom. Electr. Power Syst. **42**(4), 39–46 (2018)
2. Fu, Z.G., Liu, B.H., Wang, P.K., et al.: A comprehensive evaluation model for performance indexes of gas-fired power plants. Thermal Power Gener. **48**(3), 7–13+21 (2019)

3. Liu, S., Liu, G.C., Zhou, H.Y.: A robust parallel object tracking method for illumination variations. Mob. Netw. Appl. **24**(1), 5–17 (2019)

4. Wang, F.: Dynamic controllable flexible resource scheduling algorithm under cloud computing. Sci. Technol. Eng. **18**(3), 291–296 (2018)

5. Han, H.F., Yujuan, M.A., Huang, Y., et al.: Research on the construction of intelligent power plant of 1000 MW thermal power unit. Electr. Power **51**(10), 49–55 (2018)

6. Jianqiu, L.: Construction of real-time interactive mode-based online course live broadcast teaching platform for physical training. Int. J. Emerg. Technol. Learn. (iJET) **13**(6), 73 (2018)

7. Fu, W., Liu, S., Srivastava, G.: Optimization of big data scheduling in social networks. Entropy **21**(9), 902 (2019)

8. Liu, S., Bai, W., Zeng, N., et al.: A fast fractal based compression for MRI images. IEEE Access **7**, 62412–62420 (2019)

9. Liu, S., Fu, W.N., He, L.Q., et al.: Distribution of primary additional errors in fractal encoding method. Multimedia Tools Appl. **76**(4), 5787–5802 (2017)

10. Zhou, J.C.: Simulation of invulnerability measure index optimization correction in wireless mobile network. Comput. Simul. **36**(2), 283–286+413 (2019)

11. Fm, A., As, A., Nd, A., et al.: Modelling the influence of peers' attitudes on choice behaviour: theory and empirical application on electric vehicle preferences. Transp. Res. Part A Policy Pract. **140**(12), 278–298 (2020)

12. Bindra, A.: Electric vehicle batteries eye solid-state technology: prototypes promise lower cost, faster charging, and greater safety. IEEE Power Electron. Mag. **7**(1), 16–19 (2020)

13. Karray, S., Martín-Herrán, G., Zaccour, G.: Pricing of demand-related products: can ignoring cross-category effect be a smart choice? Int. J. Prod. Econ. **223**, 107512 (2020)

14. Kohtamki, M., Einola, S., Rabetino, R.: Exploring servitization through the paradox lens: coping practices in servitization. Int. J. Prod. Econ. **226**, 107619 (2020)

Vulnerability Evaluation Method of Big Data Storage in Mobile Education Based on Bootstrap Framework

Xi-liu Zhou[✉] and Yang-bo Wu

College of Mathematics and Computer, Xinyu University, Xinyu 338000, China

Abstract. In order to better improve the effect of mobile education and ensure the quality of teaching, a vulnerability evaluation method of big data storage in mobile education based on bootstrap framework is proposed. By mining the characteristics of mobile education big data, this paper constructs the vulnerability evaluation model of mobile learning, and constructs the quality evaluation index of education big data storage vulnerability, so as to judge whether the learners' learning effect reaches the expected level. Finally, experiments show that the method based on bootstrap framework has high effectiveness in the practical application process, and fully meets the research requirements.

Keywords: Bootstrap framework · Mobile education · Big data

1 Introduction

Today's education mode is undergoing a very significant change, and education is developing towards the direction of ubiquitous, intelligent and personalized. With the advent of the Internet plus era, online learning has gradually become the mainstream learning mode. Large scale open online course is a typical form of online learning, which is characterized by the convenience of two-way vulnerability evaluation, the richness of teaching resources, and the diversity of teaching interaction [1]. There are a lot of vulnerability assessment behaviors in mobile learning platform. In order to ensure the teaching effect, it is necessary to store and evaluate massive data effectively. Vulnerability assessment data storage and evaluation is the key to the integration of big data storage in mobile education. The establishment and formation of knowledge in the learning process depends on the development of vulnerability assessment data storage, and the effective teaching data storage is fragile Weak evaluation can improve the learning effect. The correct application of vulnerability assessment to online courses can provide learners with the opportunity to expand their learning experience and create a new e-commerce platform for mobile education. Learning vulnerability assessment can promote learners' knowledge construction. Relevant scholars have made some progress in this field. Literature [2] proposes a video based active learning evaluation method for educational resources, which uses Hadoop cloud platform and combines HDFS with the

S. Liu and X. Ma (Eds.): ADHIP 2021, LNICST 417, pp. 169–182, 2022.
https://doi.org/10.1007/978-3-030-94554-1_15

existing higher vocational education cloud platform. It effectively solves the problem of massive data storage of educational resources, and effectively improves the effectiveness of educational resources storage evaluation. However, in order to consider the impact of this method on teaching quality. In reference [3], a teaching evaluation method of Informatics based on open education resources is proposed. This paper analyzes the process of oer's use, processing and dissemination in computer science courses to improve the effectiveness of education evaluation methods, but the accuracy of teaching evaluation is not good.

In view of the above problems, this paper proposes a bootstrap framework based on the mobile education big data storage vulnerability evaluation method, combined with bootstrap framework for online learning data vulnerability evaluation. It accelerates the research and exploration of mobile education mode, and the construction of mobile education model can not only improve the learning efficiency, but also improve the learning quality [3]. It provides a new direction for educational decision-making and more possibilities for educational learning.

2 Optimization of Big Data Storage Vulnerability Evaluation Method for Mobile Education

2.1 Data Storage Vulnerability Feature Mining

Through the data left by learners' online learning, we can analyze the specific storage behavior of their learning behavior, and comprehensively process the learning browsing content, online interactive communication data and other aspects of information, so as to obtain the data storage support of mobile education model under big data, and combine with the bootstrap framework to analyze many factors that may affect the data storage vulnerability behavior Factors for deep mining [4]. The premise of mining influencing factors of mobile education data storage lies in the accurate control and application of a variety of data mining tools and algorithms, which is composed of tools and algorithms, basic theoretical data and data mining, as shown in Fig. 1.

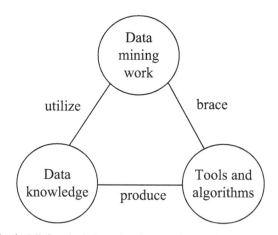

Fig. 1. Mining the influencing factors of education data storage

In the process of mining the influence factors of education data storage, we should take the data mining results as the reference basis of data vulnerability evaluation index. Through mining and collecting the basic data, we can achieve the comprehensive application of teaching data classification and interpretation evaluation, and get the impact of data mining on online learning data vulnerability evaluation behavior and model construction. The impact of construction [5]. According to the development of the industry, we should abandon the content of poor credit evaluation in the database, ensure the vulnerability evaluation of data storage based on the credit benchmark value, set corresponding multi-level vulnerability evaluation indicators according to the teaching needs, and carry out different data analysis, so as to ensure the practical application effect of the model and avoid the situation that the final stored information is not credible. The practical application of bootstrap framework can ensure the objectivity of data, and through the application of real-time dynamic analysis of mobile education big data, it can ensure the monitoring of teaching content storage effect, and establish a comprehensive and open curriculum evaluation model [6–8]. Different platforms and teaching models can be integrated into it, and the teaching content can be adjusted through practice feedback, which can improve the final teaching accuracy and scientificity.

The curriculum development of mobile education is more in line with the learning needs of learners, which is conducive to the realization of the established goals. At the same time, managers can obtain the knowledge categories they need to learn through the access records of learners, and recommend targeted information data according to their own interests, so as to meet the personalized learning needs of learners. The establishment of mobile education model and the analysis of mobile education curriculum help to form a win-win situation and promote the design and implementation process of mobile education model in real life [9, 10]. In order to ensure the storage effect of teaching data, the data mining process of learning process is further optimized, and the specific steps are as follows (Fig. 2):

As shown in the figure, setting personalized learning service tools in mobile education model can not only meet the needs of social development, but also enhance their own competitive strength. The mobile education module is the cornerstone of the whole teaching quality evaluation system. In order to widely collect data, it is divided into three sub modules, namely user registration module, user login module and questionnaire module. By entering the questionnaire module through the registration and login module, the system automatically determines the user identity and enters into different evaluation survey sub modules, which can effectively improve the learning efficiency and stimulate students' interest in learning.

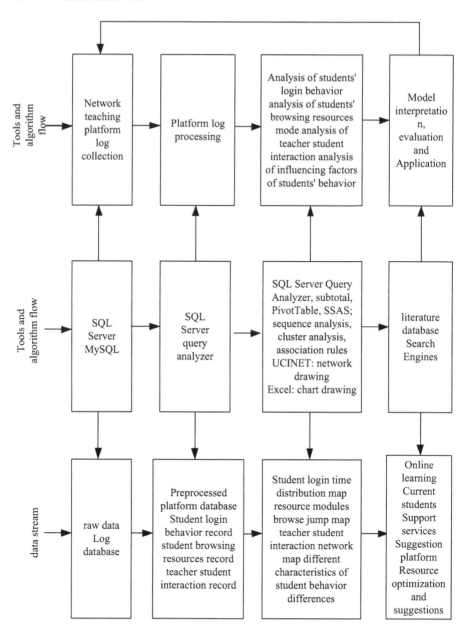

Fig. 2. Data mining process of learning process

2.2 Optimization of Data Storage Mode for Mobile Education

Three types of vulnerability assessment in mobile education data storage mainly include: data storage vulnerability assessment of learners and learning materials, data storage vulnerability assessment of learners and teachers, and data storage vulnerability assessment of learners and Learners [11]. In the research on the evaluation quality of mobile education data storage vulnerability, various evaluation vulnerability indicators are proposed [12]. This paper evaluates the storage quality of learning resources, selects the vulnerability evaluation data of learning resources as the research basis, and then quantitatively evaluates the vulnerability evaluation of learners in the learning process. Mobile teaching data activities include video viewing, page navigation, test participation and other vulnerability assessment behaviors. Vulnerability evaluation is recorded as click stream log, and each click stream is a collection of records [13]. Each vulnerability assessment activity record can be composed of learner ID, learning resource ID, vulnerability assessment time and vulnerability assessment feature vector. The vulnerability assessment feature vector contains vulnerability assessment information such as viewing documents and videos. In view of the vulnerability evaluation relationship between learners and different learning resources in the learning process, this paper constructs the learning vulnerability evaluation network model as shown in the figure below (Fig. 3).

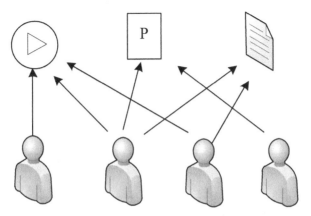

Fig. 3. Learning vulnerability assessment network

By comparing the expected test range with the test data to be evaluated, the consistency of the data is evaluated [14]. By detecting the key information such as data storage structure, file name and header file, the detection data type is determined to identify whether the data is valid. By comparing the attribute information of the test data with the key information of the existing data in the data management system, we can judge whether the data is consistent. The proportion of valid data is measured by reading the validity mark of data content, statistical data outlier interval proportion and data validity period. According to the unique attribute, whether the data is repeated or not is judged, the repeated data is associated, and the data resume is established. Through the established business association attributes, view the association of data and other data, and establish the association according to the potential relationship. Learning vulnerability evaluation network includes two kinds of nodes: learners and learning resources [15]. The connecting edge between the two kinds of nodes represents the vulnerability characteristics of learning resources. After collecting a large amount of data, in order to give full play to the data utility, we need to sort out the data and realize the transformation process from data to information, and then to knowledge. Therefore, database technology is used for data storage (Fig. 4).

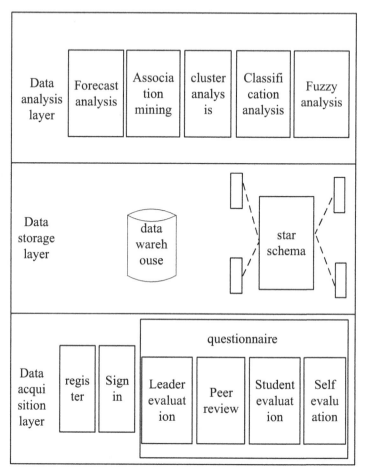

Fig. 4. Architecture of teaching data storage quality evaluation platform

The biggest advantage of the platform structure is that most of the required data can be obtained through one-step connection, and the results can be obtained quickly, which is difficult to achieve in the conventional transactional database. On the basis of data warehouse, this paper analyzes the data and obtains reasonable teaching quality evaluation results. The specific model settings are shown in the figure (Fig. 5).

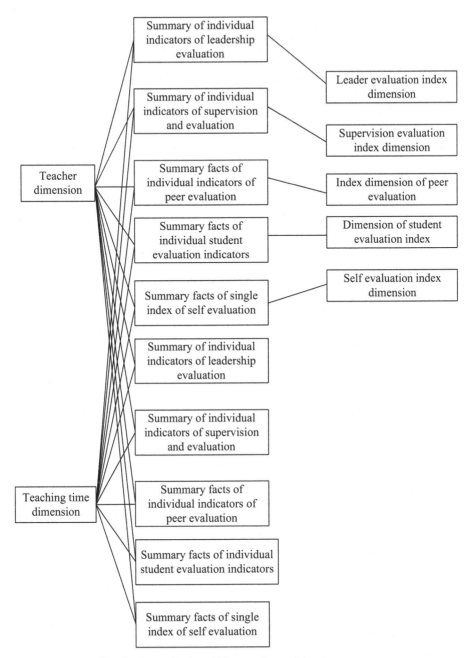

Fig. 5. Steps of vulnerability test for teaching data storage

Using cloud computing technology, running on the platform, and mainly using parallel programming model. The analysis tasks include cluster analysis, association rule mining, classification mining, prediction analysis and fuzzy analysis. There are many kinds of analysis algorithms for each type of task. These data analysis tasks will be decomposed into multiple executable parallel sub tasks on the cloud platform, and automatically decomposed and merged, so as to improve the analysis efficiency.

2.3 Data Storage Vulnerability Evaluation Algorithm

Teaching quality is not only the basis of the survival and development of colleges and universities, but also the inevitable requirement of the internationalization of higher education. Therefore, teaching quality evaluation is one of the most important points in university evaluation. With the development of information technology, colleges and universities have accumulated a large number of teaching evaluation data, such as student status management data, performance management data, personnel management data. Vulnerability evaluation is based on the construction of educational information resources, combined with the big data analysis of mobile education platform, in order to improve the quality of teaching, information-based teaching evaluation design is carried out on the sharing platform of teaching evaluation resources.

Because vulnerability assessment may occur many times, and the types and characteristics of vulnerability assessment are not the same, each edge is marked with information such as vulnerability assessment time and vulnerability assessment characteristics. In this case, the learning features are embedded into (1) to get the following results:

$$u(t) = \sigma \left(W_1^u u(t^-) + W_2^u i(t^-) + W_3^u f + W_4^u \Delta u \right) \tag{1}$$

Among them, $W_1^u, \cdots, W_n^e$ is the RNN parameter of the learner, which is obtained by training; Δu represents the time difference between the last vulnerability assessment of the learner and any learning resource and the current vulnerability assessment; F represents the vulnerability assessment feature vector; $u(t^-)$ is the embedded representation of learners before time t; I is the embedded representation of learning resources before time t. Update the learning features to get the final embedding function $u(t)$.

Embedded learning resources (2) get: 1:

$$j(t) = \sigma \left(W_1^i \dot{s}(t^-) + W_2^i u(t^-) + W_3 f + W_4^i \Delta i \right) \tag{2}$$

Among them, matrix $W_1^u, ..., W_n^e$ is the parameter of learning resource RNN, which is obtained by training. Δi represents the time difference between the last vulnerability assessment of learning resources and learners and the current vulnerability assessment; $i(t^-)$ represents the embedded representation of learning resources before time t; $u(t^-)$ represents the embedded representation of learners before time t. Finally, the learning resources are embedded in i(t). According to this, the evaluation index of learning feature vulnerability quality is proposed:

$$D(p) = j(t) \sum_{\tau \hat{I}i, Q_1} \|u(p) - i(q)\|_z \tag{3}$$

Among them, p represents a single learner, q represents a learning resource, and Q is the number of learning resources. The smaller the value of $D(p)$, the better the learning quality of the student's vulnerability assessment, and the larger the value, the worse the learning quality of the student's vulnerability assessment. In order to ensure that the final evaluation results are more accurate, it is necessary to increase the evaluation vulnerability index of the trust index on the basis of the original information; in the mobile education evaluation, the establishment of a complete credit evaluation can not only make the final result more accurate, It can also improve the credibility of the results, and through further sorting, the results of the fragile storage node selection can be obtained. The specific standards are shown in the following table (Table 1):

Table 1. Evaluation criteria for data storage

Importance of thinking	Number of attacks	Protection	Score
It's not important. The information system will not be damaged after being destroyed	The frequency is very low, basically not attacked	The safety monitoring measures are perfect, the emergency system is sound, the system spare parts are sufficient, and the implementation is very good	1
It's not very important. The security level is lower than level 2. It will cause a lower degree of loss after being damaged	The frequency is relatively low, with less than 5 attacks in the past year	The relevant safety system is basically complete, the emergency system is sound, and the implementation is good	3
It's more important. The security level belongs to level II, which may cause moderate losses after being damaged	High frequency, 5–10 attacks in the past year	The safety management and monitoring mechanism is sound, and the implementation status is general	5
Important, the security level belongs to the third level, does not have the control function, and its destruction may cause more serious losses	High frequency, attacked 10–20 times in the past year	There are basically no relevant protective measures	7
It is very important. The security level belongs to level 4, which has control function and may cause very serious losses after damage	The frequency is very high, and it has been attacked more than 20 times in the past year		9

Every link of mobile teaching involves a lot of data. Once the data is leaked or modified, it will seriously threaten the orderly operation of mobile teaching. Data security threats generally include: lack of unified standards for data, loopholes in data access settings, weak links in user and management authentication, which makes data easy to be leaked and modified, data storage center has not been established or operation management is not perfect, data can not be recovered after loss, data protection measures are weak, and data protection is not strong, which affects the development of mobile teaching Stable operation. This paper will evaluate the attack times, design standards and data security management. Let z represent the characteristics of mobile vulnerable nodes

$$z = h(x)i(y) + vD(p) \tag{4}$$

Among them, x represents the horizontal calculation component of the mobile vulnerable node feature, y represents the vertical calculation component of the mobile vulnerable node feature, $h(x)i(y)$ represents the fixed calculation method for the horizontal and vertical components, and v represents the law weakening parameter. Using formula (1), the communication channel positioning result of the mobile vulnerable node can be expressed as:

$$\varepsilon = z \cdot i + \tau(x)w(y) \tag{5}$$

Among them, i represents description accuracy, $\tau(x)$ represents horizontal positioning accuracy, and $w(y)$ represents longitudinal positioning accuracy. Through the above calculation process, the channel positioning accuracy of the new mobile vulnerable node can be fully enhanced. The normal application of the vulnerability storage evaluation method, after the completion of the above construction, also needs to calculate the stability characteristics of the data storage in the bootstrap framework according to the following formula.

$$\lambda = \mu\left(\frac{n\varepsilon}{1 - \varepsilon}\right) + 1 \tag{6}$$

Among them, μ represents the stability coefficient, and n represents the number of iterations of feature solving. Normally, n is a natural number not greater than 20. If $\lambda \leq 0.5$, the analysis result of the method is considered to be of high use value; $\lambda > 0.5$, the method is considered to have no application value. Mobile education is a complex network system, and its network topology is an important evaluation aspect of the vulnerability of data storage. The structural vulnerability of mobile education is measured from three aspects: node vulnerability, line vulnerability and maximum power supply area after cascading failure. According to the risk theory, the node vulnerability index is set as the product of the node importance and the node's comprehensive risk value. The node importance can comprehensively reflect the network characteristics and power characteristics of the node, namely:

$$V_N = \sum \lambda R_{is} \tag{7}$$

In the formula, V_N is the vulnerability index of the node; s is the importance of the data, that is, the degree of the node, and R_{is} is the risk value of the node. The data vulnerability index is defined as the product of the line betweenness and the data risk value. The line betweenness index can reflect the characteristics of the network structure and the distribution characteristics of the source flow path, namely:

$$V_L = H \sum B_i R_{is} - V_N \tag{8}$$

In the formula, V is the vulnerability index of the line, B_i is the line dielectric value and H is the line risk value. There are various calculation methods for the risk value, which are obtained by replacing the probability of failure based on state maintenance data. According to the historical statistical data to be evaluated, the evaluation formula based on the data storage health index is:

$$R = V_L a * e^{b \cdot Hi} \tag{9}$$

The vulnerability evaluation of data storage is performed based on the above algorithm to ensure the accuracy and effectiveness of the evaluation results.

3 Analysis of Results

In order to verify the actual application effect of the mobile education big data storage vulnerability evaluation method based on the Bootstrap framework, a comparative experimental study was carried out. The experimental data set includes learner/learning resource information, vulnerability evaluation information in the form of characteristics, and vulnerability evaluation Time of occurrence etc. The statistics of the two data sets are listed (Table 2).

Table 2. Experimental data set statistics

	KDD15	XTdata
Learner	7047	6371
Learning resource	98	27
Number of interactive activities	411 749	397 083
Dropouts	4066	4982

The mobile education big data storage vulnerability evaluation method based on the Bootstrap framework is compared with other static network representations as a method. The results are listed in the following table (Tables 3 and 4).

Table 3. Data storage status prediction results on the data set

Method	N = 5	N = 10	N = 20
Deep walk	0.30683	0.32102	0.41022
Struc2vec	0.13352	0.01776	0.20227
Hin2vec	0.51989	0.53759	0.55106
Method of this paper	0.55397	0.62269	0.62670

Table 4. Prediction results of data vulnerability status

Method	N = 5	N = 10	N = 20
Deep walk	0.79937	0.80377	0.80534
Struc2vec	0.79310	0.80534	0.80847
Hin2vec	0.75548	0.76766	0.75510
Method of this paper	0.95298	0.87912	0.83987

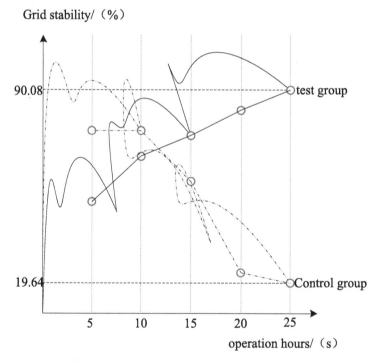

Fig. 6. Comparison of transmission grid stability

It can be seen from the comparison table that the method for evaluating the vulnerability of mobile education big data storage based on the Bootstrap framework proposed in the experiment is scalable. Furthermore, in view of the differences in the performance of learner state prediction accuracy on different data sets, the average vulnerability evaluation times of learners with different learning states in the two experimental data sets are respectively counted. The following figure reflects the comparison of the experimental group and the control group method and the stability of the transmission grid.

According to the analysis of Fig. 6, when the running time is 10s, the transmission grid stability of the experimental group is 72%, and that of the control group is 83%; When the operation time is increased to 15, the stability of the transmission grid in the experimental group is 80%, and that in the control group is 48%; In the experimental group, the stability of transmission grid increases with the running time. Based on the detection results of the above figure, compared with traditional methods, the mobile education big data storage vulnerability evaluation method proposed in this paper based on the Bootstrap framework has higher recognition accuracy in the actual application process and can better guarantee the data storage effect.

4   Conclusion and Prospect

In the context of the combination of education and the Internet, evaluating mobile education data storage vulnerability characteristics and using the generated learning vulnerability evaluation data to improve learners' learning performance and learning effects will be the inevitable trend of future education development. Bootstrap framework and computer technology conduct more in-depth mining and analysis of learner's vulnerability evaluation data, and intervene in mobile education data storage mode, which is the direction of more in-depth technology application. Establish a clear mobile education dynamic vulnerability evaluation network model, and use the evaluation results to judge whether the learner's learning effect meets expectations.

In the future, we will study how to improve the quality of education faster and further improve the accuracy of education evaluation.

References

1. Freschi, V., Delpriori, S., Lattanzi, E., et al.: Bootstrap based uncertainty propagation for data quality estimation in crowdsensing systems. IEEE Access 5(5), 1146–1155 (2019)
2. Grossman, G.D., Simon, T.N.: Student perceptions of open educational resources video-based active learning in university-level biology classes: a multi-class evaluation. J. Coll. Sci. Teach. 17(21), 45–52 (2020)
3. Ali, L., Werkes, R., Rpke, R., et al.: Der Einsatz von Open Educational Resources im Informatikunterricht Praxisbeispiel an der RWTH Aachen, vol. 32, no. 17, pp. 65–72 (2020)
4. Walton, K.: Role of campus community in open educational resources: the benefits of building a collaborative relationship with campus IT and distance education departments. Libr. Trends 69(2), 395–418 (2020)
5. Kumar, M., Jindal, S.R., Jindal, M.K., et al.: Improved recognition results of medieval handwritten Gurmukhi manuscripts using boosting and bagging methodologies. Neural Process. Lett. 50(1), 43–56 (2019)

6. Scholastica, A.J., Uhegbu, A.N.: Decades of advocacy: towards effective utilization of Open Educational Resources (OER) in universities in Nigeria: the missing link, vol. 13, no. 21, pp. 43–48 (2021)

7. Cai, K., Chee, Y.M., Gabrys, R., et al.: Correcting a single indel/edit for DNA-based data storage: linear-time encoders and order-optimality. IEEE Trans. Inf. Theory 67(6), 3438–3451 (2021)

8. Yang, G., Jan, M.A., Rehman, A.U., et al.: Interoperability and data storage in internet of multimedia things: investigating current trends, research challenges and future directions. IEEE Access 8(10), 124382–124401 (2020)

9. Otto, D.: Grosse Erwartungen: Die Rolle von Einstellungen bei der Nutzung und Verbreitung von Open Educational Resources (MedienPdagogik: Zeitschrift für Theorie und Praxis der Medienbildung). MedienPädagogik Zeitschrift für Theorie und Praxis der Medienbildung 1(13), 21–43 (2020)

10. Huang, Q., Huang, C., Huang, J., et al.: Adaptive resource prefetching with spatial-temporal and topic information for educational cloud storage systems. Knowl.-Based Syst. 181(10), 104791–1047915 (2019)

11. Warren, J.E.: Open Educational Resources (CLIPP 45): edited by Mary Francis, Chicago, IL: Association of College and Research Libraries, A Division of the American Library Association, 2021. J. Access Serv. 32(21), 1–16 (2021)

12. Liu, S., Li, Z., Zhang, Y., et al.: Introduction of key problems in long-distance learning and training. Mob. Netw. Appl. 24(1), 1–4 (2019)

13. Chang, S., Shafee, T.: Open Educational Resources and Scholarship in Law - Steven Chang and Thomas Shafee 30 July 2020 - ALLA PD event, vol. 12, no. 37, pp. 32–40 (2020)

14. Amo, D., Gómez, P., Hernández-Ibáez, L., et al.: Educational warehouse: modular, private and secure cloudable architecture system for educational data storage, analysis and access. Appl. Sci. 11(2), 806–812 (2021)

15. Amine, T.M., Amal, R., Abdelalime, S.: Storage management of educational scenarios modeled with the Recursive Entity Modelling Method. In: 2020 Fourth International Conference on Intelligent Computing in Data Sciences (ICDS) (2020)

Research on Big Data Ad Hoc Queries Technology Based on Social Network Information Cognitive Model

Yang-bo Wu[1], Xi-liu Zhou[1]([✉]), Ying Xu[1], Ming-xiu Wan[1], and Lin Ying[2]

[1] College of Mathematics and Computer, Xinyu University, Xinyu 338000, China
[2] College of Foreign Languages, Xinyu University, Xinyu 338000, China

Abstract. In order to improve the efficiency of big data ad hoc query, a big data ad hoc query technology based on social network information cognitive model is designed. Firstly, the hierarchical structure of data query is established, then the dimension measurement and query conditions are determined, and the result analysis function is proposed. On this basis, the linkage update model of distributed multi database information resources is constructed, and finally the query data is clustered and stored, so as to realize on-the-spot big data query. The experimental results show that the research of big data ad hoc query technology effectively improves the query efficiency, and improves the accuracy of data query.

Keywords: Social network information cognitive model · Ad hoc query · Big data

1 Introduction

In recent years, with the continuous development of society and the continuous improvement of business model, the business data in the enterprise has achieved explosive growth. The traditional management software of each business department has been unable to meet the growing data needs. It is urgent to seek a software that can not only deal with these huge data, but also provide decision support, provide reliable and accurate comprehensive decision support for enterprise management, and realize the integrated business intelligence of the whole enterprise. Business intelligence has built-in complete data analysis function. Without other auxiliary tools and rich it knowledge, users can realize data analysis work such as data specification, semantic mapping, real-time data query, multi-dimensional data analysis, Chinese style report compilation, visual graphic design, etc., improve the utilization rate of data, and significantly save users' investment. Similarly, business intelligence can also be closely combined with the existing business systems of enterprises to accurately analyze the massive data generated by the existing resource management system (ERP), customer relationship system (CRM), business process system (BPM), financial performance management system (FPSM), business analysis system and other core systems of enterprises. Generate clear and reliable data analysis results and embed them, so as to provide reliable data analysis means

© ICST Institute for Computer Sciences, Social Informatics and Telecommunications Engineering 2022
Published by Springer Nature Switzerland AG 2022. All Rights Reserved
S. Liu and X. Ma (Eds.): ADHIP 2021, LNICST 417, pp. 183–195, 2022.
https://doi.org/10.1007/978-3-030-94554-1_16

for users at different levels of government, finance, electric power, communication and other enterprises and institutions in various business scenarios, and significantly improve customers' existing system data analysis ability.

At present, there have been some good research results on big data query. Reference [1] presents a low latency publishing protocol for large data query process confidential data based on Manhattan metrics, classifies confidential data matrices by information gain method, extracts confidential data features from classified matrices, divides them into different data clusters according to feature differences, uses Manhattan metrics algorithm to measure the change of data cluster centers before and after, determines the low latency value according to the relative distance of cluster center changes, and judges the low latency situation of confidential data by comparing the low latency value with the selected low latency threshold size. Reference [2] presents a sampling method based on acceleration ratio and potential distribution, which supports various sampling algorithms, realizes the assurance of randomicity, performance assurance and nearness evaluation of sampling and query in a distributed environment, and is compatible with accurate query. This method can be quickly applied to the columns with a large amount of data, and has good scalability and maintainability. However, the two or more traditional methods cannot realize the impromptu query of large data and cannot be applied to social networks.

Based on this background, this study designs a big data ad hoc query technology based on social network information cognitive model. Ad-hoc Queries is a concept in the field of data warehousing that refers to queries that a user customizes to his or her current needs when using the system. Ad hoc statistics is an aggregation operation based on ad hoc query, which is a kind of statistical function of user's temporary definition of statistical indicators. Compared with fixed statistics, ad hoc statistics has the advantages of flexible use and personalized definition. This paper establishes the hierarchical structure of social network data query, determines the measurement of dimension and query conditions, and gives cognitive analysis function of social network information. This paper constructs a link updating model of distributed multi-database information resources, clusters and stores the query data, and realizes the extempore query of big data. This technology can greatly improve the efficiency of social network information processing.

2 Establishment of Data Query Hierarchy

Hierarchy is the collection of members in a dimension and the relative position between them. Hierarchy is the logical structure of organizing data, which enables us to organize and view data organically instead of facing a large number of detailed data directly. Generally speaking, drill down includes a behavior of traversing the hierarchy, that is, traversing from the high level of a one-dimensional hierarchy to the low level of detail, while roll up includes a process of generalizing the low-level detail data to the high-level data in a one-dimensional hierarchy. Hierarchy is an indispensable part of business, and it is also a basic problem in data warehouse and OLAP. There are two types of hierarchy: level based and parent child value based. This paper mainly discusses level based hierarchy, because it is more common and universal in practice. Hierarchical structure is also called hierarchical tree, which is a special type of "parent-child" connection. Children represent low-level details or father's granularity. The columns in the dimension table represent a specific level in a hierarchy. Because there may be one or more relationships between multiple members in a dimension table, a dimension table can correspond to multiple hierarchical structures. There are several types of hierarchical structures: Normal hierarchy, skip hierarchy, ragged hierarchy, ragged with skip hierarchy. In addition, according to whether the data in the dimension of hierarchy can be summarized, it can be divided into aggregation hierarchy and non aggregation hierarchy.

Normal hierarchy: For example, there are three levels of country, province, and city in the regional dimension. The level of their composition is "country, one province, one city". Without considering municipalities, this level structure is a neat and balanced level structure.

Uneven hierarchy: It means that the process of traversing from high level to the lowest level of dimension will cross one or more levels. In the above-mentioned regional dimension, if the municipality directly under the central government is added, because there is no province at the upper level of Shanghai directly under the central government, but it is directly subordinate to the whole country, then the hierarchical structure will become uneven.

Unbalanced hierarchy: It means that there will be no data of the lowest level in the process of traversing from the high level to the lowest level of dimension. That is, the branches of the hierarchy are reduced to different levels.

Unbalanced and uneven mixed hierarchies: It refers to the hierarchical structure with both spanning branches and unbalanced branches in the process of traversing from high level to the lowest level of dimension.

Aggregation hierarchy and non aggregation hierarchy: In the data warehouse, data aggregation can be defined through hierarchical structure. For example, in commodity sales, from the perspective of time dimension analysis, a hierarchical structure of "year, quarter, and month" composed of level month, quarter, and year can be established, so that monthly commodity sales can be aggregated to quarterly commodity sales, and quarterly commodity sales can be aggregated to annual commodity sales. The hierarchy with such additive facts is called aggregation hierarchy. However, there are some factual values in the data warehouse, which do not show the additivity of commodity sales, such as the GDP growth rate of a province. If the annual growth rate is added up, it is meaningless. However, if the average value is taken, a meaningful value will be obtained. Such a measure is also called semi additivity. These semi additive or non additive facts are embodied in the hierarchy to form a non aggregated hierarchy. For example, in the national level of one province and one city, the GDP growth rate of each province can not be aggregated by the GDP growth rate of the cities below it, and the GDP growth rate of the whole country can not be aggregated by the GDP growth rate of each province. The provincial GDP growth rate and the national GDP growth rate must be stored as facts in the fact table of the data warehouse.

Through the predefined hierarchical structure in the dimension, the function of data summary and drilling can be realized. For the non aggregation level, it mainly uses the data drilling function of the level to view and display the data through the up and down analysis of the level. The hierarchical modeling of data warehouse determines the description of metadata, the design and implementation of ad hoc query software, and the implementation of data drilling and aggregation. Hierarchical modeling also determines that data warehouse modeling adopts star model and snowflake model. At the same time, data aggregation also has a great impact on hierarchical modeling, which will be discussed below.

First, data aggregation [3]. As data is the normal level of aggregation, star model or snowflake model can be directly used for modeling. For the abnormal level of data aggregation, it is necessary to make the uneven branches and unbalanced branches in the level neat and balanced respectively in the process of modeling. If there are multiple levels between the child members and the direct parent, one level at a time will be processed by the same method, so as to realize the leveling of the non leveling level. Similar to the non-uniform hierarchy processing method, the non-uniform hierarchy can also be balanced by using placeholders, but the processing order is opposite to the non-uniform hierarchy. In the non-uniform hierarchy, the parent member occupies the position of the child member and iterates to the lowest level of the hierarchy, that is, the leaf node is reduced to the lowest level of the hierarchy tree.

Second, the data is not aggregated, and the non aggregated hierarchy mainly uses the data drilling function of hierarchy to view and display the data through the analysis of roll up and drill down by hierarchy. Due to the non aggregation of data, the data of the column corresponding to dimension in fact table comes from all nodes in dimension hierarchy tree. Star model is difficult to realize this kind of hierarchical modeling, and snowflake model is generally used.

3   Dimensional Measurement and Query Condition Determination

Measurement: Columns in the database that can be used for aggregate calculation of sum, average and count.

Dimension: The angle from which users observe and analyze the data. For example, time, region.

The main contents include:

1) Find the data model by searching or in the model list. After users get the data model they want, they can view dimensions or measures.
2) Configure query criteria. According to the needs of enterprises or organizations, dimensions or measures in the model are selected.
3) Set filter conditions, such as time or some metric limits.
4) In order to browse query results more clearly, users can also sort dimensions and measures separately.
5) According to different query needs, users can select list or crosstab. A list is a linear table, and a crosstab is a report with groups in both row and column directions (Fig. 1).

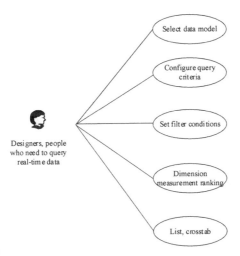

Fig. 1. Use case diagram of configuration table

4 Result Analysis Function Development

If the data is simply listed, it is impossible to find the key points and trends at a glance. This requires designers to use the idea of data visualization to show the data graphically and explore its in-depth value. The main functional steps involved are as follows:

1) Create a new calculation column [4] to calculate the function of the metric. Get the custom calculation result column.
2) Summarize and calculate the measurement. Merge multiple data into one cell, such as sum, average, count, etc.
3) Set an alert. According to some rules or opinions, highlight important data reports, prompt special situations, so as to avoid the ignorance of injury or lack of preparation conditions, so as to minimize the loss caused by danger.
4) For the data, the trend analysis chart can be drawn to facilitate observation and comparison.
5) Query results can generate graphics, which can help analysis through data visualization (Fig. 2).

Edit query data table function. As a complete and detailed report, in addition to accurate data and clear analysis results, designers also need to further improve the unit, font size, decimal point and so on, so as to complete a unambiguous and professional work result [5].

1) Designers can add units to the table data. One is to directly add currency symbols in front of the characters in the cell; the other is to add the unit name in front of the header.
2) The user can change the data alignment in the cell. Font, font size and other style settings.

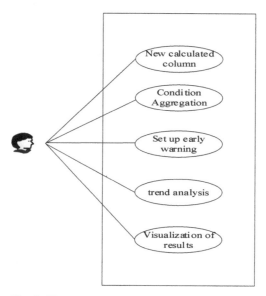

Fig. 2. Use case diagram of result analysis function

3) For the data in the table, the user can set the mantissa after the decimal point and the thousandth (Fig. 3).

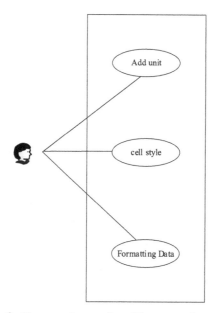

Fig. 3. Use case diagram for editing query data table

5 Construction of Distributed Multi Database Information Resources Linkage Update Model

Based on the completion of the above-mentioned basic work, a distributed multi-database information resource linkage update model is constructed. In order to better realize the linkage update of information resources, the fuzzy theory is used to process the incremental data packets in the space, extract the information characteristics collected by the sensors, and define the information collected by each sensor as a fuzzy input [6]. The fuzzy proposition judgment support degree of each information is given to obtain the spatial information optimization degree. The calculation formula for processing the observation information of the space sensor is:

$$J = \sum_b i \frac{g}{Tbv \times f} \tag{1}$$

In the formula (1), J represents the set of recognition sensors, $\sum_b i$ represents the rule inference coefficient in the data, and Tbv represents the fuzzy algorithm factor. This calculation does not perform directional analysis.

On the basis of the above calculation, the quality of information collected by different sensors in the database is evaluated, and the data with lower weights is eliminated. The calculation formula is:

$$E_b = \sqrt[w]{F} * \frac{p}{c \in v} \tag{2}$$

In formula (2), E_b represents the average of sensors in different spaces, $\sqrt[w]{F}$ represents the similarity matrix composed of sensors, and $c \in v$ represents the weight judgment factor. This calculation does not perform directional analysis.

Through the above calculation, the collected information features [7] are extracted, and the collected data is defined as the data volume of fuzzy proposition. On this basis, a distributed multi database information resource linkage update model is constructed. In order not to affect the efficiency of data update during work, the linkage update model is set to start at night, the incremental information extraction function is started within the planned time, the specified incremental information file is extracted, and the spatial data exchange format is automatically generated after extraction [8]. On this basis, the distance algorithm is used to measure the data in the distributed multi database, and the data in the database is updated by linkage. The update algorithm is introduced to synchronize the relevant data in the database to the database, and automatically update the new data, so as to complete the linkage update of distributed multi database information resources. The specific calculation is as follows:

$$K = \frac{B}{Q} \times G \times \sqrt{P_d} \tag{3}$$

In formula (3), K represents the unit value of the data, $\frac{B}{Q}$ represents the difference factor between the data, and $\sqrt{P_d}$ represents the linkage update factor. Directional analysis is not performed in this calculation.

Through the above calculation, the incremental data is imported into the database to complete the linkage update of distributed multi database information resources.

6 Query Big Data Clustering

On the basis of the above data update, the main steps of clustering the query data and information clustering are as follows:

Step 1: Users input query keywords, which is the first step in the operation of the query system in this study. Users input the keyword information of the title of the book, and the keyword information can be the title of the book, the author's name, the subject's name, the publishing house and so on [9];

Step 2: For keyword segmentation and query processing, because some users may not remember the complete information of the bibliography, so the association rules are defined for the bibliography. The keyword equivalence is defined as the smaller keyword which is most suitable for users' needs to query, so as to return multiple query results. The expression of association rules is as follows:

$$q = Q(m + s_j)/v \tag{4}$$

In formula (4), s_j represents the query record of the j-th book, m represents the book query association rule, v represents the query support, and Q represents the frequent item set of the query information.

The recommendation process of association rules is as follows (Fig. 4):

Step 3: Content clustering is used to cluster the bibliographic information returned from the previous query. Firstly, the clustering center is initialized, and the clustering algorithm is used to determine the initial clustering center. Assuming that there are n objects, n objects are divided into different clusters. The purpose is to cluster the objects with high content similarity into the same cluster as much as possible. In the object set, the mean value of all objects in each query content cluster is taken as the new cluster center. The content clustering expression is as follows:

$$M = \frac{1}{n} \sum_{i=1}^{n} s/c \tag{5}$$

In formula (5), s represents the distance between the query object and the query object, and c represents the number of cluster objects.

Then cluster the bibliographic information and calculate the edit distance to optimize the clustering result. The expression is as follows:

$$D = \frac{r_k}{1 + d/a} \tag{6}$$

In formula (6), r_k represents the k-th attribute feature value, d represents the attribute weight of the query information, and a represents the comprehensive similarity of the query information of the two books.

Step 4: Sort the bibliographic information, sort the relevance of the query content according to the results of the content clustering, and view the title of the book closest to the cluster center, that is, the most relevant information the user queries, and it will be ranked at the top of the query results;

Step 5: The query result is displayed, the result of the query content is displayed to the user, and the data is accessed [10–12].

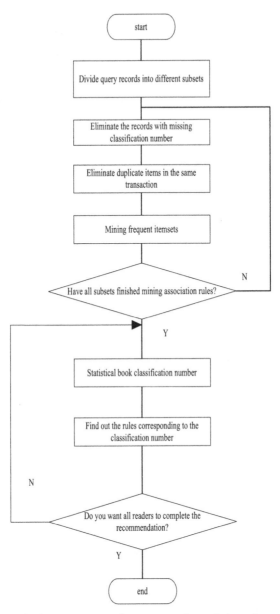

Fig. 4. Recommendation process of association rules

Step 6: Query data storage, OLAP is a kind of software technology that enables analysts, managers or executives to access information quickly, consistently and inter-actively from multiple perspectives, so as to obtain a deeper understanding of the data. The goal of OLAP is to meet the requirements of decision support or specific query and report in multi-dimensional environment. The core of OLAP technology is the concept of "dimension". Dimension is the angle from which people observe the objective world, and it is a high-level classification. Dimension generally contains hierarchical relationship, which is sometimes quite complicated. By defining multiple important attributes of an entity as multiple dimensions, users can compare data on different dimensions. Therefore, OLAP is also a collection of multidimensional data analysis tools [13].

OLAP multidimensional data analysis operations, OLAP basic multidimensional analysis operations include roll up and drill down, slice, dice, pivot, drillcross, drill through and so on. Drilling is to change the level of dimension and the granularity of analysis. It includes roll up and drill down. Upward drilling is realized by climbing up the hierarchical structure of a dimension, that is, generalizing the low-level detail data to the high-level summary data in one dimension, or reducing the dimension by dimension specification [14]; On the contrary, drill down is implemented along the hierarchy of dimensions, that is, from summary data to detailed data for observation, or by introducing additional dimensions. Slicing selects one dimension of a given cube, resulting in a subcube. The slicing operation defines a subcube by performing a selection on two or more dimensions. Rotation is the direction of transforming dimensions, that is, reordering dimensions in a table.

On this basis, for data cube calculation, OLAP tools use data cube to model data and provide multi angle analysis. Data cube calculation is a basic task of OLAP. Full precomputation (full cube materialization) or partial precomputation (partial cube materialization) of data cube can greatly reduce the response time and improve the performance of OLAP [15]. But this kind of computing is a challenge, because it requires a lot of time and storage space. The core of multidimensional data analysis is to effectively calculate the aggregation of multiple dimensional sets. In SQL terms, these aggregations are called group by. Each group can be represented by a cube, and the set of groups forms the lattice of cubes that define the data cube. The figure below shows a cube with three dimensions (Fig. 5).

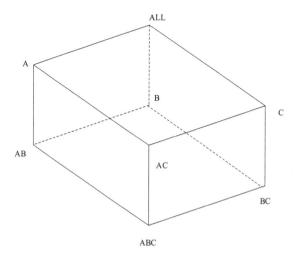

Fig. 5. Data cube

Based on the above process, it completes the big data ad hoc query and data storage.

7 Experiment

In order to verify the effectiveness of the big data ad hoc query technology based on the social network information cognitive model in this study, experiments are conducted to compare the query effects of the two methods.

7.1 Query Time Comparison

Comparing the data query time of the two methods, the comparison results are as follows:

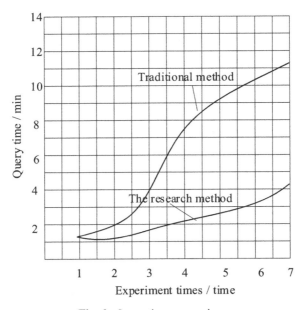

Fig. 6. Query time comparison

Through the analysis of Fig. 6, we can see that the query time of the two methods increases gradually in 7 iterations, but from the third experiment, the query time of the two methods is quite different. In the seventh experiment, the query time of the traditional method is about 11 min, in contrast, the query time of the research method is about 4 min, so we can see that the data extemporaneous query technology in this study has better application effect than the traditional query technology.

7.2 Comparison of Query Accuracy

Comparing the query accuracy of the two methods, the results are as follows:

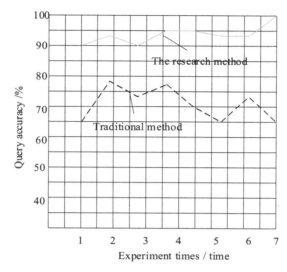

Fig. 7. Comparison of query accuracy

The results of the experiment in Fig. 7 show that the accuracy of the traditional method and the proposed method are stable throughout the experiment. In 7 iterative experiments, the range of the traditional methods is 65%–80%, and the range of the proposed methods is 90%–100%. Through the analysis of the figure above, we can find that the query technology in this study has higher accuracy and better application effect than traditional methods.

8 Conclusion and Prospect

This paper designs a big data ad hoc query technology based on social network information cognitive model, and verifies the effectiveness of the research method through experiments. Through this research, the query technology can effectively improve the efficiency and accuracy of ad hoc query, but due to the limitation of research time, the design method still has some shortcomings, and further research is needed in the follow-up research.

In the future, the research on unified query statistics of structured data and unstructured data will be further carried out, as well as predictive data mining and analysis research using historical and real-time data, so as to more fully mine the value of social network information big data and provide new technical means for the processing of network big data.

Fund Projects. Science and technology project of Jiangxi Provincial Education Department in 2018: Research on big data ad hoc query and analysis technology (Fund No. GJJ181029).

References

1. Su, X.G., Xue, J.M., Xuan, Z.Y.: Big data query process confidential data low-latency release protocol simulation. Comput. Simul. **36**(7), 363–366 (2019)

2. Qi, W., Bao, Y.B., Song, J.: Column-oriented store based sampling query process on big data. Comput. Sci. **046**(012), 13–19 (2019)
3. Liu, S., Bai, W., Zeng, N., et al.: A fast fractal based compression for MRI images. IEEE Access **7**, 62412–62420 (2019)
4. Liu, W., Zhang, T., Liu, J.: Window-based multiple continuous query algorithm for data streams. J. Supercomput. **75**(9), 5782–5807 (2019)
5. Fu, W., Liu, S., Srivastava, G.: Optimization of big data scheduling in social networks. Entropy **21**(9), 902 (2019)
6. Chen, I.C., Hsu, I.C.: Open Taiwan Government data recommendation platform using DBpedia and Semantic Web based on cloud computing. Int. J. Web Inf. Syst. **15**(2), 236–254 (2019)
7. Safari, L., Patrick, J.D.: An enhancement on Clinical Data Analytics Language (CliniDAL) by integration of free text concept search. J. Intell. Inf. Syst. **52**(1), 33–55 (2019)
8. Liu, S., Lu, M., Li, H., et al.: Prediction of gene expression patterns with generalized linear regression model. Front. Genet. **10**, 120 (2019)
9. Khan, A.W., Bangash, J.I., Ahmed, A., et al.: QDVGDD: Query-Driven Virtual Grid based Data Dissemination for wireless sensor networks using single mobile sink. Wirel. Netw. **25**(1), 241–253 (2019)
10. Watanabe, Y., Sato, K., Takada, H.: DynamicMap 2.0: a traffic data management platform leveraging clouds, edges and embedded systems. Int. J. Intell. Transp. Syst. Res. **18**(1), 77–89 (2020)
11. Cao, Y., Chen, S.W.: Extended query model for MOOC education resource metadata based on big data. Int. J. Contin. Eng. Educ. Life-Long Learn. **29**(4), 374–387 (2019)
12. Cisneros-Cabrera, S., Michailidou, A.V., Sampaio, S., et al.: Experimenting with big data computing for scaling data quality-aware query processing. Expert Syst. Appl. **178**(1), 114858 (2021)
13. Wang, H., Mu, L., et al.: Management and instant query of distributed oil and gas production dynamic data. Pet. Explor. Dev. **46**(05), 169–176 (2019)
14. Sam, T., Mauricio, S., Jonathan, B., et al.: Internet search query data improve forecasts of daily emergency department volume. J. Am. Med. Inform. Assoc. **12**, 12 (2019)
15. Jia, B., Meng, B., Zhang, W., et al.: Query rewriting and semantic annotation in semantic-based image retrieval under heterogeneous ontologies of big data. Traitement du Signal **37**(1), 101–105 (2020)

A Design of Digital High Multiple Decimation Filter in Intermediate Frequency Domain

Yaqin Tong, Wenxin Li[✉], and Fengyuan Kang

Shandong Institute of Space Electronic Technology, YanTai, China

Abstract. An important task of digital down conversion is to reduce the sampling rate. In order to carry out undistorted decimation, decimation filtering must be carried out before decimation to prevent spectrum aliasing. Based on the design idea of hierarchical cascade filter, an implementation method of 100 times decimator filter is given. The sampling factor and frequency response of each filter are analyzed in detail, and the theoretical analysis and practical logic implementation methods are given. The practical application results prove the correctness and feasibility of the design of high multiple decimator filter. This multi-stage decimation method not only enhances the stability of the system, but also greatly reduces calculation and saves resources.

Keywords: Decimator · Aliase · Filter

1 Introduction

Digital IF processing is one of the key technologies of software radio. It is an important component connecting the AD IF digital stream with the baseband data stream demodulated at the back end. The development of software radio puts forward higher and higher requirements for digital IF processing technology. Its performance index directly determines the ability of software radio signal processing. Taking a software radio radar receiver as an example, its parallel processing channel is up to 32 channels, with 55 MHz bandwidth as the step interval, and the wide band RF signal is scanned linearly. Among them, IF processing performance index of single channel is as follows:

(1) Working clock 120 MHz;
(2) AD input data bit width is 16 bits;
(3) AD sampling rate 120 M;
(4) Input data bandwidth 60 kHz;
(5) The decimator multiple is 100, and the final output rate is 1.2 M;
(6) Output data bit width 24 bits;
(7) Decimator of anti-aliasing inhibition is greater than 100 dB.

The decimator factor of the machine is as high as 100, and the anti-aliasing inhibition is more than 100 dB. Common filter types for decimator, such as CIC filter wave and half-band filter, cannot meet the requirements of anti-aliasing suppression. In the process of real - space design, the combination of CIC filter, half-band filter and polyphase filter

© ICST Institute for Computer Sciences, Social Informatics and Telecommunications Engineering 2022
Published by Springer Nature Switzerland AG 2022. All Rights Reserved
S. Liu and X. Ma (Eds.): ADHIP 2021, LNICST 417, pp. 196–202, 2022.
https://doi.org/10.1007/978-3-030-94554-1_17

must be integrated to gradually reduce the decimator rate and cascade to achieve aliasing suppression performance index [1, 2].

2 Design of Decimation Filter Bank

High multiple decimation processing is realized by the cascade combination of filters as shown in Fig. 1.

The filter bank is composed of four cascade filter groups, and the sampling rate of digital signal is reduced step by step.

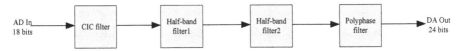

Fig. 1. High multiple decimator filter function diagram

CIC Filter of Stage 1. CIC filter structure is relatively simple, the implementation occupies less logical resources, but the frequency response transition band is large, the frequency band characteristic changes are relatively flat slow. The CIC filter can be used to achieve low-multiple logic decimator, and the signal sampling rate can be reduced in the initial step, and the timing margin of time-division multiplexing can be provided for other types of subsequent filters, so as to save logic resources.

In this design, CIC realizes 5 times decimator. To ensure decimation anti-aliasing suppression is not less than 100 dB, the frequency band amplitude response of the filter in the $[f_s/10, f_s/2]$ interval (f_s is the data sampling frequency) must be more than 100 dB relative to the main lobe peak.

Half-Band Filter of Stage 2/Stage 3. When the input signal sampling rate of digital IF equipment is reduced to 1/5, the sampling rate is changed to 24 MHz. Since the effective band width of the signal is 500 kHz, the order of FIR filter should be about 400 (24 MHz/60 kHz) if the subsequent FIR multiphase filter is directly adopted to achieve the decimator of signal bandwidth. High-order filter will take up a huge amount of multiplier resources, and the frequency response is very difficult to do sharp.

In this design, the two-stage half-band filter cascade is used to further reduce the data sampling rate by 1/4, so as to ensure that the low-order FIR multiphase filter can be adopted and the time-division multiplexing margin of the subsequent filter can be further guaranteed.

Polyphase Filter of Stage 4. After the CIC filter and the two-stage half-band filter, the signal sampling rate is reduced to 1/20 of the original AD input sampling rate, that is, 6 MHz. Because the ratio of data sampling rate to data bandwidth is relatively low, it is convenient to use low-order FIR filter to realize effective selection of data pass-band. In order to achieve the requirement of 100 times overall decimator rate, the multiphase filter structure is adopted in the 4th stage FIR filter, and the 60 kHz pass-band and 5 times decimator are realized simultaneously.

2.1 CIC Filter Design

The CIC filter achieves 5 times decimator. In order to ensure that the decimator anti-aliasing suppression is not less than 100 dB, the frequency band amplitude response of the filter should be in the range of $[f_s/10, f_s/2]$ (f_s is the data sampling frequency) and the suppression of the zero frequency point must be greater than 100 dB.

The amplitude-frequency response function of the single-stage CIC filter is shown in formula (1) [3].

$$H(jw) = \left| \frac{\sin(\frac{wDM}{2})}{\sin(\frac{w}{2})} \right| \tag{1}$$

w represents the angular rate, the design parameter D is the differential delay, and M is the decimator factor.

The stop-band attenuation performance of single-stage CIC filter is poor. As can be seen in formula (1), when $DM \gg 1$, the level of the first side lobe is $2DM/3\pi$, and the difference between the peak level and the main lobe is:

$$20\log(\frac{\frac{DM}{2DM}}{3\pi}) \tag{2}$$

In engineering applications, the multi-stage CIC cascade is generally adopted to achieve better stop-band suppression performance under the condition of high decimator rate. The design parameters of the multi-stage CIC filter in this design are shown in Table 1.

Table 1. CIC hogenauer filter parameters.

Differential delay D	6
Decimation factor M	5
Cascading order N	5

The amplitude response of the multi-stage CIC filter is shown in Fig. 2 (the horizontal axis in the figure has been normalized by $f_s/2$).

As can be seen from Fig. 2, the normalized frequency point is located at [0.2, 1] interval, the stop-band suppression is not less than 130 dB, which meets the anti-aliasing suppression requirement of 5 times decimator.

2.2 Half-Band Filter Design

The pass-band and stop-band of the half-band filter are symmetric, and the filter's coefficients are symmetric, and all the coefficients of even order terms are 0 except the center point. The actual half-band filter coefficient that needs to be stored is about 1/4 of the order, which can reduce the resources of the mass storage and multiplier and reduce the operation Quantity [4].

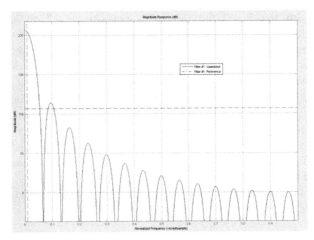

Fig. 2. Amplitude response for the CIC hogenauer filter

Table 2. Half-band filter parameters.

Order number	32
Coefficient of quantitative	16

The design parameters of the half-band filter in this design are shown in Table 2.

The amplitude response of the half-band filter is shown in Fig. 3 (the horizontal axis in the figure has been normalized with $f_s/2$, where f_s is the signal frequency input to the half-band filter).

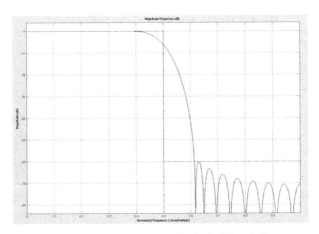

Fig. 3. Amplitude response for the half-band filter

As can be seen from Fig. 3, the stop-band suppression is at least 40 dB. Since the pre-stage CIC filter has suppressed the stop-band by nearly 100 dB, the superposition of the two also meets the requirement of aliasing suppression greater than 100 dB.

It should be noted that when the half-band filter is used for 2 times decimator, although there is no aliasing in the pass-band and stop-band, there is aliasing in the transition band. The pass-band and stop-band of the half-band filter are symmetrical. Frequency response characteristics after 2 times decimation are shown in Fig. 4. The transition band has obvious aliasing after 2 times decimator. Therefore, a higher order FIR filter must be used in the post-filter to improve the performance indicators such as pass-band, transition band and stop-band attenuation, so as to ensure that the transition band aliasing of the half-band filter is suppressed.

The second stage and the third stage half-band filter adopt the same coefficient, the total with the realization of 4 times decimator.

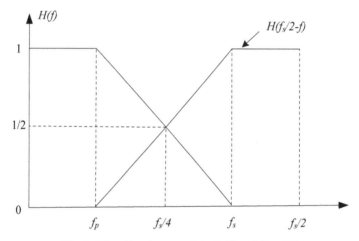

Fig. 4. The alias diagram of the halfband filter [5]

2.3 Polyphase Filter Design

Polyphase filter is a commonly used method in the process of digital down conversion decimator [6]. Compared with the ordinary FIR filter, the conversion relationship of signal sampling rate can be effectively used to remove the unnecessary operation in the process of data rate conversion, thus greatly improving the speed of operation.

Figure 5 shows the structure of FIR filter based on polyphase decomposition [7]. The filter is divided into three subbands, and each subband is extracted by M times. After decimator, it enters the subband filter for filtering. Since the decimator rate of each subband is reduced to 1/M, the subband filter can reduce the resource utilization by time division multiplexing.

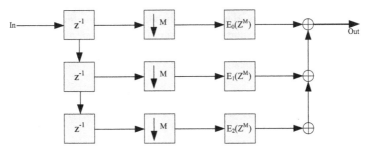

Fig. 5. The aliase diagram of the half-band filter

In the design of this paper, polyphase filter is used to achieve 5 times decimator and 60 kHz data bandwidth acquisition. After the pre-stage CIC filter and the two-stage half-band filter, the data sampling rate into the polyphase filter is 6 MHz (120 MHz/5/2/2). The design parameters of the polyphase filter are shown in Table 3.

Table 3. Polyphase decimator filter parameters.

Order number	128
Coefficient of quantitative	16

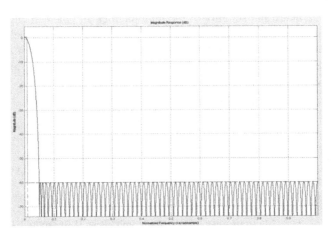

Fig. 6. The amplitude response of the polyphase decimator filter

The amplitude response of the polyphase filter is shown in Fig. 5 (where the horizontal axis has been normalized with fs/2, where fs is the signal frequency of the input multiphase filter). As can be seen from Fig. 6, the stop-band suppression is up to 60 dB, and the pass-band is 6 kHz. After 5 times of decimator, the anti-aliasing inhibition can also reach 60 dB when it is located in the stop-band. Considering the stop-band suppression of the pre-stage CIC filter and the half-band filter, the superposition performance of the cascade filter bank meets the requirement of the whole body aliasing suppression greater than 100 dB.

3   Conclusion and Outlook

The four-step cascade of CIC, half-band and polyphase filter banks proposed in this paper has the following characteristics in design:

(1) The design method of decomposing the anti-aliasing index and the decimator factor index into multiple filters reduces the design performance of each filter and improves the design realizability;

(2) As the CIC filter and half-band filter are used to effectively reduce the front-end working frequency, the order of the key FIR filter at the back end is reduced as much as possible, and the usage of logical resources is reduced. The filter transition band can be designed to be smaller, which ensures the SNR of the output signal.

The verification results in a radar receiver show that the digital IF decimator achieved by FPGA can achieve a high multiple decimator of 100 times, the signal bandwidth can reach 500 kHz, and the anti-aliasing index of the whole machine can reach 110 dB, which meets the design requirements. The experimental results show that the design method of high multiple decimator index decomposition has the characteristics of high design performance, simple use and less resource consumption, and is worthy of popularization and application in design practice.

If the high multiple decimator filter is implemented by single-stage filter, it will lead to the filter has to adopt higher order and larger quantization error, which will affect the performance index and stability of the multiplefilter. In practical engineering, the suitable design method is to adopt multi-stage filter cascade, and gradually reduce the sampling rate through the lower order filter. In the future, more efficient multi-stage filter will be considered, so as to meet the requirements of the design index and reduce the use of resources.

References

1. Yang, Y., Li, D., Shi, L.: Design of digital decimator filter for 14-bit 1.28 MS/s$\sum\triangle$ADC. J. Xidian Univ. (Nat. Sci. Ed.) **37**(2), 315–319 (2010)
2. Liu, Q., Gao, J., Huang, G., Zhu, W.: Low power CIC decimator filter with improved frequency response. J. PLA Univ. Sci. Technol. (Nat. Sci. Ed.) **12**(2), 114–119 (2011)
3. Guangshu, H.: Digital Signal Processing: Theory, Algorithm and Implementation. Tsinghua University Press, Beijing (2012)
4. Chu, S., Burrus, C.S.: Multirate filter design using comb filters. IEEE Trans. Circuits Syst. **31**, 913–924 (1984)
5. Tian, Y., Xu, W., Zhang, W.: Design of FPGA for Wireless Communication. Press of Electronics Industry (2010)
6. Sun, J., Han, T., Yu, W., Liu, R.: DDC design and FPGA implementation based on polyphase filtering principle. Space Electron. Technol. **2**, 54–58 (2014)
7. Bellanger, M.G., Bonnerot, G., Coudreus, M.: Digital filtering by polyphase network: application to sample rate alternation and filter banks. IEEE Trans. Acoust. Speech Signal Proc. ASSP **24**, 109–114 (1976)

Improved Sliding Window Kernel RLS Algorithm for Identification of Time-Varying Nonlinear Systems

Xinyu Guo[1], Menghua Jiang[1], Ying Gao[1], Shifeng Ou[1(✉)], Jindong Xu[2], and Zhuoran Cai[1]

[1] School of Optoelectronic Information Science and Technology, Yantai University, Yantai 264005, China
ousfeng@ytu.edu.cn

[2] School of Computer and Control Engineering, Yantai University, Yantai 264005, China

Abstract. The sliding window kernel recursive least squares (SW-KRLS) algorithm is one of the most widely used approach in dealing with nonlinear problems because of its simple structure, low computational complexity and high predictive accuracy. However, as data size increases, the computational efficiency of the SW-KRLS algorithm will be affected by the redundant data and the size of sliding window. In order to solve these problems, this paper proposes a variable sliding window sparse kernel recursive least squares (VSWS-KRLS) algorithm. It first uses the sliding window to constrain the size of the novelty criterion dictionary. After that, the algorithm combines the improved novelty criterion with the SW-KRLS to remove the less relevant data. In addition, mechanisms for window size adjustment are added to adjust the size of sliding window adaptively according to the system changes. The experimental results show that the proposed algorithm has better performance in identification of nonlinear system.

Keywords: Kernel recursive least squares · Nonlinear system · Variable sliding window · System identification

1 Introduction

Adaptive filtering algorithm is widely used in system identification, channel equalization, echo cancellation and other fields [1–4]. In 1950, Plackeet first proposed the recursive least squares [1] (RLS) algorithm, which has fast convergence and small prediction error. In 1960, Widrow and Hoff proposed the least mean square algorithm [5] (LMS), which is widely used because of its simple calculation and good convergence performance. However, the effect of linear filters in dealing with nonlinear problems is not ideal. In recent years, kernel methods [6] have been widely used to deal with nonlinear problems, such as support vector machines [7], kernel principal component analysis [8], Nuclear adaptive filter [9] and so on. The kernel recursive least squares [10] (KRLS) algorithm were proposed by Engel in 2004. This algorithm solves the nonlinear inseparable problem

© ICST Institute for Computer Sciences, Social Informatics and Telecommunications Engineering 2022
Published by Springer Nature Switzerland AG 2022. All Rights Reserved
S. Liu and X. Ma (Eds.): ADHIP 2021, LNICST 417, pp. 203–220, 2022.
https://doi.org/10.1007/978-3-030-94554-1_18

by projecting the input data to the Hilbert space (Reproducing Kernel Hilbert Space, RKHS), and using the positive definite kernel function that satisfies the Mercer [11] condition to calculate the inner product. The projection is shown in Fig. 1. Because the KRLS algorithm has a strong tracking ability in solving nonlinear problems, it has been successfully applied in data mining, machine learning and other fields [12–14].

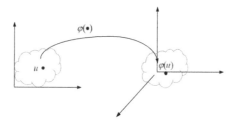

Fig. 1. Nonlinear mapping from input space to feature space.

But due to the high computational complexity of the algorithm, it is difficult to complete, so most studies mainly concentrated on the sparsification of the KRLS algorithms. In 2006, Van Vaerenbergh et al. proposed a sliding window based kernel recursive least squares algorithm [15] (SW-KRLS), which used a sliding window to limit the growing kernel matrix. This algorithm obtains lower computational complexity while retaining the advantages of KRLS. In 2010, Van Vaerenbergh et al. proposed a fixed budget recursive least squares algorithm [16] (FB-KRLS), which used the distance between the sample and the dictionary as a constraint to the sample sparsification. In 2013, Chen et al. proposed the quantized kernel recursive least squares [17] (QKRLS), which used a quantized method to reduce the input dimension. In 2018, Han et al. proposed an adaptive dynamic adjustment kernel recursive least squares method [18] (ADA-KRLS), which performs online sparsification by combining fixed budget with dynamic adaptation. In 2019, Han et al. proposed an adaptive normalized sparse quantized kernel recursive least squares [19] (ANS-QKRLS), which integrated dynamic adjustment, coherence criterion, approximate linear dependency criterion and the QKRLS algorithm for online sparsification. Zhong et al. proposed a dynamic adaptive sparse kernel recursive least squares [20] (DASKRLS) in 2020. It used approximate linear dependency (ALD) and online vector projection standards to make the data sparse, and combined with regularized maximum correlation entropy to deal with the noise impact on data. The literature [21] proposed an initial framework with forgetting factor on the basis of KRLS (FFIKRLS), which combines the ALD-KRLS algorithm with the QKRLS algorithm, and introduces a forgetting factor to sufficiently track the strongly changeable dynamic characteristics.

In summary, research based on KRLS has made good progress. However, in the face of the increasing kernel matrix size, online prediction also faces enormous challenges. In addition, KRLS has some difficulties in accommodating abrupt change. Because KRLS is based on mean square error (MSE) criterion, which is sensitive to the change. So the mutation value in the kernel matrix will affect the convergence of the algorithm. For the above problems, the variable sliding window sparse kernel recursive least squares

(VSWS-KRLS) algorithm proposes a suitable solution. In this paper, we present three major contributions:

- We reduce kernel matrix size by combining sliding window and novelty criterion with KRLS. A lot of irrelevant data in the sliding window will affect the convergence. Therefore, in order to improve calculation efficiency, we add a novelty criterion, which can raise the threshold of entering kernel matrix.
- We propose a new variable sliding window technology to enhance the capability of algorithm to track system changes. This method can adaptively adjust the window size according to the system environment.
- The algorithm has performed system identification experiments in the Wiener nonlinear systems, and has achieved good results.

The rest of this paper is arranged as follows: In Sect. 2, we introduce the KRLS algorithm. In Sect. 3, we introduce several common sparse methods. In Sect. 4, we introduce the proposed methods in this paper. In Sect. 5, we conduct simulation experiments. In Sect. 6, we present our conclusions and prospects.

2 Kernel Recursive Least Squares Algorithm

2.1 Recursive Least Squares Algorithm

The least squares [22] (LS) algorithm does not need to make assumptions about the statistical characteristics of the input signal. The RLS algorithm is a recursive extension of the LS algorithm, which can recursively update the estimated value by applying the old data. For the training data, the RLS algorithm estimates the filter coefficients ω_{i-1} by minimizing the cost function. The cost function is:

$$J(\omega) = \min_{\omega} \sum_{j=1}^{i-1} \left\| d_j - \mathbf{u}_j \omega \right\|^2 \tag{1}$$

The function is computed as:

$$\omega_{i-1} = (\mathbf{U}_{i-1} \mathbf{U}_{i-1}^T)^{-1} \mathbf{U}_{i-1} \mathbf{d}_{i-1} \tag{2}$$

where $\mathbf{U}_{i-1} = [\mathbf{u}_1, \mathbf{u}_2, ..., \mathbf{u}_{i-1}]_{L \times (i-1)}$, $\mathbf{u}_j$ is the input vector at time j, $\mathbf{d}_{i-1} = [d_1, d_2, ..., d_{i-1}]^T$, d_j is the expected response at time j. By introducing the matrix inversion lemma (Woodbury identity), the RLS algorithm can be updated:

$$r_i = 1 + \mathbf{u}_i^T \mathbf{P}_{i-1} \mathbf{u}_i \tag{3}$$

$$\mathbf{k}_i = \mathbf{P}_{i-1} \mathbf{u}_i / r_i \tag{4}$$

$$\omega_i = \omega_{i-1} + \mathbf{k}_i e_i \tag{5}$$

$$\mathbf{P}_i = \mathbf{P}_{i-1} - \mathbf{k}_i \mathbf{k}_i^T r_i \tag{6}$$

The RLS algorithm uses the inverse of the correlation matrix to whiten the input data, which improves the convergence performance of the filter.

2.2 Kernel Recursive Least Squares Algorithm

The KRLS algorithm projects the input data to the RKHS based on the RLS algorithm, and it can solve the nonlinear relationship without knowing the specific mapping form of input. Supposed that $\mathbf{X}$ represents the original space, $\mathbf{X} = [\mathbf{x}_1, \mathbf{x}_2,...,\mathbf{x}_n] \in \mathbf{R}^N$, $\mathbb{H}$ represents the Hilbert space, and the mapping φ is represented as:

$$\varphi : \mathbf{X} \rightarrow \mathbb{H}, \mathbf{x} \rightarrow \varphi(\mathbf{x}) \tag{7}$$

$\varphi(\mathbf{x})$ represents the projection of $\mathbf{x}$ in the feature space, and the nonlinear mapping is realized by the kernel function:

$$\kappa(\mathbf{x}_i, \mathbf{x}_j) = <\varphi(\mathbf{x}_i), \varphi(\mathbf{x}_j)> \tag{8}$$

where $<, >$ represents the inner product operation. As the kernel function used in this paper, the Gaussian kernel function is expressed as:

$$\kappa(\mathbf{x}_i, \mathbf{x}_j) = \exp(-\|\mathbf{x}_i - \mathbf{x}_j\|^2/2\sigma^2) \tag{9}$$

where σ represents the Gaussian kernel parameter. In KRLS algorithm, Given the expected sequence $\{d_1, d_2,...\}$ and the input sequence $\{\varphi_1, \varphi_2,...\}$, the cost function:

$$J(\boldsymbol{\omega}_i) = \min_{\boldsymbol{\omega}} \sum_{j=1}^{i} \left\|d_j - \boldsymbol{\varphi}_j^T \boldsymbol{\omega}_i\right\|^2 \tag{10}$$

where d_j and $\boldsymbol{\varphi}_j$ represent the expected sequence and the input sequence. We assume that $\boldsymbol{\Phi}_i = [\varphi(\mathbf{x}_1),...,\varphi(\mathbf{x}_i)]$, and a $n \times n$-dimensional kernel matrix is defined as:

$$\mathbf{K}_i = \boldsymbol{\Phi}_i^T \boldsymbol{\Phi}_i \tag{11}$$

when $\boldsymbol{\omega}_i = \boldsymbol{\Phi}_i a_i$, the cost function of the KRLS algorithm at time i can be obtained:

$$J = \min_{a_i} \|d_i - \mathbf{K}_i a_i\|^2 \tag{12}$$

where $a_i = [a_1, a_2,...,a_i]^T$ is the weight vector and $d_i = [d_1, d_2,...,d_i]^T$ is the expected output vector. The estimated value of α_i is as follows:

$$\hat{a}_i = \hat{\mathbf{K}}_i^{-1} d_i \tag{13}$$

The kernel matrix is expressed as:

$$\hat{\mathbf{K}}_i = \mathbf{K}_i + \lambda \mathbf{I} \tag{14}$$

where λ is the regularization factor. Unlike the RLS algorithm, the KRLS algorithm focuses on updating the kernel matrix during recursion. However, as the input data increases, the size of the kernel matrix is expanding, so the KRLS algorithm optimization problem has been transformed into the sparse problem of the kernel matrix.

3 Sparsification

The previous part makes a summary of the RLS algorithm and the KRLS, and compares the two algorithms. This part will introduce some common sparse methods.

3.1 Novelty Criterion and Approximate Linear Dependency

In the KRLS algorithm, the size of the kernel matrix will upsize linearly with the update, which will bring challenges to online prediction. Therefore, many sparse methods have been proposed, the novelty criterion [23] (NC) is a simple way to check whether the newly data is useful. In 2004, Engel et al. proposed the approximate linear correlation criterion [10] (ALD) to solve this problem. Besides, Richard et al. studied another similar method, called the coherence criterion [24]. This paper mainly introduces NC and ALD online sparse methods.

In the novelty criterion, online sparsification starts from an empty set and gradually adds samples to the central set of the dictionary according to the judgment. Assuming the current dictionary is:

$$C = \{c_j\}_{j=1}^{m_i} \tag{15}$$

where c_j is the center of time j, and m_i is the count of the set. When a new data pair $\{u_{i+1}, d_{i+1}\}$ appears, algorithm will decide whether to add u_{i+1} as a new center to the dictionary. First we calculate the shortest distance from u_{i+1} to the dictionary.

$$dis = \min_{c_j \in C} |u_{i+1} - c_j| \tag{16}$$

If the distance is less than the threshold $\delta 1$, then u_{i+1} will not be added to the dictionary. On the contrary, the algorithm continues to compare the calculated prediction error e_{i+1} with the threshold $\delta 2$. Only if e_{i+1} is greater than the threshold $\delta 2$, u_{i+1} will be added to the dictionary as a new center. The approximate linear correlation criterion can be defined as:

$$\delta_i = \left\| \sum_{n=1}^{i-1} \alpha_n \varphi(x_n) - \varphi(x_i) \right\|^2 > \nu \tag{17}$$

In the ALD criterion, α represents the coefficient vector, and ν represents the threshold. When the input data arrives, the ALD criterion will calculate the linear dependence between the input data and the dictionary data. If the value is greater than the preset threshold, the input data will be added to the dictionary, otherwise the input will be removed. Similar to the NC criterion, through the comparison of the dictionary and the input data, the ALD criterion will perform sequential sparsification in order to reduce the size of the kernel matrix.

3.2 Sliding Window Method

In order to limit the kernel matrix size, Van Vaerenbergh et al. applied the sliding window method to the KRLS algorithm [15]. In this method, the window size M is fixed, the observation matrix $\mathbf{\Phi}_i = [\varphi(x_{i-M+1}), \ldots, \varphi(x_i)]$, a kernel matrix expressed as:

$$\mathbf{K}_i = \mathbf{\Phi}_i\mathbf{\Phi}_i^T + c\mathbf{I} = \begin{bmatrix} \mathbf{K}_{i-1} & \kappa_{i-1}(\mathbf{x}_i) \\ \kappa_{i-1}(\mathbf{x}_i)^T & \kappa_{ii} + c \end{bmatrix} \tag{18}$$

where $\kappa_{i-1}(\mathbf{x}_i) = [\kappa(\mathbf{x}_{i-M+1}, \mathbf{x}_i), \ldots, \kappa(\mathbf{x}_{i-1}, \mathbf{x}_i)]^T$, $\kappa_{ii} = \kappa(\mathbf{x}_i, \mathbf{x}_i)$, c is the regularization factor. In order to keep the size of kernel matrix unchanged, the kernel matrix uses the new sample data to add new rows and new columns (19)–(20).

$$\mathbf{K} = \begin{bmatrix} \mathbf{A} & \mathbf{b} \\ \mathbf{b}^T & d \end{bmatrix}, \ \mathbf{K}^{-1} = \begin{bmatrix} \mathbf{E} & \mathbf{f} \\ \mathbf{f}^T & g \end{bmatrix} \ \Rightarrow \begin{cases} \mathbf{A}\mathbf{E} + \mathbf{b}\mathbf{f}^T = \mathbf{I} \\ \mathbf{A}\mathbf{f} + \mathbf{b}g = 0 \\ dg + \mathbf{b}^T\mathbf{f} = 1 \end{cases} \tag{19}$$

$$\mathbf{K}^{-1} = \begin{bmatrix} \mathbf{A}^{-1}(I + \mathbf{b}\mathbf{b}^T\mathbf{A}^{-1H}g & -\mathbf{A}^{-1}\mathbf{b}g \\ -(\mathbf{A}^{-1}\mathbf{b})^Tg & g \end{bmatrix} \tag{20}$$

where $\mathbf{A}$ is a kernel matrix before expansion, which is non-singular, $g = (d - \mathbf{b}^T\mathbf{A}^{-1}\mathbf{b})^{-1}$. After expansion, the kernel matrix needs to compress the expanded kernel matrix, and the oldest rows and columns are removed by (21)–(22).

$$\mathbf{K} = \begin{bmatrix} a & \mathbf{b}^T \\ \mathbf{b} & \mathbf{D} \end{bmatrix}, \ \mathbf{K}^{-1} = \begin{bmatrix} e & \mathbf{f}^T \\ \mathbf{f} & \mathbf{G} \end{bmatrix} \ \Rightarrow \begin{cases} \mathbf{b}e + \mathbf{D}\mathbf{f} = 0 \\ \mathbf{b}\mathbf{f}^T + \mathbf{D}\mathbf{G} = \mathbf{I} \end{cases} \tag{21}$$

$$\mathbf{D}^{-1} = \mathbf{G} - \mathbf{f}\mathbf{f}^T/e \tag{22}$$

The above equations can make the size of the kernel matrix fixed. The Fig. 2 is the principle of the sliding window.

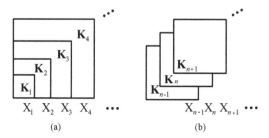

Fig. 2. (a) Kernel matrix $\mathbf{K}_i$ with growing size. (b) Kernel matrix $\mathbf{K}_i$ with a fixed size.

In summary, the advantage of the SW-KRLS method is that it can track time variation without additional computational complexity.

4 Improved Kernel Recursive Least Square Algorithm

In this section, we will introduce proposed algorithm. We begin with some relevant optimizations and introduce the improved algorithm in the end.

4.1 Sliding Window and Novelty Criterion

The SW-KRLS algorithm has low computational complexity and good tracking ability, but it lacks of judgment on the input data and directly adds the input data to the kernel matrix. Furthermore, novelty criterion can eliminate less relevant data in the input, but the size of the dictionary will increase with the number of training data. For the above problems, we use the sliding window method to limit the size of dictionary. Meanwhile, this paper applies the improved novelty criterion to the SW-KRLS algorithm, and proposes the sliding window sparse kernel recursive least squares algorithm (SWS-KRLS).

First, the NC will determine whether to add input data to the sliding window.

$$dis = \min_{\mathbf{c}_j \in C} |\mathbf{u}_{i+1} - \mathbf{c}_j| > \delta 1 \tag{23}$$

$$e_i = d_i - f(\mathbf{u}_*) = d_i - \sum_{j=1}^{i} \mathbf{a}_j \kappa(\mathbf{u}_j, \mathbf{u}_*) > \delta 2 \tag{24}$$

- If the above equations are both satisfied, it means that the input has a greater influence on the algorithm, so input can be added to the sliding window and dictionary.
- If any of the above equations is not satisfied, it means that the new input has little effect on the algorithm. Therefore, the system will not add the input to the sliding window, and the algorithm error is equal to the previous time error.

Table 1. Sliding window sparse kernel recursive least squares algorithm (SWS-KRLS).

Initialization: sliding window size M, threshold $\delta 1$, $\delta 2$, regularization coefficient c, $\mathbf{K}_0 = (1+c)\mathbf{I}$.
for j=2,3,... **do input:** $\{\mathbf{u}_j, d_j\}$
if dis>$\delta 1$, **and** e>$\delta 2$, **then**
add the input to the dictionary $C = \{c_j\}_{j=1}^{mi}$.
else
$e_n = e_{n-1}$, **start** the next iteration $\{\mathbf{u}_{j+1}, d_{j+1}\}$.
if dictionary size is greater than M
dictionary expressed as $C = \{c_{m_{i-L+1}}, c_{m_{i-L}}, \ldots, c_{m_{i-L+1}}\}$.
calculate the kernel matrix according to Eq. (18) -Eq. (22).
calculate the coefficient vector according to Eq. (14).
calculate the system error according to Eq. (24).
end for

Contrary to the standard NC, we apply sliding window method to the dictionary. When the dictionary size is greater than M, the sliding window delete the i-M th data in the dictionary to keep the dictionary size unchanged. Assuming that the size of the fixed novelty criterion dictionary is M, the definition of dictionary C is as follows:

$$C = \{\mathbf{c}_{i-M+1}, \mathbf{c}_{i-M}, \ldots, \mathbf{c}_i\} \tag{25}$$

We use the dictionary $\mathbf{C}$ to calculate the kernel matrix, on the one hand, the improved algorithm has a better steady-state error because deleting these data can increase the proportion of the closer data, on the other hand, this dictionary can decrease memory usage. The process of the algorithm is shown in Table 1.

4.2 Variable Sliding Window Method

Using a sliding window to limit the size of the kernel matrix is one of the effective sparse methods. However, if a fixed size window is used, it is difficult to achieve good parameter tracking when change occurs. When the system changes, the window size will affect tracking ability. Therefore, adaptive adjustment of the window size can better track system change. Julian [25] first applied the variable sliding window method to KRLS, and the method achieved good results. This paper improves this method by adding the Mechanisms for window size adjustment and the change detection mechanisms.

Mechanisms for Window Size Adjustment
Suppose the time required to upsize is U_m and the time required to downsize is D_m:

$$D_m = m^2 + m + O(1) \tag{26}$$

$$U_m = 5m^2 + 2mT_\kappa + 3m + O(1) \tag{27}$$

where m is the kernel matrix size, T_κ is the calculation cost of the kernel function. The total calculation time can be calculated as:

$$T_m = 6m^2 + 2mT_\kappa + 4m + O(1) \tag{28}$$

If the size of the kernel matrix is smaller than M, the total calculation time T_m will less than the allowable calculation time. Thus, it is this "residual" computation time that the algorithm can adjust the size of the kernel matrix online. Suppose the size of the kernel matrix is m, where $1 < m < M$. When the size of kernel matrix is upsized, the up range of the kernel matrix sizes is:

$$R_m^U = \left\{ \overline{m} \le M : \sum_{i=m}^{\overline{m}} U_i \le T_M \right\} \tag{29}$$

When the kernel matrix is downsized, enough time must be reserved to upsize. Therefore, the donsizing range of the kernel matrix size is:

$$R_m^D = \left\{ \overline{m} \ge 1 : \sum_{i=\overline{m}}^{m} D_i + U_{\overline{m}} \le T_M \right\} \tag{30}$$

In general, when the kernel matrix scale is m, the change range can be expressed as:

$$R_m = \left\{ R_m^U \cup R_m^D \cup R_1^U \right\} \tag{31}$$

where RU 1 is the discarded kernel matrix data. The kernel matrix can be upsized and downsized within this range. In addition, in order to improve the parameter tracking performance of the algorithm, this paper proposes a method of adjusting window size. The implementation is shown in Fig. 3.

In stage I, when no change is detected, the SW-KRLS algorithm uses a fixed window size M to update; in stage II, when the system parameter changes are detected, the window size can quickly decrease. By using a smaller window size, the algorithm can track parameter changes more sensitively: in phases III and IV, the window size is gradually restored until it is extended to the maximum allowable size L, so as to obtain higher convergence accuracy.

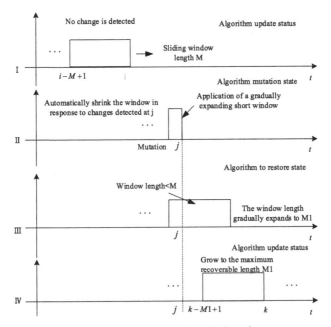

Fig. 3. Method of adjusting window size.

Change Detection Mechanisms

To adjust the window size according to system changes, the algorithms need to use change detection mechanisms, Literature [26] introduced several different methods of detecting parameter changes in adaptive filtering algorithms, which can be divided into parameter detection methods and error detection methods. However, because the parameter ω in the KRLS algorithm is replaced by α, the parameter detection method cannot be used. Therefore, this paper uses error detection to detect system change. The mean square error of adjacent time is defined as Δe:

$$\Delta e_i = e_i - e_{i-1} \tag{32}$$

In order to reduce the false alarm rate, we use the synchronous superposition averaging algorithm to update the system error difference, and this method will be verified in

Sect. 6. Suppose that the window size of the averaging algorithm is L, then the change detection function is defined as:

$$k_i = \frac{1}{L} \sum_{j=i-L+1}^{i} \Delta e_j \tag{33}$$

As defined in [26], the change threshold $\theta = 3\sigma$, where σ is the standard deviation of the background noise. When $k_i > \theta$, the algorithm judges that the system has changed. The process of variable sliding window kernel recursive least squares algorithm is shown in Table 2.

Table 2. Variable sliding window kernel recursive least squares algorithm (VSW-KRLS).

Initialization: sliding window size M, threshold θ, regularization coefficient c, $\mathbf{K}_0=(1+c)\mathbf{I}$.
for j=2,3,...**do input:** $\{\mathbf{u}_j, d_j\}$
calculate the error k_i according to Eq. (32) and Eq. (33).
if $k_i < \theta$ **then**
calculate the kernel matrix according to Eq. (18) –Eq. (22).
calculate the coefficient vector according to Eq. (14).
else
calculate the maximum changeable window size M1 .
calculate the kernel matrix according to Eq. (18)-Eq. (20).
when the size of the kernel matrix is restored to M1
calculate the kernel matrix according to Eq. (18)-Eq. (22).
end for

4.3 Improved Kernel Recursive Least Square Method

Through above analysis, novelty criterion, variable sliding window method combined with sliding window kernel recursive least squares algorithm to form our improved KRLS algorithm, which is called the sparse variable sliding window kernel recursive least square algorithm (VSWS-KRLS).

The proposed algorithm needs to judge whether to add the input to the dictionary according to the NC. Only if the distance and error are both greater than the preset threshold, the input data will be added to the dictionary and participate in the calculation of the kernel matrix. Therefore, this method can use more useful information to update the algorithm with low computational complexity.

In addition, we introduced a variable sliding window method to improve the ability to track time-varying characteristics. Because it can solve the problem that KRLS is not sensitive to outliers or time-varying characteristic environmental changes. The algorithm flow chart is shown in Fig. 4.

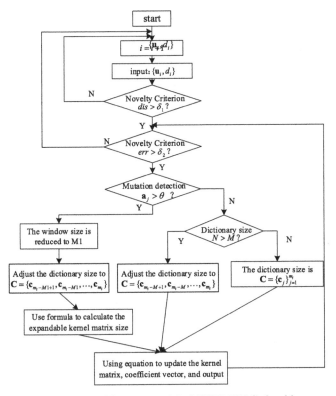

Fig. 4. The specific process of the VSWS-KRLS algorithm.

Table 3. Computational complexity of different algorithms in a single iteration.

Algorithm	Computational complexity	
ALD-KRLS	a_i	$O(l^2)$ $(l \leq i)$
	Sparse method (ALD)	$O(l^2)$ $(l \leq i)$
	P_i	$O(l^2)$ $(l \leq i)$
FB-KRLS	a_i	$O(L^2)$ $(L < i)$
	P_i	$O(L^2)$ $(L < i)$
EX-KRLS	a_i	$O(i^2)$
	P_i	$O(i^2)$
SW-KRLS	a_i	$O(L^2)$
	P_i	$O(L^2)$
VSWS-KRLS	a_i	$O(j^2)$ $(j \leq L)$
	Sparse method (NC)	$O(j^2)$ $(j \leq L)$
	Dictionary D_i	$O(j^2)$ $(j \leq L)$

4.4 Computational Complexity Analysis

In order to prove that the proposed VSWS-KRLS algorithm has lower computational complexity, this paper compares this algorithm with common KRLS algorithms (KRLS-ALD, FB-KRLS, EX-KRLS, SW-KRLS). This paper uses the weight coefficient vector α_i, the autocorrelation matrix $\mathbf{P}_i$ and the sparse method to compare the computational complexity. The results are shown in Table 3.

Suppose that at the i-th moment, the input size is i, the dictionary size is l, and the size of sliding window is L. For ALD-KRLS, the computational complexity of sparse method ALD, weight coefficient vector and autocorrelation matrix are both $O(l^2)$. In EXKRLS, the complexity of the weight coefficient vector and the autocorrelation matrix are $O(i^2)$. Both FB-KRLS and SW-KRLS limit the growth of the kernel matrix by setting a fixed size of window, so the computational complexity of their weight coefficient vector and autocorrelation matrix are both $O(L^2)$. In VSWS-KRLS, two sparse methods (NC and sliding window) reduce the computational complexity. And the computational complexity of the weight coefficient vector, sparse method and dictionary are all $O(j^2)$ (j is the size of the NC dictionary, and the dictionary size is controlled by a variable sliding window, so $j \leq L$). Therefore, compared with other algorithms, the SWS-KRLS algorithm has the smallest computational complexity in a single iteration, and as the iteration progresses, the advantage of computational complexity will become more obvious.

5 Simulation

In this section, in order to prove the effectiveness of the algorithm, we apply the algorithm to a nonlinear time-varying system for system identification. The linear channel coefficients will change at a given moment to compare the tracking ability of the algorithm. This paper uses four channels for simulation: In the first part of the simulation, the linear channel is $H_1(z) = 1 + 0.8362z^{-1} - 0.7732z^{-2} - 0.4484z^{-3}$, after receiving 500 data, it is changed into $H_2(z) = 1 - 0.8045z^{-1} + 0.9962z^{-2} + 0.4678z^{-3}$. This paper nonlinear Wiener system as the nonlinear system, and the model is shown in Fig. 5.

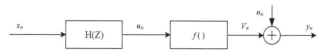

Fig. 5. Nonlinear Wiener system.

A binary signal x_n is sent through this channel, then a nonlinear function $v = \tanh(x)$ is applied to it, and v_n is the channel output. Finally, 20 dB of additive white Gaussian noise (AWGN) is added. The performance of the model is shown through the MATLAB simulation. In addition, the ALD-KRLS algorithm and the NC-KRLS algorithm are evaluated to show the reason for choosing the novelty criterion as the sparse method. In the experiment, the mean square error (MSE) is selected as the evaluation index of prediction accuracy. If the value of MSE is smaller, it means that the prediction performance is better. At the same time, this paper also uses training time as an indicator of computational complexity.

This experiment was simulated on windows10 operating system, Intel Core i3-9100 CPU 3.60 GHz, RAM 8.00 GB, and the codes are operated in MATLAB R2016a software, and the experiment results are obtained through 100 Monte Carlo experiments.

5.1 Selection of Sparse Methods

In order to choose a suitable sparse method, we compare different sparse methods, we uniformly select 20 values of $\delta 1$ in the interval of [0.1, 0.3] and 20 values of $\delta 2$ in the interval of [0.05, 0.1] for the NC. And we select 80 values of $\delta 3$ are uniformly selected in the interval of [0.05, 0.3] for the ALD. For each threshold, we use KRLS-NC and KRLS-ALD to train, and record the kernel matrix size and MSE values. Figure 6 shows the MSE values corresponding to two different sparse methods in different kernel matrix size.

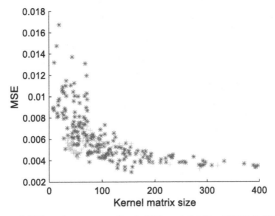

Fig. 6. Comparison of different sparse methods NC and ALD of KRLS. The "*" point comes from KRLS-NC, the " + " point comes from KRLS-ALD.

The results show that when the kernel matrix size is less than 70, the performance of ALD criterion is better than NC, and NC can actually perform better than ALD in other places. Because the sliding window size in the SW-KRLS algorithm is concentrated between 100–200, so this paper applies the NC to the SW-KRLS algorithm to solve the sparse problem.

5.2 Sliding Window Sparse Kernel Recursive Least Squares Algorithm

To further test the performance of SWS-KRLS algorithm, this section shows the influence of algorithm performance with different NC thresholds. The size of the sliding window is 150, and the threshold $\delta 1 = [0.2, 0.5, 0.7]$. The simulation result is shown in Fig. 7. It is observed that the steady-state error is improved by adding novel criterion to the

SW-KRLS algorithm. For thresholds, the greater the value of δ1 is, the better the steady-state error of the system. But the sparse methods will destroy the complex relationship between the data, it will reduce the convergence speed of the algorithm.

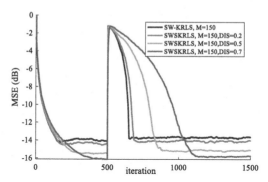

Fig. 7. MSE learning curves of SW-KRLS and SWS-KRLS with different NC thresholds.

5.3 Improved Algorithm

This paper introduces a change detection mechanism in the variable sliding window method. Different from the literature [25], the error difference after simultaneous superposition and averaging can better detect change. In a single experiment, the algorithm mean square error difference and mean square error value after using synchronous superposition averaging algorithm is shown in Fig. 8. In one iteration, the error difference is unstable, so the judgment is also inaccurate. The synchronous superposition average algorithm selects adjacent time to calculate the average value to reduce the influence of instability.

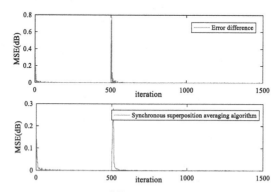

Fig. 8. Algorithm mean square error difference and mean square error value after using synchronous superposition averaging algorithm.

In order to evaluate the impact of different window lengths on the performance of the algorithm, we choose the window length M of the SW-KRLS algorithm to be 50, 100, and 150. Meanwhile, the variable sliding window method is applied in the algorithm of M = 150. The simulation results are shown in Figs. 9 and 10. Clearly, when the SW-KRLS algorithm undergoes a sudden change, the window size affects the performance of the algorithm. However, the VSW-KRLS algorithm can improve the tracking ability of the algorithm by changing the window size.

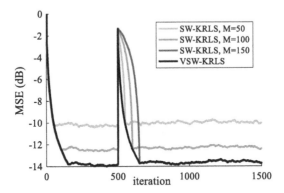

Fig. 9. MSE learning curves of VSW-KRLS and SW-KRLS with different window lengths.

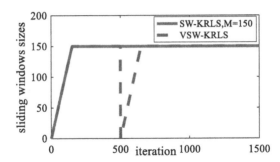

Fig. 10. Changes in the sliding window length of the VSW-KRLS and SW-KRLS.

As shown in Fig. 11, the VSWS-KRLS algorithm is compared with the SW-KRLS algorithm and SWS-KRLS algorithm. The simulation results confirm that the VSWS-KRLS algorithm has not only a fast convergence rate but also a small steady-state misalignment in comparison to SW-KRLS algorithms. Table 4 shows the simulation results of different algorithms in a nonlinear system, where, M represents the size of the sliding window, $\delta 1$, $\delta 2$ and θ represent the NC threshold and change detection threshold, and MSE represents the mean square error value. t represents the system running time of a single iteration. As can be seen, compared with other algorithms, the calculation performance of the VSWS-KRLS algorithm has been greatly improved.

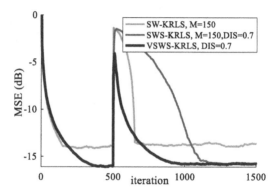

Fig. 11. MSE learning curves of SW-KRLS SWS-KRLS and VSWS-KRLS.

Table 4. Simulation results of different algorithms in nonlinear time-varying systems.

Algorithm	M	$\delta 1$	$\delta 2$	θ	MSE	t
SW-KRLS	100				−12.26 dB	0.42
	150				−13.63 dB	0.59
	200				−14.95 dB	0.73
SWS-KRLS	150	0.2	0.05		−14.21 dB	0.60
	150	0.5	0.05		−15.35 dB	0.45
	150	0.7	0.05		−15.89 dB	0.37
VSW-KRLS	150			3σ	−13.61 dB	0.60
VSWS-KRLS	150	0.7	0.05	3σ	−15.86 dB	0.36

6 Conclusion and Prospect

This paper proposed VSWS-SKRLS algorithm for identification of nonlinear systems. The model used sliding window and NC to make the dictionary and kernel matrix sparse, it could reduce the computational complexity and memory occupancy of algorithm. Moreover, online sparsification could improve prediction accuracy by reducing impact of irrelevant data, but it also influenced the tracking ability of model. To further improve the tracking ability of the model, adaptive mechanisms for sliding window size were employed. Finally, the experimental results showed that the VSWS-SKRLS algorithm had better prediction performance and low computational complexity in identification of nonlinear time-varying systems. Furthermore, our work will focus on selection the other sparse methods in the future, which has certain effects on performance and deserves further studied.

References

1. Haykin, S.: Adaptive Filter Theory, 4th edn. Prentice-Hall, Englewood Cliffs (2001)

2. Wu, C., Wang, X., Guo, Y.: Robust uncertainty control of the simplified Kalman filter for acoustic echo cancelation. Circuits Syst. Signal Process. **35**(12), 4584–4595 (2016)
3. Liu, Y., Mikhael, W.B.: A fast-converging adaptive fir technique for channel equalization. In: 2012 IEEE 55th International Midwest Symposium on Circuits and Systems (MWSCAS), pp. 828–831. IEEE, Boise (2012)
4. Mariappan, S., Rao, G.S.: Enhancing GPS receiver tracking loop performance in multipath environment using an adaptive filter algorithm. Indian J. Sci. Technol. **7**(S7), 156–164 (2014)
5. Widrow, B., Hoff, M.E.: Adaptive switching circuits. In: IRE WESCON Convention Record, vol. 4, pp. 96–104. IRE, New York (1960)
6. Kuh, A.: Adaptive least square kernel algorithms and applications. In: 2002 International Joint Conference on Neural Networks, vol. 3, pp. 2104–2107. IEEE, Honolulu (2002)
7. Williamson, R.C., Kivinen, J., Smola, J.: A online learning with kernels. Signal Process. **100**, 1–12 (2004)
8. Bai, Y., Xiao, J., Long, Y.: Kernel partial least-squares regression. In: International Joint Conference on Neural Networks, IJCNN 2006, Part of the IEEE World Congress on Computational Intelligence, WCCI 2006, pp. 1231–1238. IEEE, Vancouver (2006)
9. Liu, W., Príncipe, J.C., Haykin, S.: Kernel Adaptive Filtering: A Comprehensive Introduction. Wiley, Hoboken (2010)
10. Engel, Y., Mannor, S., Meir, R.: The kernel recursive least-squares algorithm. IEEE Trans. Signal Process. **52**(8), 2275–2285 (2004)
11. Aronszajn, N.: Theory of reproducing kernels. Trans. Am. Math. Soc. **68**(3), 337–404 (1950)
12. Wang, X., Han, M.: Multivariate time series prediction based on multiple kernel extreme learning machine. In: IEEE International Joint Conference on Neural Networks (IJCNN), pp. 198–201. IEEE, Beijing (2014)
13. Han, M., Zhang, S., Xu, M.: Multivariate chaotic time series online prediction based on improved kernel recursive least squares algorithm. IEEE Trans. Cybern. **99**, 1–13 (2018)
14. Zhou, H., Huang, J., Lu, F.: Echo state kernel recursive least squares algorithm for machine condition prediction. Mech. Syst. Signal Process. **111**, 68–86 (2018)
15. Van Vaerenbergh, S., Via, J., Santamaria, I.: A sliding-window kernel RLS algorithm and its application to nonlinear channel identification. In: 2006 IEEE International Conference on Acoustics Speech and Signal Processing Proceedings (ICASSP 2006), vol. 5, pp. 789–792. IEEE, Toulouse (2006)
16. Van Vaerenbergh, S., Santamaria, I., Liu, W.: Fixed-budget kernel recursive least-squares. In: 2010 IEEE International Conference on Acoustics Speech and Signal Processing (ICASSP 2010), pp.1882–1885. IEEE, Dallas (2010)
17. Chen, B., Zhao, S., Zhu, P.: Quantized kernel recursive least squares algorithm. IEEE Trans. Neural Netw. Learn. Syst. **29**(4), 1484–1491 (2013)
18. Han, M., Ma, J., Kanae, S.: Time series online prediction based on adaptive dynamic adjustment kernel recursive least squares algorithm. In: The 2018 Ninth International Conference on Intelligent Control and Information Processing (ICICIP), pp. 66–72. IEEE, Wanzhou (2018)
19. Han, M., Zhang, S., Xu, M.: Multivariate chaotic time series online prediction based on improved kernel recursive least squares algorithm. IEEE Trans. Cybern. **49**(4), 1160–1172 (2019)
20. Zhong, K., Ma, J., Han, M.: Online prediction of noisy time series: dynamic adaptive sparse kernel recursive least squares from sparse and adaptive tracking perspective. Eng. Appl. Artif. Intell. **91**, 103547 (2020)
21. Guo, J., Chen, H., Chen, S.: Improved Kernel recursive least squares algorithm based online prediction for nonstationary time series. IEEE Signal Process. Lett. **27**, 1365–1369 (2020)
22. Sayed, A.H.: Fundamentals of adaptive filtering. IEEE Control. Syst. **25**(4), 77–79 (2003)
23. Platt, J.: A resource - allocating network for function interpolation. Neural Comput. **3**(2), 213–225 (1991)

24. Richard, C., Bermudez, J.C.M., Honeine, P.: Online prediction of time series data with kernels. IEEE Trans. Signal Process. **57**(3), 1058–1067 (2009)
25. Julian, B.J.: Modifications to the sliding-window kernel RLS algorithm for time-varying nonlinear systems: online resizing of the kernel matrix. In: 2009 IEEE International Conference on Acoustics, Speech and Signal Processing, pp. 3389–3392. IEEE, Taipei (2009)
26. Gustafson, F.: Adaptive Filtering and Change Detection. Wiley, New York (2000)

The Prediction Method of Regional Economic Development Potential Along Railway Based on Data Mining

Hui-fang Guo[1(✉)] and Qing-mei Cao[2]

[1] School of Transportation and Municipal Engineering, Inner Mongolia Vocational and Technical College of Architecture, Hohhot 010010, China
[2] Department of Computer Technology and Information Management, Vocational and Technical College of Inner Mongolia Agricultural University, Baotou 014100, China

Abstract. The prediction accuracy and efficiency of regional economic development potential along the railway line are poor when the common methods are used to predict the economic development potential along the railway. In view of this problem, a data mining based prediction method for regional economic development potential along the railway line is designed. Select the influencing factors of regional economic development along the railway, construct the regional economic development index system, collect index data, clean, cluster, sort and standardize the index data, extract the economic development characteristics in the pre-processing index data through data mining, input the economic development characteristics to the neural network, and output the predicted value of economic development potential after training and learning, Divide the economic development potential level. The experimental results show that the design method reduces the deviation of economic development potential prediction, shortens the prediction time, improves the accuracy and efficiency of prediction.

Keywords: Data mining · Economic development potential · Data preprocessing · Economic forecast

1 Introduction

High speed railway is a modern system engineering with high technical difficulty and complexity. It is a public transportation tool with huge carrying capacity. With the construction of high-speed railway, the number of economic zones along the railway increases. Therefore, this paper studies the prediction method of regional economic development potential along the railway, puts forward reasonable decision according to the regional economic trend along the railway, and realizes the optimization of economic decision, It is of great significance.

Foreign research on regional economic development potential prediction is relatively mature. For the study area, the URL address table related to economic development potential is maintained, and the URL address is stored. When the URL address already

© ICST Institute for Computer Sciences, Social Informatics and Telecommunications Engineering 2022
Published by Springer Nature Switzerland AG 2022. All Rights Reserved
S. Liu and X. Ma (Eds.): ADHIP 2021, LNICST 417, pp. 221–238, 2022.
https://doi.org/10.1007/978-3-030-94554-1_19

exists, it is not added to the address list, otherwise it is added to the list, so as to eliminate the URL address of the Web page and get the characterization information of regional economic development potential, On the basis of completeness and scientificity, the forecast value of regional economic development potential is calculated. The domestic regional economic development potential prediction research has also made great progress. It analyzes the regional engineering cost and land cost, including technical economy, technical design, land acquisition, etc., determines the percentage of the cost in the total project investment, and counts the regional infrastructure construction cost, including all costs related to terrain preparation and platform construction, The amount of these costs varies greatly due to the different terrain characteristics. When encountering the geographical obstacles with special technical difficulties, it is necessary to build the viaduct, bridge and tunnel in the region. In this case, the infrastructure construction cost is doubled. Combined with the engineering cost, land cost and infrastructure construction cost, the regional economic development potential is predicted.

Combined with the above theories, data mining technology is introduced into the prediction of regional economic development potential along the railway. Data mining technology can convert a large number of regional economic development potential data along the railway into useful information and knowledge. Based on this, it can further divide the level of regional economic development potential along the railway and optimize the prediction performance of regional economic development potential along the railway.

2 Design of Economic Development Potential Along Railway Based on Data Mining

2.1 Extract the Influencing Factors of Regional Economic Development Along the Railway

Combined with the cost-effectiveness of the railway, the influencing factors of regional economic development along the railway are extracted. The first type of cost-benefit is infrastructure construction cost. The construction cost of new high-speed railway line is as follows: superstructure cost, which is the specific elements of railway, including guide rail, side rail along the line, signal system, catenary, electrification mechanism, communication and safety facilities. These elements usually account for 5–10% of the total investment respectively. The second type of cost-effectiveness is operation cost. The operation cost of high-speed railway service is as follows: infrastructure operation cost, including labor, energy and other material consumption required for daily operation and maintenance of guideway, wharf, station, energy supply, signal system, traffic management and safety system; Daily operation fixed cost and variable cost include train maintenance cost, electric traction device and catenary maintenance cost, power system maintenance cost, signal system maintenance cost, telecommunication system maintenance cost and other costs, of which track maintenance cost accounts for about 50%; Vehicle operation cost includes dispatching and train operation cost, vehicle and equipment cost, energy cost, sales cost and management cost. Vehicle operation cost mainly includes all labor costs such as vehicle service and driving, while vehicle and facilities mainly refer to the cost of vehicle and equipment depreciation and maintenance.

When analyzing the railway benefits, it is assumed that the service life of high-speed railway is A year, the completion time of high-speed railway infrastructure and superstructure is initial time $t = 0$, and the high-speed railway operators purchase high-speed rolling stock at the initial time. In the A time region, assuming that the ticket price C and annual passenger volume D are constant, and the project investment cost is E, including the infrastructure construction cost and the present value of rolling stock, under the condition of passenger volume D, the fixed cost remains unchanged, only the variable cost is considered, and all costs are calculated as opportunity cost, then the conditional expression of positive net present value is obtained as follows:

$$[F(D) - G(D)]e^{-(C-t)}dt - \int_0^A He^{-Ct}dt \geq E \tag{1}$$

Among them, $F(D)$ is the annual social benefit of the project, $G(D)$ is the annual operation and maintenance cost based on D, and H is the investment cost. It can be seen from formula (1) that during the operation of high-speed railway, the operation and maintenance cost, labor cost and energy consumption of track, station, signal and other real estate will be generated.

It can be seen from the cost-effectiveness of the railway that the stability of the technical safety level of the high-speed railway affects the passenger choice and the change of the traffic volume level. Under certain safety technical conditions, the reasons that affect the regional economic development along the high-speed railway are the length of the high-speed railway, the level of economic development, urban population, visitors, etc. The influence of these factors on the regional economic development along the line is dynamic and relative. In the early stage of the development of high-speed railway, every new railway section can open up the high-speed railway passenger transport market. At this time, the length of high-speed railway becomes the bottleneck of restricting the high-speed railway passenger transport volume. In the state of high-speed railway development has been saturated, the extension of high-speed railway line may not bring significant increase in passenger transport volume. On the other hand, the purpose of passengers taking high-speed railway mainly includes school, commuting, business trip, travel and other private activities. These activities are affected by the level of regional economic development to varying degrees, and have different influence structures in different regions. The size of the target market ultimately depends on the number of people with the ability to pay, who may come from their own countries or may be international tourists, The degree of its influence is directly related to the structural distribution ratio and the regional economic structure. So far, the extraction of influencing factors of regional economic development along the railway is completed.

2.2 Collection of Regional Economic Development Indicators Data Along the Railway

To screen the influencing factors of regional economic development along the railway, build the index system of regional economic development along the railway, and collect the index data. Taking the county as the main research unit, combined with the relevant indicators of county economic development and the availability of data, this paper selects per capita GDP, per capita gross industrial product, per capita investment in fixed assets, per capita fiscal revenue, per capita savings deposit balance, per capita total retail sales of social consumer goods, per capita net income of rural residents, the proportion of fiscal revenue in GDP, per capita net income of rural residents, per capita net income of rural residents, per capita total retail sales of social consumer goods This paper discusses the economic development in the railway radiation belt from nine factors such as the speed of economic development. The comprehensive development level of county nodes is represented by comprehensive quality index and measured by composite index. From the three levels of economic development, social development and urban construction, 14 indexes are selected to construct the index system of county comprehensive development level. The maximum method is used to standardize the data, and with the help of SPSS software, the county comprehensive development level is preliminarily measured. The index system is shown in Table 1.

Table 1. County economic development potential index system

Target layer	Code layer	Index layer
Development potential of county economy	Economic development	Gross Regional Product
		Per capita GDP
		Proportion of output value of secondary and tertiary industries
		Public revenue
		Investment in fixed assets
		Balance of household savings deposits
	Social development	Number of employees
		Total retail sales of consumer goods
		Urban per capita disposable income
		Business volume of Posts and Telecommunications
		Number of beds in health institutions
	Urban construction	Per capita park green area
		Road area per capita
		Green coverage rate of built up area

The basic data are from the regional statistical yearbook along the line, namely, China Urban Statistical Yearbook in 2020. It is difficult to obtain the relevant logistics data in the county, and the freight capacity is usually measured by the freight volume index. The proportion of freight shifts in County contact is less, and the travel time and passenger flow between counties can reflect the regional accessibility level more. Therefore, passenger transport is used as the contact medium. Considering that railway transportation is suitable for long-distance transportation in County, the railway passenger flow data has the characteristics of high accuracy and easy access, which can better reflect the reality of economic contact between counties. The shortest travel time of county railway passenger transport is selected as the basic data for calculating accessibility and economic connection. The data is from Ctrip. During the query process, the shortest travel time of all direct passenger trains between the two counties is selected. If there is no direct train between two cities, the website will automatically provide the shortest transit route for it, only the distance time will be counted, Transfer time and stop time of transfer are not considered.

Besides the collection source of statistical yearbook, the index data can also be collected by crawler. Taking the regional economic development along the railway as the key word, search the relevant Web pages of development potential indicators in the Internet, and join the crawler queue from the seed UPL, analyze and download the Web page, grab the UPL and get a new URL. In the real network traffic data, statistics the usage heat of various indicators, read the page on the homepage of the Web page, find other link addresses in the Web page, find the next page, set the number of access layers of different pages until all pages of the website are captured [1]. The grab page is preprocessed, the economic development of the railway area is taken as the theme content, the pages that are inconsistent with the theme are filtered out, the table label is used to repair the wrong or irregular tags, the repaired pages are stored in the HTML document, and the HTML file is selected as the root node, that is, an Internet unit, the tag tree is constructed, and the visual information of the Web page is used, Block processing Web pages, starting from the prediction demand of regional economic development along the railway, remove redundant information of Web pages, link useful information together, find text files related to the subject content, mark hypertext, and integrate Web pages [2–4]. Finally, through HTTP protocol, the browser can download the Web page, grab the effective information in the Web page, including the documents such as sound, text and image, obtain the index data of regional economic development field along the railway line, add the relevant contents of regional economic development along the railway line by custom, and collect the video, audio, database, picture, text data and other types of data in the Web page, Eliminate new URL, join new crawl queue, loop above operation. So far, the data collection of regional economic development indicators along the railway has been completed.

2.3 Preprocessing the Data of Regional Economic Development Indicators Along the Railway

According to the massive data of regional economic development indicators along the railway, preprocessing is carried out to make the indicator data accurately express the potential of regional economic development.

Cluster Processing Index Data

The distributed k-means algorithm is used to cluster the related regional economic development index data along the railway. Firstly, the original historical data of county economic development potential indicators are cleaned, and the wrong data, data noise and invalid data are processed. Then, the time series of regional economic development data along the railway are counted in the historical data, and the time series are classified and processed according to the quarterly order, so as to keep the regional economic development index data continuous, and then fill in the missing historical data. Through attribute mapping, the character data of the original data set is transformed into the numerical standardized data. The mapping formula is as follows:

$$m = \left[\frac{(m_{max} - m_{min})(n - n_{min})}{(n_{max} - n_{min})} \right] + m_{min} \tag{2}$$

Among them, m is the standardized index data after processing, m_{max} and m_{min} are the maximum and minimum values of the index data after processing, n is the original historical data of the regional economic development index along the railway, and n_{max} and n_{min} are the maximum and minimum values of the original data respectively. In the data set, k data objects are randomly selected as the initial cluster center of the regional economic development index data along the railway. By using horse distance, the initial cluster center and the remaining data objects are compared. The calculation formula of horse distance is as follows:

$$H_{ij} = \sum_{k=1}^{n} \frac{|x_{ik} - x_{jk}|}{x_{ik} + x_{jk}} \tag{3}$$

Among them, n is the original dimension of the indicator data, x_{ik} is the dimension value of the i-th and j-th indicator data of the indicator data object in the k-th data dimension, and H_{ij} is the Mahalanobis distance between the indicator data i and the indicator data j [5]. The closer the Mahalanobis distance H_{ij} is to 1 or -1, the higher the correlation degree and the closer the distance between the two index data are judged. If H_{ij} is closer to 0, the lower the correlation degree and the farther the distance are judged. The remaining index data objects are classified into the nearest initial cluster center, and then the cluster center is re selected, iterating for many times until the criterion function converges, and the k cluster centers do not change. The definition formula of criterion function J is as follows:

$$J = \sum_{r=1}^{k} |Z_r - E_r| \tag{4}$$

Among them, Z_r is the center value of cluster center r, and E_r is the average value of Z_r. After cleaning the clustered index data, delete the irrelevant records of regional economic development along the railway, including image content request, file request and spider crawler program request. When the spider initiates HTTP request, it separates the illogical spider session and records a large number of regional economic development information along the railway through HTTP header. After the index data is partitioned and clustered, the collected data is classified finely to get different local data tuples [6–8]. The refined data items after segmentation are shown in Table 2.

Table 2. Data items of county economic development potential index

Field code	Field type	Meaning
MCI	VarChar2	Gross Regional Product
MC	VarChar	Per capita GDP
SJY	VarChar	Proportion of output value of secondary and tertiary industries
SJYLX	VarChar	Public revenue
SJFL	VarChar	Investment in fixed assets
TXTZ	VarChar2	Balance of household savings deposits
ZXZBX	Date	Number of employees
ZXZBY	Date	Total retail sales of consumer goods
YSZBX	VarChar2	Urban per capita disposable income
YSZBY	Number	Business volume of Posts and Telecommunications
BLC	Number	Number of beds in health institutions
BB	Number	Per capita park green area
SYQX	VarChar2	Road area per capita
SJYSM	VarChar2	Green coverage rate of built up area

Build a distributed SQL database to represent the attribute structure of data items, and provide data support for the prediction of regional economic development potential along the railway through various refined data sets. So far, the clustering of index data is completed.

Primary and Secondary Relationship of Sort Index Data

This paper uses sprint classification algorithm to sort the primary and secondary relationship of regional economic development index data along the railway. The maximum minimum normalization formula is used to discretize the continuous numerical attribute of the index data, and the linear transformation of the index data is carried out.

$$V = \beta \frac{(L - M)}{(M - N)} \tag{5}$$

Among them, L is the index data value, M and N are the maximum and minimum values of the same attribute index data, β is the mapping interval, and V is the value after the indicator data is mapped. The neural network center is used to replace the continuous value of index data, and the data attributes are converted to discrete values, and the rules of regularity are displayed on the basis of relative attributes, and the number of values of the same attribute data is reduced [9, 10]. By using sprint classification algorithm, the paper sorts the primary and secondary relations of regional economic development index data along the railway, classifies the regional economic development level along the railway line, divides the areas along the line into multiple sub groups, takes the regions along the line with different development levels as different categories, distinguishes the economic development level of the regions along the line, and the economic development level of the same subdivision area is close to each other. The classification of index data is realized by decision tree. The attribute with the highest priority is selected in the index data. As the root, the pre processed attribute set is provided. The common character is searched from the index data, a series of sorting decisions are made, and the decision tree nodes are divided, and then the attribute data attributes are split, and the attributes are accurately associated with the sub nodes, and the attribute value segmentation data set is obtained [11]. Set the number of data set category as m, and the number of data set categories is the number of leaf node categories. Then the calculation formula of split parameter F is:

$$F = 1 - \sum_{I}^{m} p_I^2 \tag{6}$$

Where p_I is the relative frequency of data set category I. A data node is selected in the data set to represent the data of the data set. The logical judgment of the regional economic development level along the line is taken as the internal node of the decision tree, the branch result of the logical judgment is taken as the edge of the decision tree, and the data attribute is associated with the root node of the decision tree to construct the multi tree decision tree. When all the index data belong to the same class, the class label is used to define the leaf node. When the index data do not belong to the same class, the data attribute is measured according to the information entropy, and the data in the original attribute set is deleted. When the candidate set is empty, the leaf node is returned and marked as a common category [12–14]. For different types of index data, the calculation formula of information entropy Q is as follows:

$$Q = -\sum_{I=1}^{m} \frac{|C_I|}{|\xi|} \log_2 \left(\frac{|C_I|}{|\xi|} \right) \tag{7}$$

Where ξ is the given data set of decision tree and C_I is the set of data set belonging to class I objects. The data set ξ is classified according to the attribute characteristics to obtain multiple different objects. The weighted sum of the information entropy Q is carried out through the partition entropy to calculate the information gain attribute of the index data

$$A = -\sum_{I=1}^{\phi} \left(\frac{|\xi|}{|D|} \times Q \right) \tag{8}$$

Among them, A is the information gain of index data, ϕ is the number of attribute features of data set, D is the amount of information needed for data set classification, and ζ is the amount of information needed for data set partition. In the attribute set, the attribute with the highest information gain A is selected, and the leaf node is marked to get a score of the attribute with the highest information gain, and the elements of the dataset subset can meet the score. When the categories of nodes are the same, the remaining attributes can not be divided again, or the given score has no data, the class label is created, the division of decision tree is terminated, and the classification of regional economic development level along the line is completed. So far, the sorting of the primary and secondary relationship of the index data is completed, and the preprocessing of the regional economic development index data along the railway is completed.

2.4 Mining the Characteristics of Regional Economic Development Along the Railway

Through data mining, in the preprocessed regional economic development index data along the railway, mining the characteristics of regional economic development. This paper defines the semantic attention of index data, highlights the index data of regional economic development along the railway with similar semantics, and applies it to the prediction of regional economic development potential along the railway. The factors influencing semantic distance between data items are selected, which are economic development, social development and urban construction. The three factors are taken as the dynamic characteristics of the index data, so that the index data changes with time, showing different characteristics of regional economic development along the railway. Three kinds of dynamic features are combined to get multidimensional data information. Data exploration is used to reduce the dimension of index data. According to the dimension combination obtained from data exploration, multidimensional dynamic features are transformed into 2D dynamic features [15, 16]. If the characteristic dimension of index data is set as j, the formula of index data combination exploration condition G is as follows:

$$G = j \times (j-1) \times (j-2) \cdots \times 1 \tag{9}$$

In the dynamic features of two-dimensional display, the data items with mutual reference and link relationship are selected to judge the semantic distance between the data items is close, which is more meaningful for the prediction of regional economic development potential along the railway. Abstract features of index data in information space are extracted, and the types of abstract features are divided into three categories: time series, network and hierarchy. The calculation parameters of semantic distance are determined by using the structural relationship among the three categories of index data. Suppose that the dynamic characteristic object of the regional economic development index along the railway is s, and any index data object in the information space is x, then the semantic distance $d(s, x)$ between s and x is:

$$d(s, x) = w_1 f(s, x) + w_2 g(s, x) + w_3 l(s, x) \tag{10}$$

Among them, $f(s, x)$ is x 2-dimensional display of s dynamic feature dimension combination. The 2-dimensional display includes two parts: implicit intention and explicit intention. The implicit intention is used to determine the impact of index data on regional economic development potential prediction. The explicit intention is used to clarify the regional economic development potential prediction intention. $g(s, x)$ is the structural relationship of abstract feature of index data, which represents the correlation between x and s, $l(s, x)$ is the center distance after semantic representation of x and s, and w_1, w_2 and w_3 are the weights of 2-dimensional presentation, structural relationship and center distance respectively.

The semantic distance is regarded as an important parameter of the semantic attention degree of the indicator data. Through $d(s, x)$, the distance between x and s at the semantic level is reflected. In the prediction of regional economic development potential, the prior importance of different data items of the indicator data is determined, and the threshold of semantic attention degree is set to limit the set of indicator data items. The semantic attention H of index data item x to s is:

$$H = \frac{k(s, x)C}{d(s, x)} \tag{11}$$

Among them, $k(s, x)$ is the prior importance of x about s, and C is the threshold of semantic attention. The greater the semantic attention, the closer the semantic of x and s, otherwise, the less the judgment semantics is, the more the index data items with close semantics are aggregated to assist the prediction of regional economic development potential along the railway.

According to the semantically similar regional economic development index data along the railway, the association rules of regional economic development potential along the railway are mined, and the economic development mode of each region along the railway is determined according to the relationship between different attribute characteristics of the index data. In the data set, the attribute information of regional economic development index data along the railway is extracted and divided into continuous attribute set, original invariant attribute set and nominal attribute set [17]. In order to control the data semantic processing process of related attributes, the attribute sets which are similar to the economic development index data of the regions along the railway are generated by using collaborative filtering technology, By classifying the implied semantics, the corresponding concepts of attribute semantics are obtained, and the rule set for mining the characteristics of regional economic development is generated. By using HowNet knowledge base, we define the words in semantic dictionary, take the def item in HowNet as the word concept, and replace the words with similar meaning, so as to make the words have semantic relevance. At the same time, considering the semantic similarity interval between words, we expand the concept of HowNet, delete the same words in def item, and through the minimum semantic similarity and maximum semantic similarity, The semantic similarity of different index data is calculated. The calculation formula is as follows:

$$H = \left(\frac{|K_a \cap K_b|}{\min(K_a, K_b)} + \frac{|K_a \cap K_b|}{\max(K_a, K_b)} \right) \bigg/ 2 \tag{12}$$

Among them, K_a, K_b is the concept of a, b, $|K_a \cap K_b|$ is the number of two concepts with the same definition, $\min(K_a, K_b)$, $\max(K_a, K_b)$ is the number of semantic words with fewer and more semantic words, and H is the semantic similarity of index data. The value of concept similarity is between [0,1]. The smaller the similarity is, the lower the possibility of concept semantic correlation between the feature attributes mined and the prediction of regional economic development potential along the railway line is. The greater the similarity is, the closer the concept semantic is judged. Set the threshold of semantic similarity, select the feature attributes whose H is greater than the threshold, extract the index data and identify their association. The semantic similarity matrix is used to represent the semantic similarity of all index data, and the index features of deep semantic connection are mined. Combined with the semantic elements of regional economic development potential prediction along the railway, the common parts of the semantic elements of index feature attributes are analyzed, the key points of semantic connection are obtained, and the semantic information describing the characteristics of regional economic development is described, According to the semantic bias of economic development characteristics to the prediction of economic development potential, semantic processing index data mining results, the regional economic development characteristics are defined. So far, the mining of regional economic development characteristics along the railway has been completed.

2.5   Preast Economic Development Potential Along the Railway

The characteristics of regional economic development along the railway are input into BP neural network to predict the potential of regional economic development along the railway. BP neural network uses the radial basis function of Multivariable Interpolation, selects three-layer forward neural network as the typical structure of neural network, transforms the characteristic attributes of regional economic development extracted from the input layer in the middle layer, so as to make the category of characteristic attributes of regional economic development closer to the center of the network. Suppose the output value of the i-th neuron is x_i, the sample point of the j-th network center is G_j, and the middle layer neuron of the j-th network center is T_j, then the modified new network center B is:

$$B = \frac{\sum x_i}{G_j} \forall x_i \in T_j \tag{13}$$

The characteristic attributes of regional economic development along the railway line are divided into new network center B, and the network center set is used as the value range to replace the characteristic values of regional economic development along the railway line, so as to eliminate the influence of different dimensions of each dimension data on the prediction of regional economic development potential along the railway line, so as to find the change law of regional economic development potential along the railway line. In the prediction of regional economic development potential along the railway, with the increase of prediction length, the error of prediction value will be larger and larger. Therefore, BP neural network adopts the learning training of fitting error difference to ensure the prediction accuracy of nonlinear factors in the prediction of economic development potential. The learning algorithm of BP neural network is based on the error back propagation algorithm of neural network, and consists of four processes, In the first stage, the input mode is the forward propagation from the input layer to the output layer through the middle layer. In the second stage, the expected output and the actual output of the network are the error back propagation from the error signal to the input layer through the middle layer, and the connection weight of the neural network is corrected layer by layer. In the third stage, the error back propagation and the mode forward propagation are alternately repeated. In the fourth stage, the neural network tends to converge, The learning convergence process of the global error of the network tends to the minimum [18–20].

In the three-layer BP network structure, the number of nodes in the input layer is 2, the number of nodes in the hidden layer is 6, the number of nodes in the output layer is the number of output vectors, and the number of output vectors in the target value of the neural network is 1, which is the prediction result of the regional economic development potential along the railway. The characteristic attributes of regional economic development along the railway are used as training data and test data, BP neural network is trained with the training data, and the predicted value of regional economic development potential along the railway is output [21]. The predicted value of economic development potential is divided into 1–5 levels, as shown in Table 3.

Table 3. Level of regional economic development potential along the railway

Estimate	Grade	Meaning
0–20	1	Great potential for regional economic development along the railway
20–40	2	The potential of regional economic development along the railway is great
40–60	3	The potential of regional economic development along the railway line is general
60–80	4	The regional economic development potential along the railway is poor
80–100	5	The potential of regional economic development along the railway is poor

According to Table 3, the potential level of regional economic development along the railway is determined [22–24]. So far, the prediction of regional economic development potential along the railway has been completed, and the design of prediction method of regional economic development potential along the railway based on data mining has been realized.

3 Experiment and Analysis

The design method is compared with the macro-economic development prediction method based on optimized multi-dimensional grey model in reference [1] and the marine economic development prediction method based on grey prediction model in reference [5], and the accuracy and efficiency of prediction are compared.

3.1 Experimental Preparation

Taking the counties along the Lanzhou Xinjiang Railway economic belt as the experimental object, the absolute economic difference of the county is expanding year by year from 2010 to 2020, and the relative economic difference of the county is fluctuating, which can be roughly divided into two stages: 2010–2013 is the stage of narrowing the economic difference of the county, the coefficient of variation and Searle index fall below the average level, and the Searle index reaches the lowest value in 2012, In 2013, the coefficient of variation reached the lowest value, which was 0.85352013-2019, which was the stage of continuous expansion of county economic differences. The coefficient of variation and Searle index in 2019 were 1.18 and 1.32 times of 2000, respectively.

From the frequency density distribution curve of per capita GDP of each county, the county economic differences along the Lanzhou Xinjiang Railway are positively skewed, reflecting that the regional economic development level depends on a few higher level counties, and the deviation coefficient fluctuates around 2.0. Using geoda software to calculate the per capita GDP and global autocorrelation coefficient from 2010 to 2020, the statistics of per capita GDP and global autocorrelation coefficient are all positive, showing a trend of gradually increasing first, then gradually decreasing, and finally rising again. In terms of time, the per capita GDP and the global autocorrelation coefficient gradually increased from 2010 to 2013, showing the characteristics of spatial agglomeration. The index gradually decreased from 2013 to 2015, and reached the lowest value of 0.1829 in 2015. After 2016, the per capita GDP and the global autocorrelation coefficient increased slowly. Although there was a trend of gradual recovery, the per capita GDP and the global autocorrelation coefficient in 2019 were still lower than those in 2010, It shows that the economic agglomeration of the economic belt along the Lanzhou Xinjiang Railway can not return to the initial level after the shock is reduced, and the spatial agglomeration of the county economy shows signs of recession, reflecting the trend of the expansion of the county economic differences.

According to the section, the economic belt along Lanzhou Xinjiang Railway is mainly divided into Gansu section, Qinghai section and Xinjiang section. According to the spatial decomposition characteristics of Searle index, the overall difference of county economy in the economic belt along Lanzhou Xinjiang Railway is composed of the difference between Gansu section, Qinghai section and Xinjiang section and their respective internal differences. From 2010 to 2020, the differences among the three regions are small and show a trend of gradual narrowing. The Searle index gradually decreases from 0.0353 to 0.0028, and the contribution rate of the overall regional differences also decreases from 11.10% to 0.67%. On the whole, the internal difference of Gansu section is expanding year by year, and the contribution rate to the regional economic difference is the largest, reaching 71.20% in 2019. Although the internal difference of Qinghai section remains at a low level, it narrows slightly in the early stage, but expands in the later stage. The internal difference of Xinjiang section is shrinking year by year, The internal difference of Gansu section is the main contributor to the overall difference of economic belt along Lanzhou Xinjiang Railway.

3.2 Experimental Results

Prediction of Deviation Degree Experimental Results

According to the above historical data, the three methods respectively calculate the predicted value of regional economic development potential along the railway. Combined with the actual development of the region, the deviation degrees of the upper limit, lower limit and basic value of the prediction are calculated.

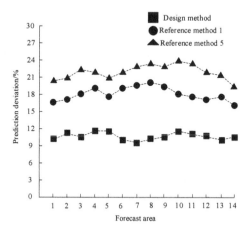

Fig. 1. Results ast upper limit deviation

It can be seen from Fig. 1 that the prediction deviation of the upper limit value of the design method is significantly smaller than that of the two groups of commonly used methods, with an average deviation of 10.4%. The average deviations of reference [1] and reference [5] are 18.7% and 22.1% respectively, and the prediction deviation of the design method is reduced by 8.3% and 11.7% respectively. The lower limit values of the regional economic development potential predicted by the three methods are counted, and the prediction deviation degree is calculated.

It can be seen from Fig. 2 that the prediction deviation of the lower limit value of the design method is also smaller than that of the other two commonly used methods, the average deviation is 10.9%, the average deviation of reference [1] and reference [5] are 18.2% and 21.1% respectively, and the prediction deviation of the design method is reduced by 7.3% and 10.2% respectively. Finally, the basic value of regional economic development potential along the railway is predicted, and the comparison results of prediction deviation are shown in Fig. 3.

As can be seen from Fig. 3, compared with the upper and lower limit values, the prediction deviation of the basic value has decreased, but the deviation of the design method is still less than that of reference [1] and reference [5], with an average deviation of 3.1%, and the average deviation of reference [1] and reference [5] is 9.2% and 14.1% respectively, and the prediction deviation of the design method has decreased by 6.1% and 11.0% respectively. Based on the comprehensive analysis of the prediction upper

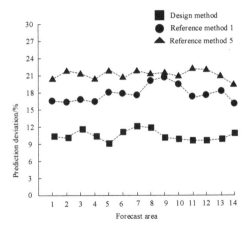

Fig. 2. Comparison result of the lower limit of prediction and deviation

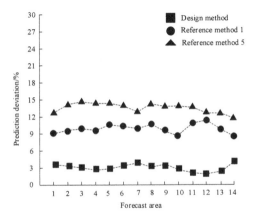

Fig. 3. Predicting comparison result of deviation of base value

limit, lower limit and overall deviation, the introduction of data mining technology into the prediction of regional economic development potential along the railway effectively reduces the impact of influencing factors on the prediction accuracy and the prediction deviation.

Predicted Time-Consuming Experimental Results
The total prediction time of regional economic development potential along the railway in 10 groups of experiments is counted. Each group of experiments is conducted 20 times respectively. The average prediction time of each group of experiments is counted to obtain the total prediction time of regional economic development potential along the railway under different methods. The experimental comparison results are shown in Table 4.

Table 4. Results of development potential (s)

Number of experiments	Design method: Forecast time	Reference [1]: forecast time	Reference [5]: forecast time
1	34.65	100.23	121.73
2	35.23	102.32	120.84
3	35.89	99.34	122.44
4	35.18	101.32	121.37
5	34.04	100.32	120.87
6	35.82	101.76	122.93
7	35.12	99.63	121.83
8	35.46	100.73	120.84
9	35.23	101.83	121.83
10	34.72	100.92	120.83

It can be seen from Table 4 that the average prediction time of the design method is 31.62 s, and the prediction time of reference [1] and reference [5] are 100.84 and 121.55 s respectively. Compared with reference [1] and reference [5], the prediction time of the design method is reduced by 69.22 s and 89.93 s respectively. To sum up, compared with the common methods, this design method reduces the deviation of the predicted value of the regional economic development potential along the railway, and the predicted result is closer to the actual development area of the regional economy. At the same time, it shortens the prediction time, and improves the prediction efficiency while ensuring the accuracy.

4 Conclusion and Prospect

In order to improve the accuracy and efficiency of the prediction of regional economic development potential along the railway, a prediction method of regional economic development potential along the railway based on data mining is proposed, which gives full play to the technical advantages of data mining. The prediction of regional economic development potential along the railway is realized by extracting influencing factors, collecting index data, pre-processing index data and mining development characteristics.

However, there are still some shortcomings in this study. In the future research, we will adopt the custom way to provide the semantic class view of the economic development characteristic attributes, and delete the attributes that will not have a related impact on the potential prediction, so as to further improve the reliability of the prediction results.

References

1. Wu, P., Qiu, S.: Prediction of macroeconomic development based on optimized multidimensional grey model. Stat. Decis. **36**(3), 42–45 (2020)
2. Li, Y., Lu, S., Yuan, X., et al.: Forecasting algorithm of macroeconomic indicators based on correlation analysis. Command Inf. Syst. Technol. **11**(1), 84–88+100 (2020)
3. Lu, B., Ming, Q., Guo, X., et al.: Current and future aspects of coupling situation of tourism technological innovation-regional economy in China. Geogr. Geo-Information Sci. **36**(2), 126–134 (2020)
4. Yang, R., Xu, T.: Development scale of Chongli Ski tourism industry under Winter Olympics: from the perspective of economic forecasting. J. Shenyang Sport Univ. **38**(6), 1–7 (2019)
5. Xu, X., Zhu, R.: The development forecast of Shanghai marine economy based on grey prediction model. Ocean Dev. Manage. **36**(10), 44–46 (2019)
6. Fan, J., Duan, H., Shu, L.: Forecast analysis of coordinated development of highwaytransportation and national economy. J. Chang'an Univ. (Philos. Soc. Sci. Ed.) **21**(3), 49–58 (2019)
7. Zhou, W., Yang, W.: Research on macroeconomic forecasting based on interval-valued financial time series data. On Economic Problems **487**(3), 35–41 (2020)
8. Wang, J.: Monitoring and forecasting economic performance with big data. Data Anal. Knowl. Discov. **4**(1), 12–25 (2020)
9. Gao, P., Li, J., Liu, S.: An introduction to key technology in artificial intelligence and big Data Driven e-Learning and e-Education. Mobile Netw. Appl. 1–4 (2021)
10. Chen, H.-J., Hu, X.-B., Deng, X.: A short-term macroeconomic Forecasting model based on GMDH. J. Sichuan Univ. (Nat. Sci. Ed.) **57**(5), 915–919 (2020)
11. Wu, R., Zhou, X.: Research on short-term prediction of local economy based on big data. J. Univ. Sci. Technol. Liaoning **43**(4), 304–308 (2020)
12. Liu, S., Liu, D., Srivastava, G., et al.: Overview and methods of correlation filter algorithms in object tracking. Complex Intell. Syst. **3**, 1–23 (2020)
13. Liu, S., Liu, X., Wang, S., Muhammad, K.: Fuzzy-aided solution for out-of-view challenge in visual tracking under IoT assisted complex environment. Neural Comput. Appl. **33**(4), 1055–1065 (2021)
14. Feng, X., Li, N., Wang, G., et al.: Development of a liver cancer risk prediction model for the general population in China: a potential tool for screening. Ann. Oncol. **30**, ix46–ix47 (2019)
15. Pei, C., Liu, Y.: The simulation of the prediction model of the economic development potential of the Coastal Area. J. Coastal Res. **112**(sp1), 211–215 (2020)
16. Hoffmann, A., Ponick, B.: Method for the prediction of the potential distribution in electrical machine windings under pulse voltage stress. IEEE Trans. Energy Convers. **36**(2), 1180–1187 (2020)
17. Bushuk, M., Msadek, R., Winton, M., et al.: Regional Arctic sea-ice prediction: potential versus operational seasonal forecast skill. Clim. Dyn. **52**(5–6), 2721–2743 (2019)
18. Sturniolo, S., Liborio, L.: Computational prediction of muon stopping sites: a novel take on the unperturbed electrostatic potential method. J. Chem. Phys. **153**(4), 044111 (2020)
19. Bai, K., Li, K., Chang, N.-B., Gao, W.: Advancing the prediction accuracy of satellite-based PM2.5 concentration mapping: a perspective of data mining through in situ PM2.5 measurements. Environ. Pollut. **254**, 113047 (2019)
20. Wang, H., Huang, Z., Zhang, D., et al.: Integrating co-clustering and interpretable machine learning for the prediction of intravenous immunoglobulin resistance in Kawasaki disease. IEEE Access **8**, 97064–97071 (2020)

21. Wang, C., Bi, J., Sai, Q., et al.: Analysis and prediction of carsharing demand based on data mining methods. Algorithms **14**(6), 179 (2021)
22. Li, H., Lu, Y., Zheng, C., Yang, M., Li, S.: Groundwater level prediction for the arid oasis of northwest china based on the Artificial Bee Colony Algorithm and a back-propagation neural network with double hidden layers. Water **11**(4), 860 (2019)
23. Trumpis, M., Chiang, C.H., Orsborn, A.L., et al.: Sufficient sampling for kriging prediction of cortical potential in rat, monkey, and human ECoG. J. Neural Eng. **18**(3), 036011 (2021). (18pp)

Data Security Transmission Algorithm of Remote Demonstration System Based on Internet of Things

Yong-sheng Zong[1,2][✉] and Guo-yan Huang[1]

[1] College of Information Science and Engineering, Yanshan University, Qinhuangdao 066004, China
[2] Qinhuangdao Vocational and Technical College, Qinhuangdao 066100, China

Abstract. Traditional remote demonstration data transmission methods neglect the evaluation of data trust, which leads to the problem of high packet loss rate in data transmission. In order to solve this problem and optimize the security of remote demonstration, a secure data transmission algorithm of remote demonstration system based on Internet of Things is proposed. Through four parts of system data fusion processing, system data compression, data trust evaluation and data encryption transmission, the design process of remote demonstration system data security transmission algorithm based on the Internet of Things is completed. The example test link is constructed, and the application effect of the Internet of Things algorithm is confirmed through three groups of experiments with different indexes. In the future use of the remote demonstration system, this algorithm can be used to realize the high security transmission of data.

Keywords: Internet of Things technology · Remote demonstration system · Data encryption · Data transmission

1 Introduction

With the development of Internet and industrial manufacturing level, more and more industrial data are stored in enterprise database, but due to the lack of effective data analysis and management tools, data can not fully play the function of equipment application guidance, and only through remote demonstration system can this function be realized. The current mainstream of remote monitoring technology is the application of Internet technology, supporting TGP, IP protocol and WWW technical specifications, well-organized software, enabling staff to access network servers quickly to all their access information and timely response [1, 2]. In the future, the development of embedded system will be more and more rapid and mature. New technology will be applied to remote demo system in the future, which is a demo system design. Embedded demo system can realize information to improve the positioning of server performance, so that each device can have Internet access and service functions, that is, each device can serve independently, thus greatly improving the quality of service range.

© ICST Institute for Computer Sciences, Social Informatics and Telecommunications Engineering 2022
Published by Springer Nature Switzerland AG 2022. All Rights Reserved
S. Liu and X. Ma (Eds.): ADHIP 2021, LNICST 417, pp. 239–253, 2022.
https://doi.org/10.1007/978-3-030-94554-1_20

At present, the use of the system gradually expanded the scope of the system's data gradually increased. Massive data, which emerged in recent years with the development of information technology, especially Internet technology, is mainly used to describe huge and unprecedented data, such as various environmental and cultural data information such as spatial data, report statistics, text, voice, image, hypertext, etc. Now, many enterprises, companies are involved in the operation of mass data processing, such as water conservancy departments, meteorological departments, such as processing data are very large. Because the mass data is very large, the reasonable compression and storage of the mass data become the supporting technology to solve the mass information storage and transmission. Because of the limitation of the technology and the complexity of the mass data, the efficiency and reliability of the remote demonstration system is low. In this paper, these problems are analyzed and improved.

In the past research, many remote demonstration system data security transfer algorithms have been proposed [3, 4], but there are some problems in the use of physical network technology to optimize it. The resolution of the Internet of Things has attracted great attention all over the world, and the research on the Internet of Things is becoming more and more in-depth. According to the definition of the International Telecommunication Union, the Internet of Things mainly deals with the interconnection of goods to goods, people to goods and people to people. The Internet of Things is a new type of real-time interactive system between virtual network and real world, which is characterized by ubiquitous data perception, wireless information transmission and intelligent information processing. The extensive application of the Internet of Things can provide convenience and bring social benefits for people in their lives and work. However, because a lot of information transmission in the Internet of Things relies on wireless networks, and a lot of devices are exposed to the public, information is vulnerable to theft and attack, and devices are vulnerable to damage, the security risks brought by the Internet of Things have been paid more and more attention. In this study, the application of this technology to optimize the traditional algorithm, and strive to improve the performance of the algorithm on the basis of the traditional algorithm.

2 Design of Data Security Transmission Algorithm for Remote Demonstration System

In this study, the Internet of things technology is applied to the design of remote demonstration system data security transmission algorithm, and the design process of system data security transmission algorithm is set as shown in Fig. 1:

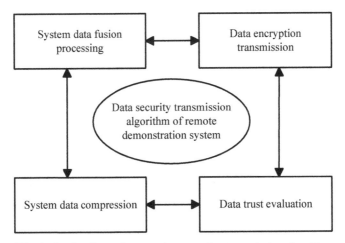

Fig. 1. Design flow of system data security transmission algorithm

In this design, the main work content is as follows:

(1) Describe the current research status of data security transmission algorithms, point out the existing problems and defects, and introduce the significance of research on privacy protection.

(2) This paper introduces the basic concepts and architecture of the Internet of Things, describes the role of the Internet of Things in data transmission and the security goal of data transmission, and finally expounds the trust evaluation theory and data encryption used in this paper in detail.

(3) This paper analyzes the existing problems in evidence merging, and proposes a new recommended merging rule, which can resist the safe data transmission algorithm of malicious defamation, and proves the performance of the algorithm through experiments.

2.1 System Data Fusion Processing

The basic idea of data fusion is that the upper nodes analyze the data transmitted by the lower nodes, filter, compress, remove the redundant information, and only send useful information to the upper nodes. This saves energy and prolongs the life of the node [5, 6]. The IOT awareness layer is made up of a lot of nodes distributed everywhere. If the nodes are densely distributed, the data between different nodes may be overlapped. If the nodes transmit all the data to the base station, it will first bring a large amount of data transmission consumption. Secondly, too much data may lead to data collision and affect the accuracy of the data. Finally, these data will be transmitted to other nodes, bringing more energy consumption. How to make full use of limited resources is the focus of wireless network research.

Through analyzing the noise characteristics of the system module, the measurement of the target parameters of the system module is described in the design of data fusion model by the following models:

$$y_i(s) = x(s) + \varepsilon_i(s) \tag{1}$$

In the above formula, $x(s)$ is used to represent the measured value of the system module during i-th measurement. $\varepsilon_i(s)$ represents the comprehensive noise generated by the system module in the i-th measurement, and meets the requirements of $C[\varepsilon_i] = 0$ and $H[\varepsilon_i] = W_i^2$. The comprehensive noise includes the noise of the system module itself and the external noise of the environment in which the system module is located. $y_i(s)$ is the overall distribution, and it is considered in the above formula, The integrated noise between the system modules is completely independent.

Set n systems module to measure the same target, the data distribution of the i-th system module and the j-th system module is V_i, V_j. V_i, V_j obeys Gauss distribution, and its PDF curve is taken as the characteristic function of the system module, denoted as $r(V_i)$, $r(V_j)$, V_i and V_j, which are the first observation values of V_i and V_j respectively.

In order to reflect the deviation between v_i and v_j, the concept of relative distance is introduced:

$$d_{ij} = abs|v_i - v_j| \tag{2}$$

Where, d_{ij} is the distance. From the expression d_{ij}, the larger d_{ij} is, the greater the difference between v_i and v_j, the lower the degree of mutual support between v_i and v_j. Therefore, a function f related to e is introduced to represent the mutual support relationship between v_i and v_j. q_{ij} itself shall meet the following two conditions:

(1) q_{ij} should be inversely proportional to the relative distance d_{ij};

(2) The value range of q_{ij} should be limited between [0, 1], that is, $q_{ij} \in [0, 1]$.

The support function obtained by the above method is simple in form and easy to implement. In the process of using, it needs to carry out accurate calculation to ensure the effect of data compression. Suppose U_1, U_2, ..., U_n homogeneous system module makes a measurement once in the measurement period t, and the sampling data matrix is obtained as follows, where $q_n(m)$ represents the m-th measurement value of node n.

$$Q = \begin{bmatrix} q_1(1) & q_1(2) & \cdots & q_1(g) \\ q_2(1) & q_2(2) & \cdots & q_2(g) \\ \cdots & \cdots & \cdots & \cdots \\ q_n(1) & q_n(1) & \cdots & q_n(g) \end{bmatrix}_{n*g} \tag{3}$$

The column data in data matrix Q is normalized and the error data is eliminated by the following formula:

$$\begin{cases} Q_i < \alpha, \text{eliminate} \\ Q_i \geq \alpha, \text{retain} \end{cases} \tag{4}$$

The optimal value of threshold α is obtained through multiple data calculations. When the data deviates from 1 to a large extent, the data is removed. Otherwise, it is retained to obtain a new column matrix, which is used to complete data fusion.

2.2 System Data Compression

Using parallel compression algorithm, the parallel compression unit of system data is designed. According to the traditional algorithm, the data parallel compression program is divided into the following functional modules: statistics module, data segmentation module, task allocation module, probability model building module, encoder, decoder, process control module and protocol module [7]. Each module is responsible for part of the functions of the program and completes the parallel compression of massive system data. The Fig. 2 shows the data compression process.

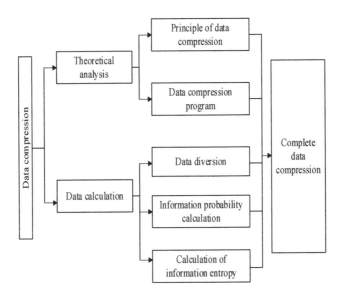

Fig. 2. Data compression process

Raw data analysis is the analysis of a large number of initial text files. Analysis needs to be done with a unicode code because it can handle both Chinese and English characters. In addition, when analyzing massive data text, there is no need to analyze the information like the header and the end of the file, just write it as it is, which is very small in proportion to the data text and is not in the scope of compressed text [8].

After analyzing the initial data stream, the frequency of the characters in the data stream should be counted. The statistical process is implemented by creating a dynamic linked list, which is used to record data and how often it occurs. As the number of characters increases, the linked list becomes larger and larger, so dynamic memory allocation techniques are used. In the statistical process, each input character also involves the data the inquiry work. To reduce the search time, the program remembers the previous character, and then when the new character is read as a memorized character, the frequency of the character is updated directly at the position specified by the memorized pointer. Otherwise, it needs to search from the beginning, find the matching character, change the character count, and if no matching character is found, reallocate the space, and insert the character into the list [9].

Probability calculation is to make statistical analysis according to the above statistical results, calculate the frequency of characters, and create a two-dimensional array to store characters and their probability. This probability will be applied to the original data segmentation. The character probability formula is as follows:

$$P_i = \frac{F_i}{\sum\limits_{1}^{n} F_i} \tag{5}$$

Among them, F_i refers to the frequency of the i character, n refers to the number of characters, and P_i refers to the probability of characters. In order to make the compression result more reasonable, the data is segmented reasonably. Data segmentation module is very important in the whole algorithm. It not only determines the task allocation of each processor, but also involves the problem of load balancing among processors. The quality of segmentation algorithm directly determines the efficiency of data transmission algorithm [10].

In the field of data transmission and data compression, information entropy is a concept used to measure the amount of information, which reflects the order degree of the system. The more orderly the system is, the lower its information entropy is; On the contrary, the higher the information entropy is. In the field of data compression, information entropy is used to measure the quality of a transmission algorithm and find a suitable compression codeword. For a data input stream L, its entropy is defined as follows [11]:

$$D(l) = -\sum_{i=1}^{n} l_i \log_2 l_i \tag{6}$$

If the input stream is k in length, the information for the entire input stream is:

$$J = k * D(l) \tag{7}$$

As can be seen, entropy is the measure of the amount of data flow information. Therefore, when we segment the input information stream, as long as the information of each part is equal, the length of the code word of each part after arithmetic coding is not much different, so that the final tasks of each processor are not different, and their synchronization problem is relatively easy to solve [12, 13]. Of course, the design of the algorithm not only considers the entropy, but also the number of processors. The above formula is substituted into the original data compression algorithm to process the data, and the processed data is used as the basis for subsequent calculation.

2.3   Data Trust Evaluation

In the above completed the system data preprocessing, in this link will be the trust of the data assessment, for the subsequent transmission of data security to provide a theoretical basis.

At present, the research of data trustworthiness evaluation model is based on a specific application, from their basic theory, it can be divided into two systems: Bayesian system and evidence theory system [14, 15]. This study will use the Bayesian system to complete the system of data trust evaluation. Bayesian estimation is one of the most commonly used and effective methods in parameter estimation. Bayesian estimates treat the parameter to be estimated as a random variable that conforms to a priori probability distribution [16]. The process of observing the sample is to transform the prior probability density into the posterior probability density, and to modify the initial estimate of the parameter by the information of the sample. In Bayesian estimation, each new observation sample makes the posterior probability density function sharper, making it form the biggest spike near the real value of the parameter to be estimated [17].

The main theory of Bayesian estimation is: let the probability distribution of a and α be $f(a, \alpha)$, where $\alpha \in R$ is the unknown random variable parameter. Sample $A_1, A_2, ..., A_n$ has been given, we need to estimate the parameter α according to the above sample.

Firstly, Bayesian estimation regards the unknown parameter α as a random variable, and considers that prior to the experiment, there is α certain understanding of α, which is called prior knowledge. This kind of prior knowledge is expressed by some probability distribution of a, and its probability density function is denoted as $e(\alpha)$. This kind of probability distribution is called "prior distribution" of α. according to the above understanding of setting unknown parameter α [18].

Then, define $f(a_1, \alpha)$, $f(a_2, \alpha)$, ..., $f(a_n, \alpha)$ when α is given, the distribution condition of system data $(A_1, A_2, ..., A_n)$, then the joint probability density function of $(\alpha, A_1, A_2, ..., A_n)$ is $e(\alpha)f(a_1, \alpha)f(a_2, \alpha)...f(a_n, \alpha)$, then the marginal probability density function of $(A_1, A_2, ..., A_n)$ is:

$$f(A_1, A_2, ..., A_n) = \int e(\alpha)f(a_1, \alpha)f(a_2, \alpha)...f(a_n, \alpha)h\alpha \qquad (8)$$

Finally, given $A_1, A_2, ..., A_n$, the conditional probability density function of α is as follows [19]:

$$f(\alpha|A_1, A_2, ..., A_n) = \frac{e(\alpha)f(a_1, \alpha)f(a_2, \alpha)...f(a_n, \alpha)}{f(A_1, A_2, ..., A_n)} \qquad (9)$$

The above formula is the "a posteriori" probability density function of α. To sum up, it can be concluded that parameter α can be estimated with prior knowledge, such as sample $A_1, A_2, ..., A_n$. It can be found that formula (7) is very similar to Bayesian formula [18]. After knowing the posterior probability distribution $f(\alpha|A_1, A_2, ..., A_n)$, the parameter E can be estimated. Using the above formula, we can get the trust evaluation results of the data in the system. According to the evaluation results, we can provide guarantee for the subsequent information encryption transmission, and choose the appropriate encryption method to process the data with low trust.

2.4 Data Encryption Transmission

According to the above data processing results, the full encryption method is used to encrypt the processed data in this study. The establishment of total homomorphism can be divided into the following steps.

At the beginning of homomorphism, it is necessary to establish a bootstrap [19, 20], which is represented by symbol. Bootstrap should be established according to the following steps:

$$(wh, oh) \leftarrow KeyGen_{\lambda}(\delta) \tag{10}$$

$$\Im_i \leftarrow Encrypt_{\lambda}(wh, \wp) \tag{11}$$

$$Decrypt_{\lambda}(oh, \cdot) \rightarrow \wp \tag{12}$$

In the above formula, $KeyGen$ represents the random generation of public key wh and private key oh by random algorithm [21]. On the basis of security parameters, public key defines a clear text space $\Re$ and ciphertext space $\aleph$. $Encrypt_{\lambda}$ means that h is calculated by a random algorithm using valid public key and clear text $\cdot \in \wp$. $Decrypt_{\lambda}$ indicates that the decrypted plaintext $\wp$ is output on the premise of private key wh and ciphertext.

In order to achieve homomorphism, we need to implement an additional process $\cdot \leftarrow Encrypt_{\lambda}(wh, \Re, \wp)$, select $\cdot$ and the generated ciphertext tuple $\wp = (\wp_1, ..., \wp_n)$ from the set of allowed functions, and input them into the circuit [22, 23].

According to the application requirements of Internet of things technology, the possibility of simple attacks in the transmission process should be minimized to ensure the security of fully homomorphic password. Then, when designing the password, we should consider that the selected fully homomorphic password mapping should meet several characteristics.

1. The key space should be large, and its sequence should have good random statistical characteristics
2. Diffusion and confusion
3. Stability

An appropriate encryption process does not guarantee the theoretical and practical security of the sequence cipher with the above three points, but if the above three points are not satisfied, then it is definitely an insecure system. Now, the application of full homomorphic cipher in cryptography is mainly two modes, one is to directly use the data itself to construct the cipher, but its complexity, security and stability need further study. Second, the combination of homomorphic cryptography and the traditional cryptography algorithm with excellent characteristics constructs a new cryptography algorithm, using a new cryptography algorithm to complete encryption. After the encryption links of the above settings are integrated, the proper key is set and the data encryption process is completed.

The content of the design is integrated into the traditional algorithm. So far, the secure data transmission algorithm of remote demonstration system based on Internet of Things is designed.

3 Analysis of Experimental Demonstration

3.1 Experimental Environment Setting

In view of the data security transmission algorithm of the remote demonstration system based on the Internet of things proposed in this study, the experimental link is constructed to analyze and verify its use effect. The experimental links in this experiment are shown in Table 1.

Table 1. Experimental environment list

Parameter serial number	Parameter	Model
1	Memory	10 GB
2	Operating environment	Intel
3	Database	M4 gmp Crypto++ HElib
4	Programing language	C++
5	Operating system	Windows 10
6	JDK	1.7 and above
7	Other requirements	Interior design function

In this experimental platform, the experimental database of a remote demonstration system is built. In the process of building the experimental database, a variety of basic databases are used. The specific structure is shown in Fig. 3.

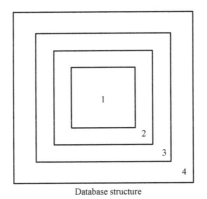

Database structure

Fig. 3. Structure of experimental database

The relationship between the four related library functions is shown in Fig. 2. One is the GNU version of macro preprocessor; 2. In this experiment, it is used as a mathematical operation library for high precision operation; 3 is a library software that can guarantee multiple floating point calculation; 4 is a C language library which can ensure the correctness of any high-order complex operation. There are 20 groups of data to be transmitted in this data, and the specific data volume and data type composition are as follows (Table 2).

Table 2. Experimental data

Serial number of experimental data group	Data volume/item	Data type
CSSJ-01	896	Data
CSSJ-02	616	Written words
CSSJ-03	882	Image
CSSJ-04	1203	Audio frequency
CSSJ-05	1798	Image
CSSJ-06	541	Written words
CSSJ-07	898	Written words
CSSJ-08	1913	Written words
CSSJ-09	1566	Written words
CSSJ-10	640	Image
CSSJ-11	1695	Image
CSSJ-12	1512	Audio frequency
CSSJ-13	1845	Data
CSSJ-14	1870	Image
CSSJ-15	1132	Audio frequency
CSSJ-16	1772	Audio frequency
CSSJ-17	896	Data
CSSJ-18	616	Data
CSSJ-19	882	Data
CSSJ-20	1203	Written words

The above experimental data is imported into the database, which is used as the data source and carrier in the process of the experiment, and the performance of the proposed Internet of things algorithm is analyzed.

3.2 Setting of Experimental Scheme

In order to make a more detailed analysis of the methods in this study, two kinds of expectations in current use are selected for comparison. At the same time, choose the appropriate experimental indicators to compare the use effect of Internet of things algorithm and traditional algorithm. In this experiment, the experimental comparison index is set as the packet loss rate of data transmission, the number of attacks on data transmission and the amount of data that cannot be read after data processing.

The calculation formula of data transmission packet loss rate is as follows:

$$R = \frac{F_i}{F_{all}} \cdot 100\% \tag{13}$$

In the above formula, F_i represents the data packet successfully sent to the target system; F_{all} is the packet that is trying to be sent to the target system. The above formula is used to calculate the packet loss rate of data transmission.

In this experiment, most of the experimental comparison indicators are statistical indicators. In the process of the experiment, we must ensure the accuracy of data calculation, so as not to affect the authenticity of the experimental results (Table 3).

3.3 Analysis of Experimental Results

Table 3. Packet loss rate of data transmission

Serial number of experimental data group	The packet loss rate of the proposed algorithm/%	Packet loss rate of algorithm in reference [3]/%	Packet loss rate of algorithm in reference [4]/%
CSSJ-01	2.16	5.91	5.08
CSSJ-02	2.63	5.52	5.18
CSSJ-03	2.09	5.66	5.08
CSSJ-04	2.87	5.79	5.33
CSSJ-05	2.61	5.02	5.31
CSSJ-06	2.68	5.91	5.22
CSSJ-07	2.82	5.59	5.56
CSSJ-08	2.02	5.22	5.79
CSSJ-09	2.22	5.03	5.69
CSSJ-10	2.86	5.39	5.19
CSSJ-11	2.48	5.62	5.47
CSSJ-12	2.95	5.21	5.28
CSSJ-13	3.00	5.87	5.15
CSSJ-14	2.35	5.17	5.71
CSSJ-15	2.36	5.43	5.41
CSSJ-16	2.79	5.83	5.29
CSSJ-17	2.11	5.34	5.44
CSSJ-18	2.01	5.13	5.71
CSSJ-19	2.47	5.22	5.73
CSSJ-20	2.16	5.92	5.08

From the above experimental results, it can be seen that the packet loss rate of system data transmission is low after the Internet of things algorithm is used, which indicates that the Internet of things algorithm can effectively control the data loss problem in the process of data transmission, and ensure that the data receiving capacity of the data receiving node can meet the current data receiving capacity requirements. Compared with the Internet of things algorithm, the packet loss rate of the traditional algorithm is relatively high. In the process of using this algorithm for massive data transmission, there will be a large number of data missing. For the remote demonstration system, a large number of data missing undoubtedly has a direct impact on the use effect of the system. According to the above experimental results, the Internet of things algorithm can effectively guarantee or improve the use effect of the system (Table 4).

Table 4. Attack times of data transmission

Serial number of experimental data group	Attack times of the algorithm proposed in this paper/per time	Attack times of algorithm data transmission in reference [3]/per time	Attack times of algorithm data transmission in reference [4]/per time
CSSJ-01	3	8	7
CSSJ-02	2	7	10
CSSJ-03	3	10	9
CSSJ-04	1	10	8
CSSJ-05	3	5	10
CSSJ-06	1	9	9
CSSJ-07	1	10	10
CSSJ-08	4	6	9
CSSJ-09	4	6	5
CSSJ-10	3	7	7
CSSJ-11	2	5	9
CSSJ-12	3	5	6
CSSJ-13	1	9	7
CSSJ-14	4	5	9
CSSJ-15	4	8	7
CSSJ-16	4	10	6
CSSJ-17	4	8	10
CSSJ-18	4	8	6
CSSJ-19	2	5	6
CSSJ-20	3	8	7

From the above experimental results, it can be seen that the system data transmission is more stable after the application of the IOT algorithm, and in different types of system data, the IOT algorithm can control the transmission security within a fixed interval, so as to avoid the high number of packet transmission attacks affecting the system operation effect. Compared with the Internet of Things algorithm, the traditional algorithm does not avoid the number of attacks on data, and in the process of data attack, it does not provide adequate protection for data packets, resulting in poor transmission security of data packets. In different types of packets, the traditional algorithm has different effects, which shows that the applicability of this algorithm is low. Based on the above experimental results, it can be concluded that the Internet of Things algorithm is better than the traditional algorithm (Table 5).

Table 5. Abnormal data volume after data transmission

Serial number of experimental data group	The algorithm proposed in this paper abnormal data volume/piece	Number of abnormal data in [3] algorithm/piece	Number of abnormal data in [4] algorithm/piece
CSSJ-01	29	100	118
CSSJ-02	22	109	104
CSSJ-03	28	90	103
CSSJ-04	22	95	109
CSSJ-05	20	110	101
CSSJ-06	28	102	114
CSSJ-07	29	107	109
CSSJ-08	29	101	115
CSSJ-09	25	96	118
CSSJ-10	23	105	120
CSSJ-11	21	102	110
CSSJ-12	29	92	104
CSSJ-13	30	104	106
CSSJ-14	26	97	111
CSSJ-15	29	95	110
CSSJ-16	26	94	114
CSSJ-17	28	90	107
CSSJ-18	27	97	100
CSSJ-19	30	107	110
CSSJ-20	29	100	118

After analyzing the above experimental results, it is found that the abnormal situation of the data transmitted by the IOT algorithm has been alleviated, and the reliability and authenticity of the data transmitted are ensured to some extent. Compared with the Internet of Things algorithm, the traditional algorithm has a relatively poor effect. After completing the system data transmission, it does not reduce the abnormal situation of data transmission, but improves in part of the data. Analysis of the above experimental results can be found that the Internet of Things algorithm based on Internet of Things technology to complete the data fusion, the data of abnormal problems have inhibition. Based on the above analysis results, it can be concluded that the IOT algorithm is effective.

All the experimental results obtained in this experiment show that the physical network algorithm is better than the traditional algorithm, and it can be used to complete the system data transmission in the future.

4   Conclusions and Prospects

With the development of Internet of things, system data security protection has become an urgent problem to be solved. Data fusion can eliminate redundancy, reduce the amount of data transmission, and then reduce energy consumption. It has become a challenge to secure data transmission while rationalizing the use of data. In this paper, a new data fusion method is proposed, and the effect of the algorithm is analyzed.

In this paper, a secure transmission algorithm of remote demonstration system based on Internet of Things technology is proposed. The bottleneck problem of mass data transmission in physical network is analyzed from three aspects: the defect of SOAP protocol, the defect of application layer and the defect of transport layer, and an improved scheme is put forward to solve each bottleneck problem. Compared with the traditional algorithm, it has some improvements in energy consumption and security, but there are still some deficiencies due to the time and condition constraints, which will be taken as the next step in the research work.

(1) The proposed trust model does not take into account the handling of new nodes.

(2) There is no actual attack test for the proposed data fusion algorithm, only theoretical analysis is carried out, and the next step is to test the algorithm.

(3) This study only considers the integrity of the data, not the robustness of the data. The next step will be to study the algorithm considering the robustness of the data.

Fund Projects. "The National Key Research and Development Program of China" and project number (2016yfb0800700).

References

1. Yang, C.Y., Liu, J.Y.: Data fusion algorithm based on weighted D-S evidence theory in Internet of Things. J. Guilin Univ. Technol. **39**(03), 731–736 (2019)
2. Tu, Y.F., Su, Q.J., Yang, G.: An encryption transmission scheme for industrial control system. J. Electron. Inf. Technol. **42**(02), 348–354 (2020)
3. Shi, L.L., Li, J.Z.: Research and design of secure data transmission mechanism in heterogeneous network. Microelectron. Comput. **36**(11), 84–88+94 (2019)
4. Geng, J.: Simulation of software secret dataloss prevention transmission based on mobile gateway. Comput. Simul. **36**(11), 151–155 (2019)
5. Xu, X.S., Jin, Y., Zeng, Z., et al.: Hierarchical lightweight high-throughput blockchain for industrial Internet data security. Comput. Integr. Manuf. Syst. **25**(12), 3258–3266 (2019)
6. Liu, S., Liu, X.Y., Wang, S., et al.: Fuzzy-aided solution for out-of-view challenge in visual tracking under IoT assisted complex environment. Neural Comput. Appl. **33**(4), 1055–1065 (2021)

7. Fang, W.W., Liu, M.G., Wang, Y.P., et al.: A distributed elastic net regression algorithm for private data analytics in Internet of Things. J. Electron. Inf. Technol. **42**(10), 2403–2411 (2020)
8. Liu, S., Li, Z., Zhang, Y., Cheng, X.: Introduction of key problems in long-distance learning and training. Mobile Netw. Appl. **24**(1), 1–4 (2018). https://doi.org/10.1007/s11036-018-1136-6
9. Gao, P., Li, J., Liu, S.: An introduction to key technology in artificial intelligence and big Data Driven e-Learning and e-Education. Mobile Netw. Appl. **26**, 2123–2126 (2021). https://doi.org/10.1007/s11036-021-01777-7
10. Yu, J.G., Zhang, H., Shu, L., et al.: Data sharing model for Internet of Things based on blockchain. J. Chin. Comput. Syst. **40**(11), 2324–2329 (2019)
11. Zhang, G.H., Yang, Y.H., Zhang, D.W., et al.: Secure routing mechanism based on trust against packet dropping attack in Internet of Things. Comput. Sci. **46**(06), 153–161 (2019)
12. Zhang, T., Zhang, J., Chen, K., et al.: Low-latency data transmission method based on Lora node cooperation. J. Northwest Univ. (Nat. Sci. Ed.) **49**(01), 88–92 (2019)
13. Yuan, M.L., Long, Y., Li, L.: Data transmission algorithms for mobile WSN networks based on piecewise retransmit link awareness mechanism. J. Electron. Measure. Instrum. **33**(12), 50–57 (2019)
14. Harbi, Y., Aliouat, Z., Refoufi, A., et al.: Enhanced authentication and key management scheme for securing data transmission in the Internet of Things. Ad hoc Netw. **94**(Nov.), 101948.1–101948.13 (2019)
15. Xu, J., Tao, F., Liu, Y., et al.: Data transmission method for sensor devices in internet of things based on multivariate analysis. Measurement **157**(25), 107536–107544 (2020)
16. Sujitha, B., Parvathy, V.S., Lydia, E.L., et al.: Optimal deep learning based image compression technique for data transmission on industrial Internet of things applications. Trans. Emerg. Telecommun. Technol. **2020**(6), e3976 (2020)
17. Hasan, M.K., Shahjalal, M., Chowdhury, M.Z., et al.: Real-time healthcare data transmission for remote patient monitoring in patch-based hybrid OCC/BLE networks. Sensors **19**(5), 1208–1215 (2019)
18. Li, X., Zhao, N., Jin, R., et al.: Internet of Things to network smart devices for ecosystem monitoring. Sci. Bull. **64**(17), 1234–1245 (2019)
19. Saracevic, M.H., Adamovic, S.Z., Miskovic, V.A., et al.: Data encryption for Internet of things applications based on Catalan objects and two combinatorial structures. IEEE Trans. Reliab. **14**(9), 1–12 (2020)
20. Duan, R., Guo, L.: Application of blockchain for Internet of things: a bibliometric analysis. Math. Probl. Eng. **21**(6), 1–16 (2021)
21. Gao, Y., Xian, H., Yu, A.: Secure data deduplication for Internet-of-things sensor networks based on threshold dynamic adjustment. Int. J. Distrib. Sens. Netw. **16**(3), 155–162 (2020)
22. Ma, H., Zhang, Z.: A new private information encryption method in internet of things under cloud computing environment. Wirel. Commun. Mob. Comput. **225**(6), 1–9 (2020)
23. Jiang, W., Yang, Z., Zhou, Z., et al.: Lightweight data security protection method for AMI in power Internet of things. Math. Probl. Eng. **15**(5), 1–9 (2020)

Safe Storage Algorithm of Spreadsheet Data Based on Internet of Things Technology

Yong-sheng Zong[1,2(✉)] and Guo-yan Huang[1]

[1] College of Information Science and Engineering, Yanshan University,
Qinhuangdao 066004, China
[2] Qinhuangdao Vocational and Technical College, Qinhuangdao 066100, China

Abstract. The traditional data security storage algorithm of spreadsheet is mainly designed for single user scenario. When encountering the problem of multi-user, the traditional method will appear the problem of high delay and high repetition rate. In order to solve this problem, a new secure data storage algorithm is proposed in this paper. The intelligent gateway of Internet of Things is designed, the core board, the expansion board and the flash memory are designed, the multi-copy association model is established, and the association of the same source data is realized by copy cataloging. Finally, lossless coding is used to compress the data of spreadsheet, and the secure data storage algorithm based on Internet of Things is realized. In the simulation experiment, the traditional data security storage algorithm and the designed algorithm are used respectively. Experimental results show that the proposed algorithm can effectively reduce the latency, and the repetition rate of data storage is very low.

Keyword: Internet of Things technology · Electronic forms · Data security storage · Algorithm design

1 Introduction

With the rapid development of computer, human society is moving towards comprehensive digitalization and informationization. Every day, a large amount of spreadsheet data is stored in the network environment. As user data is shared in the network environment, operations such as sharing, downloading or forwarding to other storage service providers by authorized users or storage service providers will inevitably result in multiple data copy files, which will make widely disseminated user data no longer subject to the constraints and controls of the data owner, such as access time periods, which are not under control [1, 2]. When a users data reaches its storage period and needs to be deleted, if there is no effective mechanism for deleting multiple copies of data, the other copies of the user's data will not be effectively deleted, which not only causes great waste of storage space, but also leads to the abuse of user data and privacy leakage and other problems. Data management is out of control and long-term storage has no definite deletion, which seriously affects social stability and endangers national security.

In the traditional safe storage algorithm of spreadsheet data, the network architecture and mathematical modeling are all aimed at the single user scenario [3]. Although the multi-user problem can be transformed into a single user problem by dividing the storage node capacity, in practice, the data of multiple users may have interference in the transmission process, In the process of allocating storage capacity can not be reasonably allocated, resulting in long delay, so this paper designs a safe storage algorithm of spreadsheet data based on Internet of things technology.

2 Secure Storage Algorithm of Spreadsheet Data Based on Internet of Things Technology

2.1 Design Intelligent Gateway of Internet of Things

The intelligent gateway in this paper is mainly composed of the hardware and communication equipment of the Arduino platform based on the technology of the Internet of Things. In the hardware structure of the Arduino platform, the top is the Arduino EthernetShield expansion board, which is mainly used to maintain the communication between the Ethernet and the Internet of Things in the data storage of the spreadsheet, and the middle is the Arduino GSM Shield expansion board, which is mainly responsible for sending short messages and the bottom is the Arduino UNO core board, and is responsible for the data analysis and processing of the whole intelligent gateway. These two expansion boards and a core board, can be called the system intelligent gateway Arduino hardware devices, plug and play with the characteristics of "electronic building blocks", so the algorithm can easily be combined together. The three hardware boards are stacked together in the form of stacked blocks, as shown in Fig. 1:

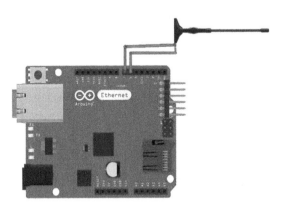

Fig. 1. Gateway hardware device in algorithm

In Fig. 1, the core board is actually a microcontroller board, which provides the most basic information data analysis and processing work for the storage algorithm. The processor model is ATMega328P, and the input voltage range is 6–20 V. It has 14 digital input/output pins, 6 PWM outputs, 3 communication interfaces, and 8 analog inputs.

The crystal oscillator value is 16 MHz, It also has USB interface to connect with other hardware drivers, and has a power jack [4]. In addition, batteries or some power adapters can be used to provide energy.

The function of GSM expansion board is relatively simple, mainly providing the function of sending short messages. It is equipped with a four band modem chip, which has four working bands. It is connected with the upper layer equipment through GPRS. The data transmission rate of uplink and downlink is 85.6 kbps. The communication protocol supports TCP/UDP and HTTP protocols. Compared with the GSM expansion board, the Ethernet shield expansion board is more powerful. It has Ethernet interface and integrated data storage function. The front-end page files of the system are stored in the SD card to avoid occupying ROM resources and improve the network performance of the gateway. In the expansion board, there are 14 digital IO pins, of which 4 are PWM outputs, and 6 Arduino pins are reserved. Pin10–13 is used for SPI, pin4 is used for SD card and pint is used for W5100 interrupt (bridge) by default, and the size of flash ROM is 32 KB, of which 0.5 kb is used for bootloader.

In order to ensure the performance of the gateway, high-end SOC chips are used in the design. In this paper, Arduino platform products are used in the design of intelligent gateway, which can save cost and shorten the delay in the storage process [5, 6].

In the data security storage algorithm, it needs to save the continuous spreadsheet data, and the storage capacity of the algorithm needs to be guaranteed. Therefore, flash memory can be used as storage support. Flash memory is a kind of safe and fast storage element. With the advantages of small size and low cost, it has gradually replaced other semiconductor storage elements and become the mainstream data program carrier [7, 8]. In the storage algorithm of this paper, four Flash memories of H27U4G8F2DTR-BC are used to form a storage array with a total capacity of 4T. The pin diagram of H27U4G8F2DTR-BC is shown in Fig. 2

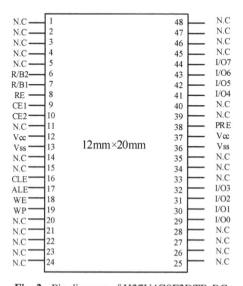

Fig. 2. Pin diagram of H27U4G8F2DTR-BC

The H27U4G8F2DTR-BC in Fig. 2 is 1t × 8-bit NAND Flash is packaged in 48 pin TSOP with a size of 12 mm × 20 mm. The function description of each pin is shown in Table 1.

Table 1. Pin function description

Pin type	Pin name	Function description
Data input/output	I/O0-I/O7	Input command, address and data, and output data during read operation
Command latch	CLE	Enter the active path of the command in the control command register
Address latch	ALE	Input controls the active path of the address in the internal address register
Selected films	CE-CE2	When the device is busy, the device will not return to write operation
Read enable	RE	Re controls serial data output and drives data to I/0 bus when valid
Write/erase	WE	The input control is written to the I/0 port, and the command, address and data are locked on the rising edge of we
Power on read enable	PRE	When connected to VCC, it controls the power on auto read operation function
Power input	Vcc	Device power input
Place of reference	Vss	–
Empty pin	N.C	No internal connection

The design of storage array directly affects the performance of data storage system. Large capacity storage array can store more ship navigation information data, which is an important guarantee for ship navigation safety.

2.2 Establish Data Multi Copy Association Model

Aiming at the problem of data multi copy Association, this paper proposes a data multi copy association model and deletion feedback mechanism, which can meet the needs of data multi copy Association and deletion caused by the operation of authorized users or storage service providers such as sharing, transferring or downloading. A data replica is an exact copy of the source data. Usually, the generation of data replica is to avoid data loss events caused by hardware failure. With the advent of the era of big data and the development of cloud storage technology, the generation of data replica is also due to the consideration of data access speed, data disaster tolerance and reliability. In mrad scheme, the data that are copied and identical with the source data is called data copy. The data copy from the same source data is also called homologous data copy, which is called data copy for short. In order to describe the data multi replica association model proposed in this paper, the concepts of child replica and parent replica are proposed. Similar to the tree structure, the parent copy is the parent node in the tree node, and the child copy is the child node in the tree node. Each parent copy can produce multiple child copies, and each child copy has only one corresponding parent copy. When a child replica is generated, the information about its parent replica will be recorded accordingly. This information is mainly used for the feedback mechanism when multiple replicas are deleted. As shown in Fig. 3:

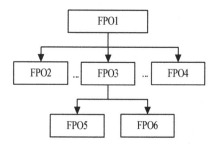

Fig. 3. Tree structure of data replica

In Fig. 3, the parent copy of FPO_3 is FPO_1, and the child copies of FPO_3 are FPO_5 and FPO_6. that is, FPO_3 is copied from FPO, while FPO_5 and FPO_6 are copied from FPO_3. With the continuous derivation of data copies, it is easy to cause copy flooding, which not only wastes storage space, but also has some security risks; And the separation of ownership and management of data copy greatly increases the difficulty of data copy management. In order to improve the multi-user and high concurrency ability of storage algorithm, the read and write operations of information data in the storage process should be separated. The storage is divided into main storage and slave storage. The main storage is responsible for data writing service. On this basis, the slave storage is added to be responsible for data reading. The working process of read-write separation is shown in Fig. 4

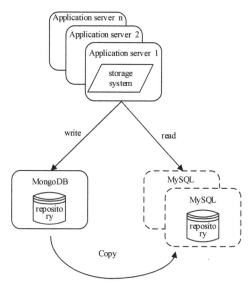

Fig. 4. Read write separation process

The purpose of adding read database is to reduce the pressure of the main database, and the read-write separation technology is applied in the database to reduce the delay phenomenon in the replication process [9, 10]. In the algorithm designed in this paper, the massive spreadsheet data in the storage array is mainly stored in the main database, and the main data stored in the database includes basic information such as users, projects and equipment [11, 12]. After a long time of data synchronization between the main database and the slave database, there will be data redundancy, which will slow down the system performance, so it is necessary to compress the data. In order to achieve data compression, the layer coding algorithm is designed. The compression flow chart is shown in Fig. 5:

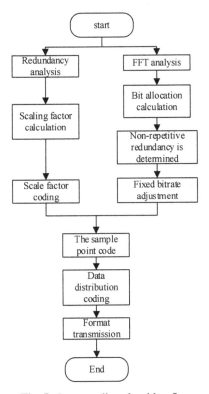

Fig. 5. Layer coding algorithm flow

In this algorithm, input the information data sample i in the database, and calculate the sample point vector $X[i]$ and window vector $C[i]$:

$$Z = X[i]C[i] \tag{1}$$

Window vector $C[i]$ can be calculated by coding standard

$$C[i] = \cos\left[\frac{(2i + 1)(k - 16) \times \pi}{64}\right] \tag{2}$$

In the above formula, the value range of i is 0–31, the value range of k is 0–63, and the calculation formula of sample vector $X[i]$ is as follows [13]:

$$X[i] = X_1[i] + X_2[i] + \ldots + X_n[i] \tag{3}$$

Each sub vector in formula (3) is a vector of 64 pieces of information data. After a series of operations of layer coding algorithm, the redundant data in the database is compressed. In the storage algorithm designed in this paper, one of the important performance indicators is whether it has the high concurrency ability of multiple users [14, 15]. It requires the algorithm to respond to a large number of users' storage requests of spreadsheet data in a short time in the storage process. If the high concurrency performance is poor, the user interaction of spreadsheet data will have a large delay, which affects the real-time update of spreadsheet data.

2.3 Lossless Coding and Compression of Spreadsheet Data

At present, the encoding method of spreadsheet data is mainly to explore the spatial correlation of data for compression [16]. The number of frequency bands is large, and the correlation between multiple frequency bands is high. The current encoding method can not eliminate the spectral redundancy, which will reduce the encoding efficiency and affect the frame storage efficiency in the process of storage. In this paper, the lossless coding scheme is improved. The complex prediction operation is transferred from the coding position to the decoding position, and the down sampling and channel coding operations are carried out in the coding position. In the coding scheme, the prediction operation is transferred to reduce the complexity of the coding end. In the decoding end, the information points of each frequency band can be selected by using the information of the frequency band that has been reconstructed before [17, 18]. There are many information points of the frequency band in the information points. In order to achieve the convenience of coding and compression, it is necessary to classify them. In the process of classification, if the pixels are of the same type, it is necessary to detect whether they have the same spectral correlation and choose the adaptive prediction method according to their correlation. Firstly, the sub-copy of the spreadsheet data needs to be down-sampled and bit-by-bit coded [19]. The more such subcopies are decoded, the more current data can be used to obtain edge information. The encoding result for the following sample is shown in Fig. 6:

1	9	3	11
13	5	15	7
4	12	2	10
16	8	14	6

Fig. 6. Schematic diagram of sampling number results

After decoding successfully, the resolution of the sub-copy of the current sample will be improved, which shows that the resolution is progressive and the precision of data storage is also improved. After the coding, we need to compress the data of the spreadsheet, extract the important node in the data of the spreadsheet, and get the sparse measured value after the coefficient transform, then store and transmit the measured value after compressing and processing, and then get the original signal after the sparse reconstruction [20, 21].

In the process of compression, suppose that x is a one-dimensional sparse signal and its size is described by $N \times 1$. When there are K and non-zero values in x, then its sparsity can be described as K. Φ represents a two-dimensional measurement matrix and its size is $M \times N$. Then the one-dimensional measurement signal can be expressed as:

$$y = \Phi x \tag{4}$$

The above equation is a system of underdetermined equations, so it needs to be transformed into an optimization problem [22]:

$$\hat{x} = \arg \min \|x\|_0 \quad \Phi x = y \tag{5}$$

This process is the perceptual reconstruction in the compression process, and it is also a constraint condition [23]. Its minimum 0 norm is a NP problem, so we need to transform the above optimization problem, that is, to transform the 0 norm problem into the 1 norm problem. In this way, through the given measurement matrix and measurement value y, the signal approximation value of the original spreadsheet data can be further calculated to complete the lossless coding compression of the spreadsheet data.

3   Simulation Experiment

3.1   Experimental Environment

The distributed storage of spreadsheet data using Internet of things technology can effectively reduce the data transmission delay in IIoT on the basis of guaranteeing the data security. The network architecture and mathematical modeling in the traditional data security storage algorithm are all aimed at the scenario of a single user. Although the multi-user problem can be transformed into a single-user problem by dividing the storage node capacity, in the actual situation, there may be interference in the transmission process of multi-user data, and it is impossible to reasonably allocate the storage capacity during the allocation process. The algorithm designed in this paper is optimized for this problem, so it needs to design simulation experiments to verify. Firstly, the network architecture of simulation experiment is built. In IIoT environment, production service puts forward a lot of requirements for delay. Taking industrial monitoring as an example, the camera first captures images of the normal production process and then uploads them to the cloud center for storage. Finally, during the production process, the camera continuously captures images of the production process and compares them with the normal production process of the cloud data center at regular intervals. If there is any deviation, adjustments will be made in a timely manner. In the SD-CFIIoT architecture, fog computing is introduced to reduce latency, distributed storage is introduced to

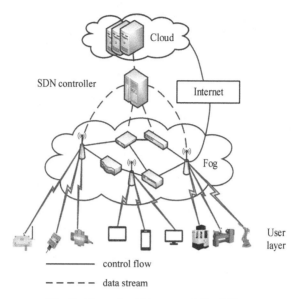

Fig. 7. Network architecture of SD-CFIIoT

enhance data security, and SDN technology is introduced to network management in multi-user environment. The SD-CFIIoT architecture is shown in Fig. 7:

The architecture consists of four layers: cloud computing layer, SDN control layer, fog computing layer and infrastructure layer. The main components of IIoT industrial monitoring application basic equipment layer include camera, electronic eye, monitor and other data acquisition equipment, as well as manipulator, intelligent lathe, alarm and other execution equipment. Data acquisition equipment after the acquisition to the video or image data uploaded to the fog calculating layer distributed storage system and the cloud for storage, after the acquisition device again after the monitoring information collected, compare the intermittent and already stored data, to update or store information according to the feedback information to perform the adjustment of the production process. Fog computing layer is mainly composed of router, base station, switch and other edge network equipment. OpenFlow protocol is running on these fog equipment, which can realize the communication function with SDN controller. In industrial control applications, the mist device receives the infrastructure layer of information collected and stored locally, due to limited storage capacity, fog between devices can form a distributed storage system at the same time, improve storage capacity, at the same time, the distributed storage can improve the safety of production when the information is stored, to prevent equipment being attacked by a single fog caused production secrets. The fog computing layer can not only store the information received, but also actively cache the production information close to the production process of the nearby factories or workshops from the cloud computing layer to provide a more comprehensive service. It can also upload its own stored non-confidential information to the cloud computing layer for global information sharing to provide services for other users. By storing image data in the fog computing layer, the response delay of industrial monitoring applications can be reduced, and the security of data stored in the fog computing layer can be guaranteed by distributed storage. The SDN controller layer is mainly composed of SDN controllers. The SDN controller centrally controls the SD-CFIIOT network, and obtains the global information of the network, including the storage capacity, communication ability and security of the device, through information interaction with the basic equipment layer, fog computing layer and cloud computing layer.

The cloud computing layer mainly refers to the remote cloud server cluster. Use of its powerful storage capacity for cloud server cluster IIoT environment in various business and equipment to provide services, in the business of industrial monitoring, its main task is according to the infrastructure layer set up production process database, the data collected camera equipment after the collected data, characteristic of information transmission to the cloud layer can be information, The cloud computing layer can rely on powerful computing power to quickly feedback the results of the comparison.

In IIoT monitoring applications, multiple users after the collected data, send data to the center of the fog calculating layer distributed storage system node, at the same time, users also upload data security requirement, how to center node according to the requirement of each user for data security, coupled with the fog calculating layer equipment storage capacity, safety, and communication ability, Reasonable distribution of all users' data to the fog computing layer devices and cloud computing layer for storage is a problem to be solved in SD-CFIIoT architecture. In this section, the total delay of data transmission in the case of multiple users is modeled, and the delay optimization problem under security constraints is proposed, and an algorithm is proposed to solve it. In the SD-CFIIoT architecture, the SD-CFIIoT network composed of M users and m fog computing devices is considered. The corresponding abstract structure of network topology is shown in Fig. 8:

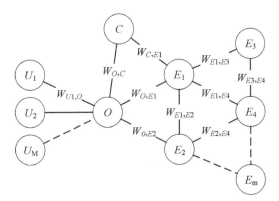

Fig. 8. SD-CFIIoT architecture device undirected diagram

In Fig. 8, the user set can be expressed as $U = \{U_1, U_2, ..., U_M\}$, which means there are a total of M users in the system. In the fog node set $E = \{E_1, E_2, ..., E_m\}$ in the figure above, there are m fog nodes, and the main fog nodes of the 9RZ-type storage system in generation O, where C represents the cloud server. The value w on the edge represents the data transfer rate between nodes. In the above figure, the data transmission rate between each node is shown in Table 2:

Table 2. Transfer rate table between storage nodes

W	E_1	E_2	E_3	E_4	E_5	E_6	E_7	E_8	O	C
E_1	∞	20	50	35	–	–	–	–	84	–
E_2	20	∞	60	–	40	37	.	–	92	–
E_3	50	60	∞	24	–	43	–	57	–	–
E_4	35	–	24	∞	–	–	–	38	–	–
E_5	–	40	–	–	∞	18	46	–	–	–
E_6	–	37	43	–	18	∞	55	70	–	–
E_7	–	–	–	–	46	55	∞	33	–	–
E_8	–	–	57	38	–	70	33	∞	–	–
O	87	92	–	–	–	–	–	–	∞	7
C	–	–	–	–	–	–	–	–	7	∞

In Table 2, the transmission rate of each node to itself is set to, indicating that there is no transmission delay when the node transmits data to itself, and - means that the two nodes are not directly connected. And take the above equipment architecture as the model, build the simulation experiment environment, and set the parameters in detail.

3.2   Parameter Settings

The above experimental environment was built in the simulation software MATLAB, and the relevant environmental parameters were set as shown in Table 3:

Table 3. Simulation parameters

The parameter name	Numerical	Unit
Total channel bandwidth	5	MHz
Subscriber transmitted power	100	mW
Background noise	– 100	dBm
Path loss factor	2	–
The number of fog nodes	8	a
Maximum number of iterations	1000	Times

Parameters of fog nodes in the simulation software are designed, as shown in Table 4:

Table 4. Fog node parameter table

Node	Attack arrival rate λ_i (times/min)	Storage capacity Z_i/MB
E_1	0.23	160
E_2	0.15	270
E_3	0.37	374
E_4	0.82	682
E_5	0.64	262
E_6	0.51	98
E_7	0.43	753
E_8	0.52	246

In order to verify that the SD-CFIIoT architecture can effectively reduce the time delay of multi-user total data transmission while ensuring the security of IIoT data, In this section, the delay performance of SD-CFIIoT architecture, cloud computing architecture and Single Fog Node (SFN) is compared when the probability of an attack Node being broken is different. In the simulation, the number of users is set to 10, and the relevant parameters of users are set as shown in Table 5:

Table 5. User parameters table

User	The channel gain H_i	Data security requirements ss^i
U_1	0.0027	0.86
U_2	0.0074	0.41
U_3	0.0019	0.73
U_4	0.0035	0.55
U_5	0.0062	0.62
U_6	0.0063	0.46
U_7	0.0054	0.18
U_8	0.0033	0.51
U_9	0.0098	0.38
U_{10}	0.0062	0.74

In the experimental environment of the above formula, the secure storage algorithm of spreadsheet data based on the Internet of Things technology and the traditional algorithm designed in this paper are respectively used for testing, and statistical analysis is conducted on the transmission experiment of total data.

3.3 Experimental Results and Analysis

In the above experimental environment, the influence of the number of users on experimental performance is obtained, as shown in Fig. 9:

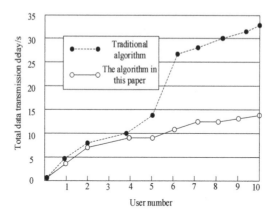

Fig. 9. The influence of the number of users on delay performance

From Fig. 9, it can be seen that with the increase of the number of users, the transmission delay of the two data security storage algorithms is increasing. When the number of users is small, the delay of this algorithm is lower than that of the traditional algorithm, but the reduction amplitude has no obvious advantage. With the increase of the number of users, the advantages of the delay performance are reflected. It can be seen that the total transmission delay of traditional data storage algorithm increases with the increase of the number of users. This is because with the increase of the number of users, the amount of data transmitted will increase, and the mutual interference between users will increase, which leads to the rapid deterioration of the delay of data transmission from the basic equipment layer equipment to the SDN controller, the increase of total transmission delay is accelerated. At the same time, if the single fog node in the traditional algorithm can not meet the needs of any user, it can not adopt this storage architecture, so its application scope is very limited; In the architecture algorithm designed in this paper, although the delay of SDN controller to receive data will increase, all users' data are allocated to a large number of fog computing equipment and Cloud Computing Center for distributed storage, which reduces the transmission delay of data in the allocation stage. Therefore, the increase of transmission delay with the increase of the number of users can basically keep stable. The paper verifies that the electronic table data security storage algorithm based on Internet of things technology can effectively reduce the delay in the case of multi-user.

Two algorithms are used to analyze the deduplication rate of spreadsheet data during secure storage. The result is shown in Fig. 10.

According to Fig. 10, the trend of repetition rates for each method of data storage is non-linear and fluctuates up and down. The repetition rate of spreadsheet data storage is always lower than 10% under the proposed algorithm, which shows that this technique can delete the same data and achieve high quality data storage.

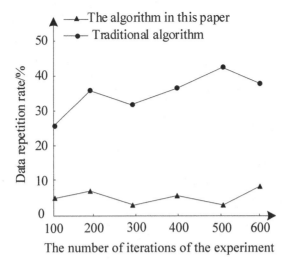

Fig. 10. Comparison of data store repetition rate of different methods

4 Conclusions and Prospects

Thanks to the rapid development of Internet technology and communication technology, the Internet of things industry has entered a stage of rapid development. The Internet of things has been widely and deeply applied in many industries, especially in the industrial field. The network architecture has the problems of high data transmission delay and heavy load of cloud computing center, which is difficult to meet the low delay requirements of massive services in iiot. This paper designs the intelligent gateway of Internet of things based on storage algorithm, designs the core board, expansion board, flash memory and other devices, establishes the data multi copy association model, realizes the association of homologous data copies through the copy directory, and finally carries on the lossless coding compression for the spreadsheet data, Complete the design of electronic form data security storage algorithm based on Internet of things technology.

However, when the intelligent gateway designed in this study realizes the functions of information collection, information input, information output, centralized control, remote control, linkage control and so on, its stability cannot be guaranteed due to the interference of relevant signals, which affects the safe storage of the data in the spreadsheet. In the future research, it is necessary to analyze the stability of the IOT intelligent gateway to avoid data loss and tampering during the remote operation of spreadsheet data.

Fund Projects. "The National Key Research and Development Program of China" and project number (2016yfb0800700).

References

1. Ma, Z.X., Fu, W.P., Li, Y., et al.: Design and implementation of substation intelligent safety management and control system based on internet of things technology. Electron. Measure. Technol. **42**(23), 6–14 (2019)
2. Cao, P.: Remote data storage and analysis platform based on industrial Internet of Things technology. Mod. Ind. Econ. Informationization **10**(04), 61–62 (2020)
3. Liu, H.N., Li, D.S.: A big data storage system architecture and data security placement mechanism. J. Chongqing Inst. Technol. **33**(08), 170–177 (2019)
4. Zhang, Q., Sun, C.: Fault-tolerant strategy optimization of medical data storage system based on erasure coding and decoding algorithm. New Gener. Inf. Technol. **02**(20), 47–52 (2019)
5. Liu, S., Li, Z., Zhang, Y., Cheng, X.: Introduction of key problems in long-distance learning and training. Mobile Netw. Appl. **24**(1), 1–4 (2018). https://doi.org/10.1007/s11036-018-1136-6
6. Gao, P., Li, J., Liu, S.: An introduction to key technology in artificial intelligence and big Data Driven e-Learning and e-Education. Mobile Netw. Appl. **24**, 1–4 (2021). https://doi.org/10.1007/s11036-021-01777-7
7. Xu, X.S., Jin, Y., Zeng, Z., et al.: Hierarchical lightweight high-throughput blockchain for industrial Internet data security. Comput. Integr. Manuf. Syst. **25**(12), 3258–3266 (2019)
8. Sheng, H., Zhou, Y., Ma, J.G., et al.: A secure sharing solution for massive heterogeneous data of traditional Chinese medicine based on alliance chain. World Sci. Technol. Modernization Tradit. Chin. Med. **21**(08), 1662–1669 (2019)
9. Ren, T.Y., Wang, X.H., Guo, G.X., et al.: Design of power Internet of Things data security system based on multiple authentication and lightweight password. J. Nanjing Univ. Posts Telecommun. (Nat. Sci.) **40**(06), 12–19 (2020)
10. Liu, S., LiuX, Y., Wang, S., et al.: Fuzzy-aided solution for out-of-view challenge in visual tracking under IoT assisted complex environment. Neural Comput. Appl. **33**(4), 1055–1065 (2021)
11. Zhu, S.S., Li, C.Q., Huang, R.J., et al.: Secure sharing model and mechanism of medical data based on block chain. Comput. Technol. Dev. **30**(10), 123–130 (2020)
12. Fu, Z.Z., Wang, L.M., Tang, D., et al.: HBase secondary ciphertext indexing method based on homomorphic encryption. Netinfo Secur. **20**(04), 55–64 (2020)
13. Yang, S.H., He, J.R.: Application of data encryption technology in computer network data security. J. Yanan Univ. (Nat. Sci. Ed.) **40**(01), 78–82 (2021)
14. Sinha, K., Priya, A., Paul, P.: K-RSA: secure data storage technique for multimedia in cloud data server. J. Intell. Fuzzy Syst. **39**(2), 1–18 (2020)
15. Rafique, A., Landuyt, D.V., Beni, E.H., et al.: CryptDICE: distributed data protection system for secure cloud data storage and computation. Inf. Syst. **96**(10), 45–56 (2020)
16. Wang, X., Yang, B., Xia, Z., et al.: A secure data sharing scheme with cheating detection based on Chaum-Pedersen protocol for cloud storage. Front. Inf. Technol. Electron. Eng. **20**(6), 787–800 (2019)
17. Yang, C., Tao, X., et al.: Secure data transfer and deletion from counting bloom filter in cloud computing. Chin. J. Electron. **29**(02), 79–86 (2020)
18. Meng, W., Ge, J., Jiang, T.: Secure data deduplication with reliable data deletion in cloud. Int. J. Found. Comput. Sci. **30**(4), 551–570 (2019)
19. Yang, G., Yang, J., Lu, Z., et al.: A combined HMM–PCNN model in the contourlet domain for image data compression. PLoS ONE **15**(8), e0236089 (2020)
20. Chen, Y.H.: New variable-length data compression scheme for solution representation of meta-heuristics. Comput. Oper. Res. **131**(3), 105256 (2021)

21. Kredens, K.V., Martins, J.V., Dordal, O.B., et al.: Vertical lossless genomic data compression tools for assembled genomes: a systematic literature review. PLoS ONE, **15**(5) e0232942 (2020)
22. Uthayakumar, J., Vengattaraman, T., Dhavachelvan, P.: A new lossless neighborhood indexing sequence (NIS) algorithm for data compression in wireless sensor networks. Ad Hoc Netw. **83**(FEB.), 149–157 (2019)
23. Justin, A., Benjamin, W.: Nuisance hardened data compression for fast likelihood-free inference. Mon. Not. R. Astron. Soc. **4**, 4–11 (2019)

Precision Adaptive Control Method for Automatic Switching Device of Power Grid Standby Power Supply

Cai-yun Di[1,2(✉)]

[1] State Grid Jibei Electric Power Company Limited Skills Training Center,
Baoding 071051, China
zz20221145@yeah.net
[2] Baoding Technical College of Electric Power, Baoding 071051, China

Abstract. In order to improve the effect of power grid standby power supply automatic switching device precision adaptive control, design a power grid standby power supply automatic switching device precision adaptive control method. Firstly, the inverter is modeled, then the instantaneous power is calculated when the standby power is switched automatically, and the mathematical model of doubly-fed induction generator is established. The experimental results show that the proposed control method can ensure the stability of power supply and voltage after control.

Keywords: Power grid · Standby power supply · Automatic switching device · Precision · Adaptive control · Voltage

1 Introduction

With the development of smart grid, the demand for power supply quality and reliability is becoming higher and higher. In particular, some customers have high demand for electricity, strict production processes and high level of automation. A sudden outage, even if it lasts only a few minutes, can shut down the entire production line, while resuming production takes a long time and is complex to operate. A sudden blackout will also bring a lot of economic losses to enterprises. People's lives have a tremendous impact, so that the national economy suffers huge losses.

The automatic standby power supply device is an important measure to ensure the continuity of power supply and improve the reliability of power supply. Using this device can simplify the configuration of power grid wiring and relay protection device to save equipment investment, reduce operational losses, and it has the advantages of simple structure, low cost, wide application, so it is widely used in power systems.

However, these devices can not meet the requirements of substation automation and unmanned shift, so a precision adaptive control method is designed for the standby power supply automatic switching device. Firstly, the inverter model is constructed, and

S. Liu and X. Ma (Eds.): ADHIP 2021, LNICST 417, pp. 270–283, 2022.
https://doi.org/10.1007/978-3-030-94554-1_22

then the instantaneous power during the automatic switching of standby power supply is calculated. Based on this, the mathematical model of doubly fed asynchronous generator is constructed to realize the droop control and adaptive state tracking control of the automatic switching device of power grid standby power supply. Finally, the accuracy and effectiveness of the adaptive control method are verified by simulation experiments.

2 Calculation Model of Instantaneous Power During Automatic Switching of Standby Power Supply

2.1 Inverter Modeling

The grid system consists of a number of inverter-based distributed generation units that rely on micropower sources for energy, such as fuel cells, micro gas turbines, DC storage, etc., while voltage source inverters are commonly used interface modules [1–3].

Specifically, the power distribution controller generates the amplitude and frequency of the inverter output voltage according to the droop characteristics. The voltage controller simulates the current vector of the reference filter inductance of a conventional synchronous generator, and the current controller synthesizes the instruction voltage vector by a pulse width modulation module. The voltage and current control loop must be able to provide sufficient damping to the filter composed of the filter and the coupled inductor. The coupling inductance determines the output impedance of the inverter, thus minimizing the coupling between active and reactive power [4]. Since the inverter of the micropower supply is controlled by a pulse signal, there is only one and only one upper or lower bridge arm on the same bridge arm, and they are alternately on each other, their switching function S_n can be defined, as shown in the formula below:

$$\begin{cases} S_n = 1 & (n = a, b, c) \\ S_n = -1 & (n = a, b, c) \end{cases} \tag{1}$$

In formula (1), n represents the conduction parameter.

The midpoint of the DC bus filter capacitor is the reference point of the system shown, so the inverter bridge output phase voltage of the three-phase inverter can be indicated by the following formula:

$$\begin{bmatrix} V_a \\ V_b \\ V_c \end{bmatrix} = \begin{bmatrix} S_a \\ S_b \\ S_c \end{bmatrix} \cdot \frac{U_{dc}}{2} \tag{2}$$

Let the phase voltage of three-phase filter capacitor C_f and the current of three-phase filter inductor L_f be the state variables of the system. According to Kirchhoff's voltage and current law, the voltage and current equations are established for the phase voltage of three-phase filter capacitor C_f and three-phase filter inductor L_f respectively.

$$
\begin{cases}
C_f \frac{d}{dt} V_{oa} = i_a - i_{oa} \\
C_f \frac{d}{dt} V_{ob} = i_b - i_{ob} \\
C_f \frac{d}{dt} V_{oc} = i_c - i_{oc} \\
L_f \frac{d}{dt} i_a = V_a - V_{oa} - R_f i_a \\
L_f \frac{d}{dt} i_b = V_b - V_{ob} - R_f i_b \\
L_f \frac{d}{dt} i_c = V_c - V_{oc} - R_f i_c
\end{cases}
\tag{3}
$$

In the formula (3), V_a, V_b and V_c are the inverter bridge output voltage of the three-phase inverter; i_a, i_b and i_c are the current of the three-phase filter inductor L_f; V_{oa}, V_{ob} and V_{oc} are the phase voltage of the three-phase filter capacitor C_f; i_{oa}, i_{ob} and i_{oc} are the current of the three-phase load.

Convert the above formula to a state space matrix [5, 6], which is represented as:

$$
\begin{bmatrix}
\dot{V}_{oa} \\
\dot{V}_{ob} \\
\dot{V}_{oc} \\
\dot{i}_a \\
\dot{i}_b \\
\dot{i}_c
\end{bmatrix}
= A \cdot
\begin{bmatrix}
V_{oa} \\
V_{ob} \\
V_{oc} \\
i_a \\
i_b \\
i_c
\end{bmatrix}
+ B \cdot
\begin{bmatrix}
V_a \\
V_b \\
V_c \\
i_{oa} \\
i_{ob} \\
i_{oc}
\end{bmatrix}
\tag{4}
$$

Modeling the inverter based on the above process can provide a basis for the adaptive control of automatic standby power supply switching device.

2.2 Calculation of Instantaneous Power in Automatic Switching of Backup Power

Based on the different topological structure of substation, different automatic bus switching models can be intelligently identified and generated by acquiring the real-time state of power network. The main function is to determine the current mode of operation of the system and to decide which model to use. The system combines the network analysis and the running state of the switch to judge the running mode of the current system, so as to decide which automatic switching mode to adopt. Therefore, the operation mode of the system must be accurately judged to prevent the misoperation and rejection due to misjudgment [7].

In order to reduce the maintenance workload and error rate effectively, the automatic generation algorithm of backup automatic switching model is developed in the main monitoring station of power network.

The design of the output inductance parameters is the core of the instantaneous power calculation. Firstly, a parameter ε_L of the fundamental voltage drop of the inductance is introduced, which is defined as the percentage of the fundamental voltage drop on the inductance L and the rated phase voltage U_g of the grid system, namely:

$$
\varepsilon_L = \frac{U_L}{U_d} \times 100\% = \frac{\omega L I_n}{U_g} \times 100\%
\tag{5}
$$

In formula (5), ω is the angular frequency of the fundamental voltage of the power grid, and I_n is the rated output current of the power grid. Since the system rated voltage is certain, the capacity of the power grid determines the rated current.

The vector relationship between the output voltage, the grid voltage and the inductance voltage. When connected to the grid, the minimum and maximum effective values of the output voltage are as follows:

$$U_{Imin} = U_g - U_L = (1 - \varepsilon_L)U_d \tag{6}$$

If an unplanned isolated island occurs and the power grid fails, there is no power exchange between the grid and the main grid, and the load is supplied separately at this time, then the effective output voltage of the converter under the isolated grid is:

$$U_I' = U_L + U_{pcc}' = \varepsilon_L U_g + U_{FCC}' \tag{7}$$

In practical engineering, the ripple current on inductor L also has certain requirements under the premise of satisfying the requirements of inductor fundamental wave voltage drop [8, 9].

When the duty cycle reaches 50%, the current ripple is greatest, so there are:

$$\Delta_{max} = \frac{T}{2} \times \frac{di}{dt} = \frac{TU_{dc}}{4L} = \frac{U_{dc}}{4Lf_{sw}} \tag{8}$$

From the above formula, the ripple current and the DC side voltage from the above formula, the ripple current and DC side voltage U_{dc}, inductance L and switching frequency f_{sw} are related. The DC side voltage is proportional to the ripple current, while the inductance is inversely proportional to the switching frequency.

Define a ripple coefficient parameter η_L, which is one half of the maximum peak ripple current as a percentage of the rated current I_n of the grid system, namely:

$$\eta_L = \frac{\Delta i_{max}}{2I_n} \times 100\% \tag{9}$$

In order to apply the droop control strategy, the instantaneous active and reactive power output from the micropower interface must first be calculated, using the two-axis theory [10], the instantaneous active and reactive components p and q injected are expressed as follows:

$$p = v_{od}i_{od} + v_{oq}i_{oq} \tag{10}$$
$$q = v_{od}i_{oq} - v_{oq}i_{od}$$

In order to have enough time span to separate power and current control loop and achieve high power quality injection, the average active power, reactive power and the corresponding fundamental component are affected by the control action. The instantaneous power component is used to obtain the active power p and reactive power Q related to the fundamental component through the lowpass filter shown in the following formula.

$$P = \frac{w_c}{s + w_c}p, Q = \frac{w_c}{s + w_c}q \tag{11}$$

In formula (11), w_c is the cut-off frequency of the filter.

3 Precision Adaptive Control of Automatic Switching Device for Standby Power Supply

3.1 Mathematical Model Construction of Doubly Fed Induction Generator

Double-fed asynchronous generator is a rising technical direction, mainly used in some large variable-speed wind turbine. In doubly fed asynchronous motors, a voltage-type transducer transmits power at differential frequencies to H-phase rotors, so the motor receives energy from both stator and rotor ends. The generator model introduced in this section ignores the hysteresis loss, eddy current loss and molten iron loss, and only considers the fundamental current component of stator and rotor, not the harmonics. The positive directions of the voltage, current and electromagnetic torque on the stator side of the motor are specified in accordance with the generators' practice; and the positive directions of the voltage, current and electromagnetic torque on the rotor side of the motor are specified in accordance with the electrification practice [11, 12]. According to the equivalent circuit diagram of the double-feed generator, the basic equations are as follows:

$$
\begin{aligned}
\dot{U}_1 &= -\dot{E}_1 - \dot{I}_1(R_1 + jX_1) \\
\frac{\dot{U}_2'}{s} &= -\dot{E}_2' + \dot{I}_2'\left(\frac{\dot{R}_2'}{s} + jX_1'\right) \\
\dot{E}_1 &= \dot{E}_2' = -\dot{I}_m(R_m + jX_m) \\
\dot{I}_1 &= \dot{I}_2' - \dot{I}_m
\end{aligned}
\tag{12}
$$

In the formula (12), R_1 and X_1 are the equivalent resistance and equivalent reactance of the stator winding; $\dot{E}_1$, $\dot{U}_1$ and $\dot{I}_1$ are the induced electromotive force, voltage and current of the stator side winding respectively; $\dot{R}_2'$ and X_m are the equivalent resistance and equivalent reactance of the rotor winding respectively converted to the value of the stator side; $\dot{E}_1$ and $\dot{E}_2'$ are the value of the induced voltage and current of the rotor winding converted to the value of the stator side respectively; $\dot{U}_2'$ is the value of the voltage of the rotor side excitation source converted to the value of the stator side respectively; R_m is the excitation resistance.

Double-fed wind turbines are a multivariable, strongly dependent, hierarchical system. In order to simplify the analysis and calculation, it is generally assumed that:

1. Stator windings of H phase with symmetrical distribution and spatial difference of 120 degrees, which ignore the influence of harmonics and magnetic saturation, generate magnetokinetic potential with sinusoidal distribution along air gap [13–15];
2. H phase winding of the rotor is of symmetrical structure;
3. Ignoring the influence of temperature, ignoring the iron loss and the skin effect.

3.2 Sag Control

After the above foundation treatment, the sag is controlled. It can be seen from the basic concept and flexible structure of power grid that the variable operation mode of power grid and the power supply service of high power quality depend on its perfect control system.

Master-Slave Control and Master-Slave Control are the control methods based on control strategy for isolated island power network. The reference voltage and frequency are provided by one or a limited number of micro-power sources, which can support the voltage and frequency of the power grid system [16]. At present, the master-slave control strategy still has some disadvantages: the fluctuation of load is balanced by the master power, so the master power is required to have certain capacity; the slave power is controlled by the master power, so the master power is too dependent on the master power; the master-slave control relies on communication, so it is restricted by the cost, complexity and reliability of communication.

Peer to Peer control, Peer to Peer control, means that there is no subordinate relationship between distributed generation and a distributed generation in the system use droop control method to regulate active and reactive power. At present, the problems of the equivalent control method are: the steady-state error can not be zero because of the instantaneous change of load; the harmonic distribution caused by non-linear load can not be coordinated; the droop characteristic method can not solve the problem that the reactive power of the system is affected by the line impedance; for the system with linear and non-linear load, the equivalent control method will not be applicable when the control mode changes due to the change of topology structure [17].

Multi-agent technology, with the rapid development of artificial intelligence, multi-agent system has been widely used. Multi-agent technology is suitable for decentralized and complex control of power network. Therefore, it can be applied to power system reconfiguration, power supply restoration, power market transaction and energy optimization and intelligent management. But at present, the application of multi-agent technology in power grid is mainly focused on the coordination of market transactions, power supply recovery and energy management, etc. In order to make multi-agent technology play a greater role in power system control systems, there is still a lot of research work to be done.

In order to ensure the normal operation of different kinds of micro-power supply, the power control method of the inverter interface of micro-power supply will use similar parallel power control method, even using similar method to the droop control of synchronous generator. This control method has the following advantages:

The micro-power supply in the power grid can be installed in any position, and will not be disturbed by some uncontrollable factors [18].

In the case of active power and reactive power balance of power grid, any micro-power supply can be integrated into or separated from the power grid, and this will not have a great impact on the power grid system.

When any micropower supply in the grid fails, the grid system can maintain the faulty micropower supply without stopping operation, and can guarantee the uninterrupted and reliable power supply to the grid load.

When the communication fault occurs between the micropower control unit and the central control unit of the power grid, the power grid system can still operate normally to ensure the reliability of the system.

Different from the traditional sag control method applied to synchronous generators, it is difficult to apply the sag control method to micro-power inverter. The self-characteristic of synchronous generator accords well with the droop characteristic, that is, when the input power of the prime mover is invariable, the increase or decrease of the load will make the speed of synchronous generator decrease or increase automatically to achieve balance. Unlike synchronous generators, micropower inverters cannot be balanced by regulating the generation of micropower, but rely on their own control strategies to achieve this balance [19, 20] (Fig. 1).

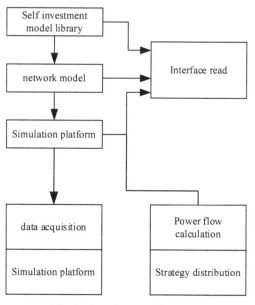

Fig. 1. Sagging control content

To sum up, the active power output of micropower can be regulated by controlling the output frequency, and the reactive power output can be regulated by controlling the output voltage amplitude of micropower [21]. Therefore, some researchers proposed using the droop control method to control the micro-power inverter. In order to realize the power distribution function in the parallel inverter system, the traditional droop control characteristics are introduced. The droop characteristics based on voltage frequency and output voltage amplitude are shown as follows:

$$w = w^* - mP \tag{13}$$

$$V = V^* - nQ \tag{14}$$

In the formulae (13) and (14), w^* and V^* shall be the rated frequency and voltage setting points; m and n shall be the static sag gain.

Under the sag control strategy, the active and reactive power output of the inverter unit is detected, then the rated frequency and voltage are obtained, and finally the active and reactive power output is reversely adjusted.

It can be seen from the sag characteristics that when the active power of the micropower output is larger, the output active power of the micropower output can be reduced by reducing the output frequency, whereas, when the active power of the micropower output is smaller, the output active power of the micropower output can be improved by increasing the output frequency; when the reactive power of the micropower output is larger, the output reactive power of the micropower can be reduced by reducing the output voltage amplitude, and vice versa, the output reactive power of the micropower output can be improved by increasing the output voltage amplitude. Through the adjustment of the above droop control strategy, the micropower in the power grid can realize the reasonable distribution of its active and reactive power.

3.3 Adaptive State Tracking Control Implementation

Before the 1960s, adaptive control was mainly used in aviation and space applications. Adaptive control focuses on the attitude control of aircraft, and many kinds of flight control schemes are proposed and realized. After the mid-1960s, adaptive control began to be gradually extended to the field of industrial control. Since the 1970s, adaptive control has been widely used in many fields. Now, adaptive control has been used in various fields maturely. The most commonly used definition of adaptive control is that an adaptive control system is one in which, when operating conditions are uncertain or change over time, it is able, during the control period, to implement effective controls based on information accumulated from online observations of reachable inputs and outputs, and to modify the parameters and control effects of the system structure so that the system is in a prescribed (and generally near-optimal) state.

Based on the idea of self-adaptive control, it is applied to the automatic switching device control of standby power supply. Uncertainties exist widely in practical control systems, and switched systems are no exception. Adaptive control is an effective method to deal with uncertain parameters, because the parameters of control are adjusted continuously according to control effect or on-line identification. Adaptive control of switched systems is a new research direction in five years, which has attracted more and more attention from the academic circles at home and abroad. With the development of research, the research on adaptive control of linear switched systems has gained some results. Consider nonlinear switched systems [22]:

$$\dot{x} = f_{\sigma 1}(x) + \theta_{\sigma 1} f_{\sigma 2}(x) + B_\sigma u \tag{15}$$

In formula (15), $f_{\sigma 1}(x)$ represents the state vector of the system, $\theta_{\sigma 1}$ is the matrix of unknown constants, $B_\sigma u$ is the matrix of known inputs, and $f_{\sigma 2}(x)$ is a known continuous function. The control objective of this paper is to design switching signals for switching systems, and to design adaptive controllers and adaptive laws for all subsystems so that the signals of closed-loop systems are bounded while $\lim_{t \to \infty} x(t) = 0$.

Considering that all subsystems can not be stabilized by designing state feedback adaptive controller, it is necessary to keep the stabilizable subsystem active all the time to achieve the goal of control. For each $i \in \Lambda$, the matrix K_{i1}^* satisfies [23, 24]:

$$\frac{\partial V_i}{\partial x}\big(f_i(x) + B_i K_{i1}^* \Phi_i(x)\big) + \sum_{j \neq 1} \beta_{ij}(x)\big(V_j(x) - V_1(x)\big) < 0 \tag{16}$$

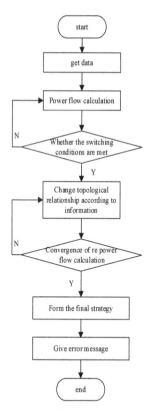

Fig. 2. Control the algorithm process

In the formula (16), $V_j(x)$ is positively definite and radially unbounded (Fig. 2).

The hyperstability theory is rooted in the theory of dissipation and passivity, which states that if the system is passive (strictly passive) with a finite input of external energy, the state of the system itself is bounded (bounded and convergent). The adaptive system can be regarded as the feedback interconnection structure of the control error system and the adaptive controller. In the design process, the hyperstability theory is applied to the system. In other words, the controller structure is chosen to make the control error system passive, and the adaptive law is chosen to ensure that the energy of the control error system is limited.

The basic idea is that the boundedness and convergence of the state are obtained by the boundedness and convergence of the energy of the system in the study of the traditional non-switched systems. This section focuses on a single subsystem to consider the problem, by studying the energy changes of the subsystem to obtain the hyperstability of the switching system. For each subsystem, the energy storage function is affected by the external energy input during the activation period, and the energy change is restricted by the state evolution of other subsystems during the non-activation period. Connecting the activation and non-activation periods, we can get the energy change of the subsystem in the continuous time domain, so we can study the properties of the system from the energy point of view. Consider linear switching systems:

$$\dot{x} = A_\sigma x + B_\sigma u$$
$$y = C_\sigma x + D_\sigma u$$

(17)

In formula (17), $A_\sigma x$ is the output of the system, $C_\sigma x$ is the input of the system, $B_\sigma u$ is the state matrix, and Q is the input matrix.

Let $V_i(x)$ be the energy storage function of the i subsystem. If the switched system is dissipative, it has:

$$V_i(x(t)) - V_i(x(s)) \leq \int_s^t \omega_i^i(u(\tau), y(\tau))d\tau$$

(18)

In the formula (18), $V_i(x(s))$ represents the self-energy supply rate when the i subsystem is activated, and $d\tau$ represents the interactive supply rate.

Based on the above process, the adaptive control of power supply automatic switching device precision is completed.

4 Simulation Experiment

In order to verify the validity of the precision adaptive control method of the standby power supply automatic switching device, the experiment was carried out and compared with the traditional control method. The simulation parameters applied in this experiment are shown in Table 1:

Table 1. Experimental parameters

Serial number	Parameter	Value
1	Active power capacity of power grid	20 kW
2	Rated voltage of power grid	220 V
3	Grid voltage angle frequency	$50 \times 2\pi$ rad/s
4	DC side voltage	700 V
5	Switching frequency	10 kHz

4.1 Comparison of Voltage Stability After Control

The fastest single opening and closing time of the power grid standby power supply automatic switching device is 225 ms, and the device is used in the substation, with large working current and arc extinguishing time of more than 65 ms. In addition, judging the loss of voltage by detecting AC, the "standby automatic switching" can complete the power supply switching within 512 ms at the fastest; By optimizing the device structure and changing the application scenario, the single bounce time of solid-state permanent magnet structure is 20 ms, the arc extinguishing time is 8 ms, the calculated DC zero crossing time is 6 ms, and the starting and closing time is 20 ms. The proposed method is compared with the traditional method. The results of voltage stability comparison between the proposed method and the traditional method are shown in Fig. 3.

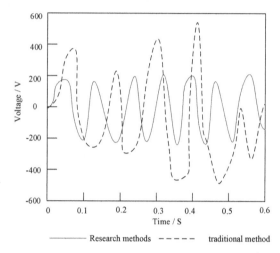

Fig. 3. Comparison of voltage stability after control

It can be seen from the analysis of Fig. 3 that the voltage stability of the adaptive control method in this study is better, and the overall value is more stable, while the voltage stability of the traditional method is obviously poor, and the fluctuation is more obvious.

4.2 Current Comparison After Control

The current comparison results of the two methods are shown in Fig. 4.

Based on Fig. 4, it can be seen that after the control of this research method, the current value is relatively stable, which can make a smooth transition and meet the good power supply quality of the load, which is more stable than the traditional method. It can be proved that the control effect of the adaptive control method is good.

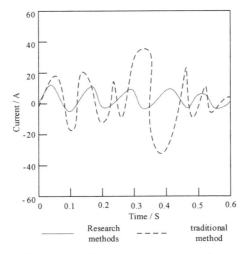

Fig. 4. Control the rear current comparison

5 Conclusion

With the acceleration of the construction of "one strong and three excellent", it is the responsibility of power grid enterprises to continuously improve the power supply reliability of power grid. Reasonable power grid structure and operation mode are the basis to ensure the safe and stable operation of power system. Whether relay protection devices can play an effective role is closely related to the power grid structure and operation mode, which must be considered as a whole. Therefore, according to the actual situation of power grid, the precision adaptive control method of automatic switching device of power grid standby power supply is designed, which is a powerful guarantee to improve the reliability of power supply.

With the development of smart grid technology, a large number of energy storage and distributed generation will be connected to the distribution network; In case of "N-1" fault under equipment maintenance mode, the power transfer mode between lines will be more flexible after the standby automatic switching action. How to establish the optimal configuration model of standby automatic switching for intelligent distribution network compatible with distributed generation and energy storage is the focus of the next step.

References

1. Zou, Z.X., Buticchi, G., Liserre, M.: Grid identification and adaptive voltage control in a smart transformer-fed grid. IEEE Trans. Power Electron. **34**(3), 2327–2338 (2019)
2. Jalalabadi, E., Salehizadeh, M.R., Kian, A.R.: Optimal control of the power ramp rate with flicker mitigation for directly grid connected wind turbines. Simulation **96**(2), 141–150 (2020)
3. Zhang, Y., Wei, W.: Decentralised coordination control strategy of the PV generator, storage battery and hydrogen production unit in islanded AC microgrid. IET Renew. Power Gener. **14**(6), 1053–1062 (2020)

4. Qian, Q., Xie, S., Xu, J., Bian, S., Zhong, N.: Passivity-based output admittance shaping of the converter-side current-controlled grid-tied inverter to improve the robustness to the grid impedance. IET Power Electron. **13**(10), 1956–1965 (2020)
5. Chen, X., Wang, X., Jian, J., Tan, Z., Li, Y., Crossley, P.: Novel islanding detection method for inverter-based distributed generators based on adaptive reactive power control. J. Eng. **2019**(17), 3890–3894 (2019)
6. Mondal, P., Tripathy, P., Adda, R., Saha, U.K.: Development of an adaptive control strategy for the three-phase grid side converter with wide range of parametric and load uncertainties. IET Power Electron. **13**(12), 2399–2412 (2020)
7. Giri, A.K., Arya, S.R., Maurya, R., Chittibabu, B.: Control of VSC for enhancement of power quality in off-grid distributed power generation. IET Renew. Power Gener. **14**(5), 771–778 (2020)
8. Ancha, S.K., Padhy, B.: Adaptive droop control strategy for autonomous power sharing and DC voltage control in wind farm-MTDC grids. IET Renew. Power Gener. **13**(16), 3180–3190 (2019)
9. Shukl, P., Singh, B.: Grid integration of three-phase single-stage PV system using adaptive laguerre filter based control algorithm under nonideal distribution system. IEEE Trans. Ind. Appl. **55**(6), 6193–6202 (2019)
10. Liu, S., Liu, D., Srivastava, G., Połap, D., Woźniak, M.: Overview and methods of correlation filter algorithms in object tracking. Compl. Intell. Syst. **7**(4), 1895–1917 (2020)
11. Abdulrahman, I., Belkacemi, R., Radman, G.: Power oscillations damping using wide-area-based solar plant considering adaptive time-delay compensation. Energy Syst. **12**(2), 459–489 (2019)
12. Liu, S., Liu, X., Wang, S., Muhammad, K.: Fuzzy-aided solution for out-of-view challenge in visual tracking under IoT assisted complex environment. Neural Comput. Appl. **33**(4), 1055–1065 (2021)
13. Li, Z., et al.: Adaptive power point tracking control of PV system for primary frequency regulation of AC microgrid with high PV integration. IEEE Trans. Power Syst. **36**(4), 3129–3141 (2021)
14. Gupta, Y., Chatterjee, K., Doolla, S.: A simple control scheme for improving reactive power sharing in islanded microgrid. IEEE Trans. Power Syst. **35**(4), 3158–3169 (2020)
15. Santhoshi, B.K., Mohanasundaram, K., Kumar, L.A.: Ann-based dynamic control and energy management of inverter and battery in a grid-tied hybrid renewable power system fed through switched z-source converter. Electr. Eng. **103**(5), 2285–2301 (2021)
16. Zhao, B.Y., Zhao, Z.G., Li, Y., Wang, R.Z., Taylor, R.A.: An adaptive PID control method to improve the power tracking performance of solar photovoltaic air-conditioning systems. Renew. Sustain. Energy Rev. **113**, 109250 (2019)
17. Liu, Z., Miao, S., Fan, Z., Liu, J., Tu, Q.: Improved power flow control strategy of the hybrid AC/DC microgrid based on VSM. IET Gener. Transm. Distrib. **13**(1), 81–91 (2019)
18. Pandey, S.K., Singh, B., Modi, G.: Frequency adaptive complex coefficient filter based control for grid integrated PV system. IET Generat. Transm. Distrib. **14**(19), 4141–4151 (2020)
19. Offringa, A.R., Mertens, F., Van, D., Veenboer, B., Gehlot, B.K., Koopmans, L., et al.: Precision requirements for interferometric gridding in 21-cm power spectrum analysis. Astron. Astrophys. **631**, A12 (2019)
20. Gong, P., Lu, Z., Lin, J., Lv, Z.: Distributed secondary control based on cluster consensus of inhibitory coupling with power limit for isolated multi-microgrid. IET Gener. Transm. Distrib. **13**(18), 4114–4122 (2019)
21. Shah, P., Singh, B.: Adaptive observer based control for roof-top solar PV system. IEEE Trans. Power Electron. **35**(9), 9402–9415 (2019)

22. Hu, C., He, H., Jiang, H.: Edge-based adaptive distributed method for synchronization of intermittently coupled spatiotemporal networks. In: IEEE Transactions on Automatic Control, pp. 1–1 (2021)
23. Nguyen, A.T., Lee, D.C.: Sensorless control of DFIG wind turbine systems based on SOGI and rotor position correction. IEEE Trans. Power Electron. **36**(5), 5486–5495 (2021)
24. Yin, Y., Vazquez, S., Alcaide, A.M., Liu, J., Franquelo, L.G.: Observer-based sliding mode control for grid-connected power converters under unbalanced grid conditions. In: IEEE Transactions on Industrial Electronics, pp. 1–1 (2021)

Fast Recommendation Method of Personalized Tourism Big Data Information Based on Improved Clustering Algorithm

Yi-lin Feng[✉], He-qing Zhang, and Cai-ting Peng

Management (Tourism) School of Guangzhou University, Guangzhou 510006, China

Abstract. The conventional tourism big data information recommendation method does not reduce the search scope of the database, resulting in a long running time of the algorithm. Therefore, based on the improved clustering algorithm, a fast personalized tourism big data information recommendation method is designed. The improved clustering algorithm is designed and the mathematical model of clustering algorithm is established. The semantic similarity of tourism information is calculated. Based on the improved clustering algorithm, the retrieval range of database is reduced, and the classification model of tourist attractions is established, so as to improve the speed of tourism big data information recommendation method. In the experiment of testing the improved clustering algorithm and test information recommendation method, the experimental data of the control group are better than the three control groups. Therefore, the fast recommendation method of personalized tourism big data information based on the improved clustering algorithm is better than the three conventional methods.

Keywords: Digital technology · Civil engineering · CAD drawing · Teaching assistant system

1 Introduction

With the rapid development of social networks, a large number of tourism related communities and websites have emerged as carriers, resulting in a huge scale of tourism information and data. After traveling, tourists like to publish their travel experience on the Internet. These travel experiences are shared online in the form of text, pictures and videos. Travelers from all over the world combine their first-hand information to form a huge tourism knowledge base. These tourism data are not only huge in quantity, but also various in types, such as picture, video, text, audio, geographical location and so on. The structure of these tourism data has strong randomness. Because of user personality, preferences and other factors, it has a very irregular data structure. Internet tourism data has formed the characteristics of big data and cross media tourism big data. Cross media tourism big data contains great application value and social benefits, and its related search, storage and knowledge mining become the research focus, and become a hot issue in the field of information retrieval and data mining [1].

© ICST Institute for Computer Sciences, Social Informatics and Telecommunications Engineering 2022
Published by Springer Nature Switzerland AG 2022. All Rights Reserved
S. Liu and X. Ma (Eds.): ADHIP 2021, LNICST 417, pp. 284–296, 2022.
https://doi.org/10.1007/978-3-030-94554-1_23

Based on the global cooperative information optimization algorithm, the paper summarizes many travel information, including text mining, Geographic Image Mining, travel route mining, and so on, and labels in detail, so that it can be applied in various scenes [2]. This information recommendation method needs to make detailed statistics and analysis of the data in the network, which takes a long time. The paper [3] through a recommendation model which combines content and relationship, uses support vector machine to train the data set of landmark scenic spots, and obtains a picture retrieval algorithm with high accuracy. This information recommendation method mainly focuses on the calculation of picture information, and has no good processing ability for text information. The literature [4] compares the past tourism information comprehensively with the personalized adaptive hybrid recommendation algorithm, and obtains a better semantic model of big data. The accuracy of push information is optimized. This method depends on the past data. If the error rate of existing data is large, it is difficult to guarantee the accuracy of the information itself.

This paper designs an improved clustering algorithm based on the above literature, and designs a fast recommendation method for personalized tourism big data information. The improved clustering algorithm is designed and the mathematical model of clustering algorithm is established. Calculate the semantic similarity of tourism information, reduce the retrieval range of the database based on the improved clustering algorithm, and establish the classification model of tourist attractions, so as to improve the running speed of tourism big data information recommendation method.

2 Design of a Fast Recommendation Method for Personalized Tourism Big Data Information Based on Improved Clustering Algorithm

2.1 Improved Clustering Algorithm

Fuzzy clustering algorithm can not be directly applied to the data set with missing data, so this paper proposes a completion algorithm which can be applied to the missing value filling. Firstly, the geometric structure of the data set with missing values is analyzed. Taking a simple two-dimensional data set as an example, the two-dimensional data set can be reflected in the plane coordinate system. For the two-dimensional data set X with missing values, for example, $N_k = (N_1, N_n)$ represents the missing point in the ordinate, which is represented by a straight line passing through point x_{k1} parallel to the ordinate in the plane, $N_k = (N_n, N_2)$ represents the missing point in the abscissa, which is represented by a straight line passing through point N_{k1} parallel to the abscissa in the plane, and $X_k = (X_{n1}, X_{n2})$ is meaningless in the clustering algorithm [5]. Therefore, the data set should not contain all attribute missing data groups. Therefore, for two two-dimensional datasets with missing values, the black dot represents the data points that are not missing, the straight line represents the data with missing attributes like $N_k = (N_{k1}, N_n)$, and the circle represents the clustering boundary of the corresponding complete dataset. If the cluster number $c = 2$ of each two-dimensional dataset with missing values is known, the missing values can be reasonably estimated. With the increase of data volume and dimension, missing data can be estimated more accurately.

From this, four missing data filling methods based on improved fuzzy clustering algorithm can be obtained, namely: complete data strategy, partial distance strategy, optimized completion strategy and best model strategy [6]. Among them, the optimized completion strategy is the best way to fill in missing data. Assume that the data sample set is $N_n = \{N_1, N_2, \cdots, N_n\}$, where N_k is the k-th w-dimensional data vector in the data set N, and satisfies $1 \leq k \leq n$. Among them, N_j is the jth value in the N_k vector, and $1 \leq j \leq s, 1 \leq k \leq n$. The data set N contains missing values, which can be represented by N_n, let:

$$N_M = \{N_j = N_n | 1 \leq j \leq s, 1 \leq k \leq n\} \tag{1}$$

The N_M in the formula is extremely missing the data set [7]. For example, when $k1 = 3B, k2 = 4CC$, the data set is set as:

$$N = \left\{ \begin{bmatrix} 1 \\ N_{k1} \\ 7 \end{bmatrix} \begin{bmatrix} 2 \\ 4 \\ 8 \end{bmatrix} \begin{bmatrix} 3 \\ 5 \\ 9 \end{bmatrix} \begin{bmatrix} N_{k2} \\ 6 \\ N_{k3} \end{bmatrix} \right\} \tag{2}$$

At this time, $N_n = \{n_1, n_2, n_3\}$ is the mathematical model of the clustering algorithm.

2.2 Semantic Similarity Calculation of Tourism Information

Firstly, the tourism information is preprocessed, and a vector composed of multiple search keywords is used to represent the tourism information similarity vector model to improve the accuracy of retrieval. On this basis, the travel information in the database is classified into temporal semantics, and its temporal semantic feature vector is extracted [8]. At this point, a vector model should be established first, so that the vector can represent tourism information in this feature space, and all the temporal semantic sentences related to tourism information in the database are extracted to establish this vector model. Each feature in the vector model has its feature value, which is the weight of the feature. Each temporal semantic keyword is one of the characteristic dimensions. Assuming that there are i total temporal semantic keywords in the database, the characteristic dimension in the database is also i. For a tourism information F_i, the frequency of temporal semantic keyword F_j in the database can be used to calculate the weight of temporal semantic keyword, and the similarity is used as its continuous measurement index [9]. The formula for calculating the frequency of tense semantic keywords is as follows:

$$F_{ij} = \frac{\delta_{ij}}{\lambda_j} \tag{3}$$

Among them, F_{ij} represents the frequency of occurrence of tense semantic keywords, δ_{ij} represents the number of occurrences of tense semantic keywords in travel information, and λ_{ij} represents the number of occurrences of keywords with the most frequent occurrence of tense semantic keywords in travel information. This formula is mainly used to calculate the relative frequency of a tense semantic keyword in travel information. The temporal semantic keywords appearing in the travel information are converted into a vector in the order of frequency, the content of the travel information is extracted, attributes and attribute values are extracted, and the temporal semantic vector describing the content of the travel information is described. Each travel information has different temporal semantic characteristics, so when calculating travel information similarity, the problem of conceptual attributes needs to be considered first. Suppose that the concept M has an instance m. At this time, the instance m can be represented as $m_y = S_{IN}[T_p]$, where $p = (p_1, p_2, \cdots, p_m)$ is the same in the instance n, $n_y = S_{IN}[T_q]$, $q = (q_1, q_2, \cdots, q_n)$ [10]. At this time, the similarity of instance n and m can be calculated. First, the attribute vector of instance m and n is a common attribute vector through the above method, and then the similarity of tourism information is calculated according to the attribute value, and the attribute values of the two instances are compared. And similarity, the obtained formula is shown below.

$$Sim_T(p, q) = \sum_{t=1}^{n} \frac{\delta_{ij} + \lambda_j}{2} \cdot Sim_T(p_1, q_1) \tag{4}$$

Among them, δ_{ij} and λ_j are the weight coefficients of attributes p_1 and q_1 in each vector, which are preset parameters. They are usually the statistical values obtained after preprocessing the tourism information. The final weight value of the tourism information is determined by this statistical value, and its value range is [0.1]. By substituting the above similarity into the semantics of tourism information, the similarity between the tourism information and the standard semantics can be calculated.

2.3 Reducing the Retrieval Range of Database Based on Improved Clustering Algorithm

The most important step of classifying text data is to retrieve the text data features. The mathematical model of text data classification established above will train any parameter in the model, which is also called text data preprocessing. In the process of preprocessing, the most difficult problem to be solved is the feature retrieval of text data, which can be solved efficiently by coding. In the process of coding, the way of improving the parameter connection coefficient and its behavior track in clustering model is studied by the form of reverse link. Although it is likely to ignore some sample data in calculating behavior trajectory, when learning efficiency is high to a certain extent, the number of missing samples can be ignored compared with the whole sample set. In the text data classification mathematical model based on improved clustering algorithm, corresponding parameters need to be established in advance. The change of the parameters selected after training is generally between 0.003–0.04. The parameter selection in this paper is 0.008. When the weight value attribute no longer changes, the maximum value of training parameter fitting can be achieved, that is, the retrieval of text data characteristics is realized. According to the different characteristics of the retrieved text data, the text data can be classified. Assuming that the data text of the input layer is $h_n = \{h_0, h_1, h_2, \cdots, h_n\}$, the text data of the input layer is converted to the hidden layer in turn, so that the data becomes $g_n = \{g_0, g_1, g_2, \cdots, g_n\}$, then, through the training model of the improved clustering algorithm, the hidden layer data is converted to the data text of the output layer through calculation. The calculation process is as follows:

$$\begin{cases} \delta_i = f(H_x y_i + H_y k_i + b_g) \\ \beta_i = H_z k_i + b_x + 1 \end{cases} \tag{5}$$

Among them, H_x is the conversion function from the input layer to the hidden layer, H_y is the conversion function within the hidden layer, H_z is the function from the hidden layer to the output layer, b_g is the deflection vector from the input layer to the hidden layer, and b_x is the deflection vector from the hidden layer to the output layer. Through formula (5), the update mode of input layer, hidden layer and output layer can be obtained, so that the state information of input layer and output layer at the previous time can be obtained by training hidden layer, and then the input and output text information at the current time can be obtained by combining with improved clustering algorithm, so as to achieve the purpose of extracting text information classification features. The improved clustering algorithm is used to train the objective function and redefine the changed function after passing through the hidden layer

$$\psi_x = -\sum_{i=1}^{n} X_i \log(x_i) \tag{6}$$

Among them, ψ_x represents the total number of text information to be classified, X_i represents the product of probability distribution and prediction value of each category after text information classification, and x_i represents the probability distribution value of each category after text information classification. The above function is a mathematical model of text data classification established by improving clustering algorithm.

2.4 Establishing the Classification Model of Tourist Attractions

Before extracting the classification features of tourist attractions, it is necessary to calculate the classification statistics of tourist attractions, and recognize the effect of automatic statistics of various tourist attractions in the classification process. In this process, there are the following parameters that will inevitably be affected. This paper uses the standard frequency, initial frequency, optimal frequency and base variable frequency to determine the classification basis of a scenic spot relative to the classification model of tourist attractions.

The first is about the change of standard frequency. In the process of eliminating standard frequency, signal overload often occurs. In order to eliminate the influence of this kind of signal, filter eliminator is usually used to reduce the signal frequency. If the short-time framing coefficient of a signal is set to $g(x)$, the adaptive function of the signal can be obtained

$$g(x) = \sum_{i=1}^{n} H_n(x) \cdot (x + g) \tag{7}$$

Where n is the number of frames of the signal; g represents the time parameter of signal elimination, and satisfies $k \in [0, n]$ generally, the value of k is between 60 Hz and 200 Hz;

$H_n(x)$ is the change range of signal windowing range function. The initial frequency is usually used to reflect the frequency frame of a tourist spot data sample, which plays a very important role in the expression of tourist spot data. When calculating the initial frequency, we can contact the classification model of tourist attractions through the specific tourism data information, and get the weighted sum of squares of the sampling points by calculating the energy mean

$$U_N = \sum_{i=1}^{N} H_i^2(\sigma_i) \tag{8}$$

Where U_N is the value of the initial frequency; N is the frame length of a tourist attraction data; i is the number of frames of this segment of scenic spot information; H_i is the average level of scenic spot information; σ_i is the electric energy calculation parameter of this section of scenic spot information. Since the average value of scenic spot information will be taken in the process of calculation, the data is an integer not less than zero. In daily calculation, the specific parameters of energy summation can be obtained in an open form.

The optimal frequency is usually due to the short-term continuous change of the scenic spot information in a certain period of time, resulting in a short-term zero energy phenomenon. This phenomenon can obviously lead to large differences between the data of various scenic spots, and this kind of phenomenon usually exists in the scenic spot information with large frequency fluctuation. The method of extracting the optimal frequency is very simple, which can be directly determined by the change of symbol

$$T_\beta = \sum_{i=1}^{n} |f_N(\beta) - \text{sgn}(\beta_0 - 1)| \tag{9}$$

Where, T_β is the frame length of the scenic spot information data with short-time zero crossing; n is the number of frames of this segment of scenic spot information; $f_N(\beta)$ is the frequency coefficient of the scenic spot information. Where sgn is usually a sign function, when $\beta_0 > 0$, the sign function value is 1, when $\beta_0 < 0$, the sign function value is 0. Most of the base variable frequency is based on the analysis of the function of tourist attractions. If the computing power of the classification algorithm of tourist attractions can be linearly and positively correlated with the information of tourist attractions, the frequency of the information of tourist attractions is low; If the computing power of the scenic spot algorithm can not be correlated with the scenic spot information data, the frequency of the scenic spot information data is higher. When a frequency filter is constructed based on this concept, it can be concluded that the frequency calculation formula of the scenic spot information data is shown in (4)

$$g(a) = x_0 \cdot \ln(1 + \frac{g(b)}{x_n}) \tag{10}$$

Where $g(a)$ is the function value of frequency filter; $g(b)$ is the calculation function of frequency; x_0 and x_n represent the initial and final terms of the function values respectively. By taking the logarithm analysis of the function value, the cosine transform parameters of the tourist attractions information data can be obtained. By accurately extracting the above four types of tourist attractions information features, we can establish a tourist attractions classification model.

2.5 Design Big Data Matching Algorithm for Tourist Attractions

To design a big data matching algorithm related to tourist attractions and tourist user information, we need to integrate the above model, and the specific algorithm structure is shown in Fig. 1.

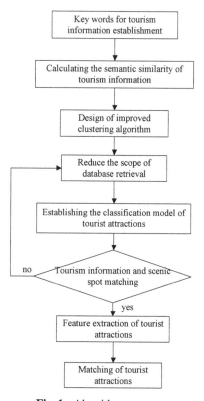

Fig. 1. Algorithm structure

As shown in Fig. 1, the classification model of tourist attractions and the similarity calculation method of semantic features of tourism information are established, and the two are matched to form the above big data matching algorithm of tourist attractions.

Taking the classification features in tourist attractions as the fourth-order dynamic error of the dynamic system, and returning the state of the system to zero, we can substitute formula (11) into the control equation to calculate its fault-tolerant approximation.

$$\lim_{x \to +\infty} a_x = b_i \tag{11}$$

Where, a_x is the node whose state is close to 0 in X system fault types; b_i is the deviation fault parameter of the system in the fault node. Through the above fuzzy logic approximation, the error of the basis vector function is calculated. In this process, the following conditions need to be ensured.

$$\frac{|s(a) - h(b)|}{2} \leq \lambda_n \tag{12}$$

Where $s(a)$ and $h(b)$ represent the approximation error of the basis vector function and the weight vector function; Represents the parametric properties of smooth functions. Based on the above formula, we can get the big data matching model of tourist attractions.

3   Experimental Study

3.1   Experimental Preparation

The main purpose of this experiment is to test the performance of the improved clustering algorithm designed above, and compare the data processing ability of the recommended method and the conventional method. Before the test, the environmental background of the experiment is explained. The test environment includes hardware environment and software environment. The details of hardware environment are shown in Table 1.

Table 1. Experimental test hardware environment

Hardware equipment	Parameter	Software required	Describe
Server under test	CPU (3.40 GHz), Memory is more than 4 G, and the total hard disk is more than 600 MB	Operating system	Windows Server2008&ubuntu 12.04
Network environment	Ethernet, or 100 m high speed network	Network protocol	IPV4
Test client	CPU (3.40 GHz), Memory is more than 4 G, and the total hard disk is more than 300 MB	Browser	Internet Explorer, Google browser, and Firefox

The software testing environment mainly includes the following professional test software, including software such as load runner, QTP, etc. The details are shown in Table 2.

Based on the above test environment, the data set is constructed. The big data of Tianchi tour guide competition is used in the data set, and several sub databases used in the algorithm test are constructed. The data set contains 1 million users' search information, 5 million scenic spots information and 30 million users' evaluation data. Delete the unqualified data, and establish a data set with 100000 user data, 1 million scenic spot data and 10 million evaluation information as the database of this experiment.

Table 2. Experimental test software environment

Name	Describe	Parameter
Operating system	Used to run the software	Include Windows xp/Windows 7/windows Server
LoadRunner	Test tools	Test model performance
QTP	Test tools	Test model security
JMeter	Test tools	Test the compressive strength of the model

3.2 Test of Improved Clustering Algorithm

For the evaluation of the improved clustering algorithm, we need to calculate the ability of data processing, the accuracy of information collection and the coverage of information search. The processing performance is used to evaluate the processing efficiency of the algorithm, which mainly measures the execution time of the algorithm, including the data preprocessing time. Accuracy is used to measure the proportion of correct items in the test set. The so-called correct items refer to the items that have appeared in the recommendation list and test set.

$$Precision = \frac{\sum_{i=1}^{n} {}_{u \in U} |R(u) \cap T(u)|}{\sum_{i=1}^{n} {}_{u \in U} |R(u)|} \tag{13}$$

Where *Precision* is the accuracy of travel information collection; $R(u)$ represents the travel information recommended by the user; U represents all user data sets; $T(u)$ represents the set of travel information that users like in the database.

$$Coverage = \frac{|U_{u \in U} R(u)|}{|I_i|} \tag{14}$$

Where $U_{u \in U}$ is the set of all user information in the database; *Coverage* is the coverage rate of information search; $R(u)$ represents the travel information recommended by the user; I_i is a collection of all travel information. Coverage is usually used to measure whether a recommender system has a strong ability to discover some unpopular items. It is generally calculated by the proportion of all recommended items and the probability distribution of all recommended items. The larger the proportion is, the more average the probability dispersion is. According to the above evaluation index, we can get the processing time of the algorithm, and then judge its performance. The algorithm is compared with the three conventional algorithms. The algorithm in this paper is taken as the experimental group, and the three conventional algorithms are taken as the control group. The optimization degree of the algorithm is judged as shown in Fig. 2.

By comparing the algorithms in Fig. 2, we can get the data results of running time. The initial running time of the experimental group is 8 s, which can reach 22 s when the number of users is 6 W. The running time of control group 1 was 4–58 s, that of control group 2 was 10–36 s, and that of control group 3 was 15–49 s. Thus, although

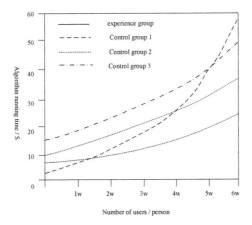

Fig. 2. Algorithm execution time test

the initial running time of the experimental group is less than that of the control group 1, the overall running time of the experimental group is less than that of the three control groups. With the increase of the number of users, the running speed of the experimental group will show a trend of rapid increase. Therefore, the improved clustering algorithm is better than the three conventional algorithms, and the running time is less than the conventional algorithm.

3.3 Actual Retrieval Effect Test

In the process of personalized tourism big data information search, the method designed in this paper makes use of the computing resources of big data platform, distributes the image feature extraction and visual vocabulary tree training process of consuming resources on the cloud computing platform, and provides users with search cloud service, which speeds up the search speed and improves users' experience of searching tourism information, To provide convenient tourism services for users. In order to test the performance of the tourism information search service designed in this paper and the three conventional methods, four data sets are designed to calculate the search time under different concurrent users and calculate the search speed. The test results are shown in Fig. 3.

As shown in Fig. 3, the retrieval time of the four data sets can be summarized as the data information shown in Table 3.

As shown in Table 3, in the four data sets, the retrieval time of the experimental group was less than that of the three control groups. It can be seen that the personalized tourism big data information fast recommendation method based on improved clustering algorithm designed in this paper has better retrieval efficiency and can quickly get the recommendation results.

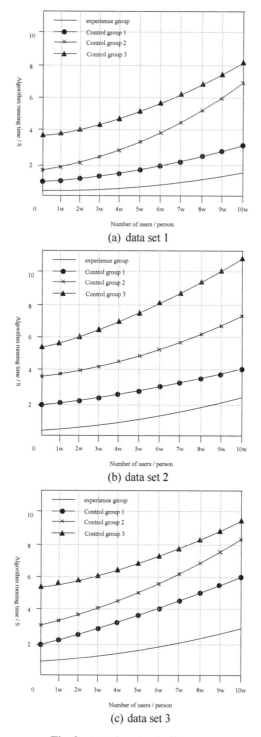

(a) data set 1

(b) data set 2

(c) data set 3

Fig. 3. Actual retrieval effect test

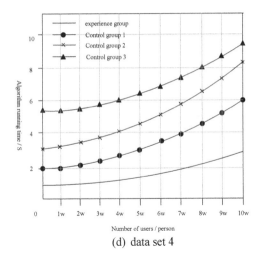

(d) data set 4

Fig. 3. continued

Table 3. Experimental results

Data set		Experience group	Control group 1	Control group 2	Control group 3
1	Maximum	0.36 s	1.98 s	1.86 s	3.76 s
	Minimum value	1.75 s	3.26 s	7.02 s	8.14 s
2	Maximum	0.36 s	2.01 s	3.67 s	5.35 s
	Minimum value	2.67 s	4.02 s	7.42 s	10.92 s
3	Maximum	1.09 s	2.00 s	3.14 s	5.47 s
	Minimum value	3.07 s	4.00 s	8.62 s	9.26 s
4	Maximum	0.84 s	2.02 s	3.26 s	5.65 s
	Minimum value	3.04 s	4.07 s	8.37 s	9.29 s

4   Conclusion

The research work of this paper mainly focuses on the problem of fast search and active push for big data personalized tourism. By fusing the travel information of tourists and the recorded scenic spot information, the travel information is matched with the scenic spot information on the semantic level to realize the rapid recommendation of personalized tourism big data information. However, this method has some shortcomings in tourism big data matching, such as the matching quantity and characteristics of tourism big data, which still need further research and improvement.

Fund Projects. National Planning Office of Philosophy and Social Science Foundation of China: Research on Cultural Heritage Conservation and the Activation of South China Historical Trail (NO: 19FSHB007).

References

1. Ma, R.X., Guo, F.Q., Liu, Z.J., et al.: Collaborative filtering recommendation algorithm for fusion context information and kernel density estimation. Comput. Technol. Dev. **31**(4), 34–39 (2021)
2. Tong, L.Y., Zhang, B.: Personalized recommendation algorithm based on global collaboration information. Value Eng. **40**(11), 233–234 (2021)
3. Wang, Y., Liu, L.: Construction of a classification model of network software test data based on feature expansion. Electron. Design Eng. **29**(8), 29–32, 37 (2021)
4. Liu, Y., Wei, M.: Personalized adaptive network hybrid information recommendation. Comput. Simulat. **38**(4), 399–402, 416 (2021)
5. Liu, S., Liu, D., Muhammad, K., Ding, W.: Effective template update mechanism in visual tracking with background clutter. Neurocomputing **458**, 615–625 (2020). https://doi.org/10.1016/j.neucom.2019.12.143
6. Liu, S., Liu, X., Wang, S., Muhammad, K.: Fuzzy-aided solution for out-of-view challenge in visual tracking under IoT assisted complex environment. Neural Comput. Appl. **33**(4), 1055–1065 (2021)
7. Wang, Z., Wang, S., Du, H.: Improved fuzzy C-means clustering algorithm based on density-sensitive distance. Comput. Eng. **47**(5), 88–96, 103 (2021)
8. Gao, P., Li, J., Liu, S.: An introduction to key technology in artificial intelligence and big data driven e-Learning and e-Education. Mob. Netw. Appl. **26**(5), 2123–2126 (2021). https://doi.org/10.1007/s11036-021-01777-7
9. Sun, Q., Chen, H., Li, C.: Clustering algorithm of big data based on improved artificial bee colony algorithm and MapReduce. Appl. Res. Comput. **37**(6), 1707–1710, 1764 (2020)
10. Feng, J., Yao, Y.: An optimization clustering algorithm based on multi-population genetic simulated annealing algorithm. Comput. Simulat. **37**(9), 226–230 (2020)

Research on Monitoring System of Ocean Observation Buoy Based on Multi-sensor

Xing-kui Yan[1,2,3(✉)] and Huan-Yu Zhao[1,2,3]

[1] Institute of Oceanographic Instrumentation, Qilu University of Technology (Shandong Academy of Sciences), Qingdao, China

[2] Shandong Provincial Key Laboratory of Ocean Environmental Monitoring Technology, Qingdao, China

[3] National Engineering and Technological Research Center of Marine Monitoring Equipment, Qingdao, China

Abstract. With the continuous development of my country's marine three-dimensional monitoring business, it has put forward higher requirements for the stability and reliability of marine monitoring buoy communication equipment. In order to meet the above requirements, this study designed a multi-sensor-based marine observation buoy monitoring system. Optimize the system hardware structure by optimizing the multi-sensor function framework, and then improve the system software operation process to improve the system positioning accuracy and precision. Experimental results show that the multi-sensor-based marine observation buoy monitoring system is highly effective in practical applications and can provide a better reference for the three-dimensional monitoring business of my country's marine environment.

Keywords: Multi-sensor · Ocean observation · Buoy monitoring · Positioning accuracy · Precision

1 Introduction

With the development of economy, human activities on the sea are becoming more and more frequent, while marine environmental problems have become increasingly prominent. In addition to being affected by various marine environmental disasters such as marine pollution and marine red tides, my country's coastal areas are also affected by disasters caused by typhoons, ocean waves, storm surges, sea ice and other dynamic phenomena, making it a country with more serious marine disasters in the world one. Vigorously developing the cutting-edge technology of ocean monitoring can not only improve the original innovation ability of the national ocean monitoring technology, but also lay a solid technical foundation for the sustainable development of other ocean technologies [1].

Buoy monitoring is the main method of ocean observation and an important part of the marine environment three-dimensional monitoring network. It has the characteristics

S. Liu and X. Ma (Eds.): ADHIP 2021, LNICST 417, pp. 297–312, 2022.
https://doi.org/10.1007/978-3-030-94554-1_24

of strong resistance to harsh environments, large capacity, long in-place time and long life, and strong ability to resist man-made damage. During the process of severe weather such as storm surge and typhoon, it can obtain valuable hydrometeorological data and provide data support for the study of severe weather process. It has become the national marine monitoring, marine military rights protection, environmental protection, disaster reduction and prevention, petroleum and biological resources important carriers such as exploration and development, harbor construction, fishery fishing, and marine aquaculture projects cannot be replaced by other ocean observation methods [2]. Therefore, the design and analysis of the marine buoy monitoring system can better promote the development of marine monitoring technology, promote the development and utilization of the ocean, and contribute to economic development and social progress.

In the process of marine buoy monitoring, many projects require far-sea, long-term data monitoring, which requires a lot of manpower and material resources [3]. The data communication function of the multi-sensor can solve the problem of ocean monitoring data communication. Not only can the monitoring data be easily obtained through the ground station, but also the two-way communication function can be used to remotely control the equipment thousands of miles away.

Based on the above research background, this paper designs an ocean observation buoy monitoring system based on multi-sensor, in order to provide better help for Marine environment three-dimensional monitoring business in China.

2 Design of the Monitoring System of Ocean Observation Buoy

2.1 Hardware Configuration of Marine Observation Buoy Monitoring System

Ocean monitoring technology is a comprehensive high-tech formed by integrating computers, information and sensors, databases, remote communications and other disciplines. It integrates the development results of multiple disciplines and represents the frontier of high-tech development. It follows the development of related disciplines and technologies. Development and rapid development [4]. A monitoring system is designed for the multi-parameter ocean buoy, including the buoy system, the shore station receiving system and the upper computer, etc. The schematic diagram is shown in Fig. 1.

The buoy management unit is mainly divided into two parts: the data acquisition system and the buoy body [5]. The data acquisition system includes acquisition main control service module, acquisition array element, cable array and sensor communication terminal. The acquisition main control service module and battery are placed in the buoy body cabin, and its structure is shown in Fig. 2.

The traditional monitoring system refers to a computer system with data collection, monitoring, and control functions. The various production parameters collected in marine environment monitoring and common industrial monitoring systems are essentially the same, but the application environment is different. In the industrial production site, the monitoring system usually connects the monitoring hosts distributed geographically by wire [6]. In the marine environment monitoring system, since the monitoring values of hydrological elements, chemical elements, and algae density should be obtained directly

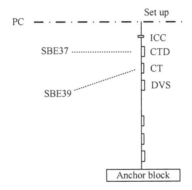

Fig. 1. Sensor detection structure of marine buoy monitoring system

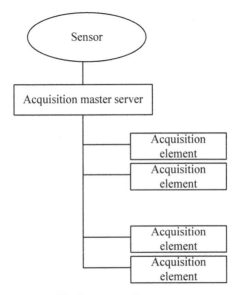

Fig. 2. Buoy unit structure

from seawater, the monitoring equipment must work in the seawater environment, and marine stations, as one of the monitoring equipment, are usually located on the coast.

In order to improve the scientificity and accuracy of the monitoring results, the monitoring coverage area of the marine monitoring system should be as large as possible, so the distance between the monitoring points is measured in kilometers. It is obviously impractical to connect buoys with a large distance in the ocean and to connect the buoys with the stations on the shore in a wired manner [7]. It can be seen that the wired connection method usually used in the monitoring system cannot meet the needs of the marine monitoring environment [8]. Therefore, this design will select the GPRS wireless communication method widely used in remote monitoring system equipment.

The configuration of the system hardware structure is shown in Fig. 3.

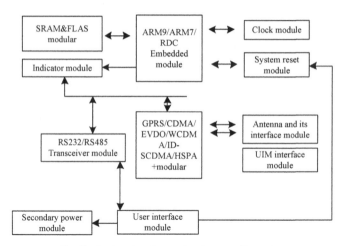

Fig. 3. System hardware structure configuration

In the system client/server, abbreviated as the C/S structure of the database system, the data processing task is divided into two parts: one part runs on the client side, and the other part runs on the server side [9]. There can be multiple division schemes. A commonly used scheme is: the client is responsible for application processing, and the database server performs the core functions of the DBMS. There are multiple hardware in the system, and the communication interface of each hardware is not the same. The specific interface between each module and the microprocessor STM32F407ZGT6 is shown in Table 1.

Table 1. Function description of each module

Modular	Model	Function description
Magnetic sensor	MicroMag3	Used to detect ships close to buoys
Infrared sensor	LH1778	It is used to detect human body near buoy
Stepper motor drive	THB7128	Used to drive stepping motor
Microprocessor	STM32F407ZGT6	Used for data acquisition, analysis and system control
TF Card		Used to store data
Electronic compass	HMC5883L	Used to output angle information and correct camera position
Camera	OV2640	It is used to collect the image around the buoy
Relay circuit	JZC-32F-012	Switch of sound light alarm
Power management circuit	TSPIH100	Used to detect output voltage and current of battery

In the C/S structure, client software and server software can run on one computer, but most of them run on different computers in the network. Client-side software generally runs on a PC, and server-side software can run on various types of computers from PCs to mainframes. The database server separates the data processing tasks to run on the client and the server, thus making full use of the server's high-performance database processing capabilities and the client's flexible data representation capabilities [10]. Usually, only query requests are sent from the client to the database server, and only the query results are sent back from the database server to the client. There is no need to transmit the entire file, which greatly reduces the amount of data transmission on the network [11]. It is given by the voltage of the CS pin. In this design, the CS pin is directly connected to the A/D mapping pin PA7 of the microprocessor, and the voltage value is obtained by sampling the voltage of the CS pin, and then calculating according to formula 1 to obtain the current of the power supply Value, the formula is as follows:

$$\text{Iout} = \frac{V_{cs} \times k}{R_{cs}} \tag{1}$$

Among them, k is the ratio of the output current to the sensing current, which is a fixed value, generally 500, and R_{cs} is the external resistor of the pin CS, that is, R34 in the circuit diagram. In order to achieve higher monitoring accuracy, R_{cs} needs to select a resistor with an accuracy of 1% or higher, and the value of V_{cs} ranges from 0 to 4 V. TPS1H100-Q1 also has a programmable current limit function. When the current exceeds the set value, the TPS1H100-Q1 will automatically disconnect the power supply. The current value setting can be calculated according to the formula, the formula is as follows:

$$R_{cl} = \frac{V_{cl} \times k_{cl}}{I_{out}} \tag{2}$$

The hardware structure mainly includes client and server. In this model, the client must install applications and tools, making the client too large and burdensome, and difficult to install, maintain, upgrade, and release the system, thereby affecting efficiency [12]. The signal generated by each sensor is automatically processed by the instrument. Its main contents include: the main body of the buoy, the mooring system, the protection device, the sensor, the data acquisition, storage and transmission module, the design and selection of the power supply and distribution module; Time is continuously recorded, unattended [13]. In view of the powerful features of LabVIEW software, flexible programming, and friendly man-machine interface, we choose it as a front-end development tool, while SQL Server has powerful data management functions. As a network database, the system functions that the monitoring system can achieve are shown in Fig. 4.

The marine environment monitoring buoy system requires the monitoring center to realize real-time connection with each data information collection buoy. Due to the large number of buoy placement points, the system requirements can meet the needs of burst data transmission, and GPRS technology can well meet the needs of burst data transmission; because the system adopts a mature TCP/IP communication architecture, it has a good With extended performance, a monitoring center can easily support the communication access of thousands of meteorological collection points [14, 15]. Although it

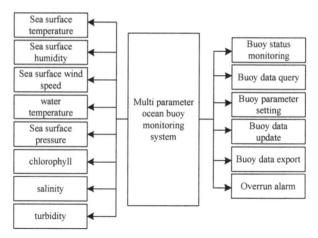

Fig. 4. Functional structure of marine buoy monitoring system

has two-way communication with a data center thousands of miles away, GPRS data transmission equipment only needs to communicate with nearby mobile base stations when it is working. Its overall power consumption is equivalent to that of an ordinary GSM mobile phone, with an average power consumption of only 200 ms. About watts, much smaller than traditional digital radio.

2.2 Software Process Optimization of Marine Buoy Monitoring System

Through the use of interrupt program design to control the operation of each subroutine, and then realize the function of each submodule. The main program of the system base station and the mobile station is basically the same. When the system starts to work, it will initialize the various hardware subsystems, and at the same time enter the waiting interrupt loop link. The interrupt request turns off the standby state, executes the interrupt program, and returns to the original standby low-power state after the interrupt program is executed. After the interrupt program is executed, the interrupt flag will be cleared, the interrupt will be exited, and the next interrupt will be entered after entering the low-power mode [16]. In summary, the main program completes data collection, communication, storage, remote transmission and other program operation functions by interrupting the work. The main program flow chart is shown as in Fig. 5.

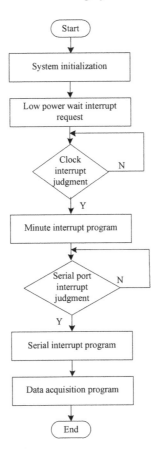

Fig. 5. Optimization of the running process of the main program of the system

According to the requirements of real-time automatic monitoring of the system, this paper proposes the following design schemes accordingly: It has the advantages of high degree of automatic integration, high positioning accuracy, safe and reliable data transmission, and improved operation efficiency. The system is mainly composed of three subsystem modules: buoy data acquisition module, data communication module and data processing analysis module. Its structure and operation principle are shown in Fig. 6.

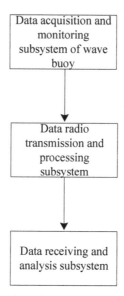

Fig. 6. Operating principle of the system structure

A data compression technology is proposed. As shown in the figure, the communication data is firstly compressed losslessly. In order to prevent data loss, the sender regularly queries the receiver and finds that the lost data is retransmitted in time to improve the reliability of data communication. Based on this, the buoy monitoring data compression process is further optimized, as shown in Fig. 7.

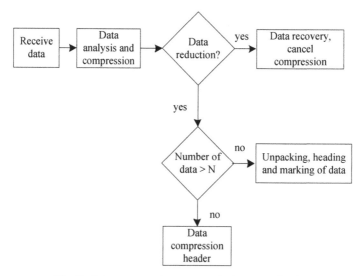

Fig. 7. Data compression process for buoy monitoring

The data conversion protocol is a protocol that converts collected data into data suitable for transmission by the sensor terminal. The conversion method is as follows: The collected data is divided into a group of 3 bytes in order, so that each group has 24 bits of data, and then each 6 bits of data is divided into 1 part in order, so that each group gets 4 parts, each part the value range of is 0 to 64. Add 48 to this value, and its range is: 48–112. This value range corresponds to the visible character part of the ASCII table, so that the collected data can ensure the correct transmission in the sensor communication terminal. The data conversion diagram is shown in Fig. 8.

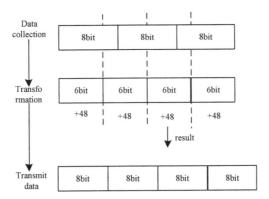

Fig. 8. Conversion of collected data to transmission data

The central station monitoring software needs to realize the functions of sending commands, receiving data, auditing data, storing data, displaying real-time data, querying historical data, generating messages, managing and monitoring the system, as shown in Fig. 9.

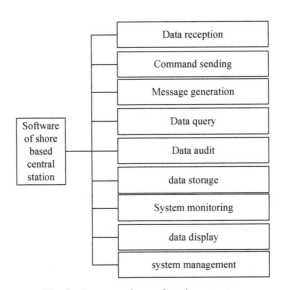

Fig. 9. System software function structure

The software needs to have the two most important functions. One is data communication, that is, sending commands to the hydrological monitoring buoy and reading data. The second is data storage, which stores all observation data in the database. Therefore, the shore-based central station software has two interfaces: communication interface and database interface. In addition to data communication and data storage, the software should also have functions such as data conversion, message generation, system management and monitoring. Data conversion is to convert and calculate the original data in accordance with relevant specifications to obtain the required result data; messages are data files generated in accordance with the specified format, used to send data to relevant application systems or relevant departments; system management and monitoring are for Centralize management and configuration of all on-site observation station equipment in the entire system, and monitor the operating status of the system. The overall structure of the shore-based central station monitoring software is shown in Fig. 10.

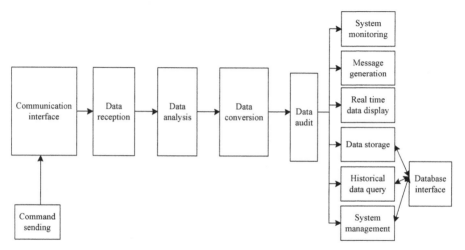

Fig. 10. The overall structure of the monitoring software

The BDS buoy wave measurement system is mainly composed of two parts: a reference station and a rover station. Relatively independent base station and mobile station make data intercommunication work through the radio, and the hardware circuit turns on the base station and mobile station power through the power control interface according to the predetermined sampling time interval. The mobile station performs differential processing between the carrier signal and observation data collected by the station and the carrier signal and data transmitted by the base station to obtain high-precision positioning coordinates of the mobile station in real time and store them in SRAM. The mobile station then sends the real-time high-precision vertical displacement of the mobile station to the shore station receiving system in the form of a message through the radio module. The fixed part of the whole week of the unknown number solution is the key to the whole process. As long as the observation value that satisfies the maintaining phase is tracked by more than four satellites and the unknown number is fixed, each epoch can be processed in real time. The mobile station can achieve centimeter-level positioning. The working data link network is shown in Fig. 11.

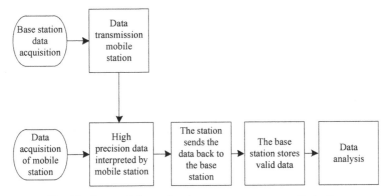

Fig. 11. System information management process optimization

When the buoy works offshore, the buoy communicates with the base station through the radio, and the communication module transmits the collected raw data through the serial port. The data processing module is the core of the program, which is responsible for filtering and correcting the original data collected by the buoy. Draw a two-dimensional graph of the wave surface movement, which is convenient for users to understand and analyze the wave graph. This software can save relevant data in the background and store it in the form of an array. Buoy data processing and ocean wave inversion programs can read relevant data in the background respectively. The signal is processed to obtain the vertical displacement of the wave, and the high-precision wave surface recording data is fitted to improve the operating efficiency of the system.

2.3 Realization of Marine Observation Buoy Monitoring

Combining the expected sea surface monitoring environment and actual conditions, in order to limit the influence of the normal operation of the system, creatively proposed the following basic requirements for the hardware: First, the floating buoy model on the sea must withstand unfavorable factors such as tsunamis, strong winds and waves, and low temperature., And strictly guarantee the tightness of the model at the same time. Secondly, the buoy will be used in the offshore waters, and it will monitor the fluctuation of sea waves for a long time in the area where it is put into use. Therefore, the system has strict requirements for low power consumption and long-term operation of the power supply. Real-time reception of buoy monitoring data, so the independently designed communication module must be stable. The satellite positioning buoy uses carrier phase difference. The principle of carrier phase difference is composed of a reference station and one or more mobile stations. The reference station transmits the three-dimensional raw data of the measured base station to the mobile station. The observation data of the mobile station and the reference station form at least four double-difference unknown equations; the unknown error amount can be eliminated in the solution process, which can reduce the orbit error and the atmosphere Errors and so on can then accurately measure the three-dimensional coordinates of the mobile station.

$$\varphi_j^p = \frac{\sqrt{(X^p - X_j)^2 + (Y^p - Y_i)^2 + (Z^p - Z_j)^2}}{cV_{t,} + cV_t - N_j^P\lambda - (V_{iw})_j^P - (V_{tap})_i^P} \tag{3}$$

According to the signal of the infrared sensor, the microprocessor judges whether a person is approaching by driving the camera to the place where the person and the ship are approaching. After detecting that someone is approaching, the system will start the stepping motor. After reaching the designated position, the system will start the camera to take pictures. After completion, the image data will be sent to the shore receiving system through the wireless data transmission radio. After the data is sent, it becomes an automatic alarm. Process, automatically enter the start position. Suppose the geometric distance from the satellite p to the receiver is:

$$\rho_j^p = \sqrt{(X^p - X_j)^2 + (Y^p - Y_j)^2 + (Z^p - Z_j)^2} \tag{4}$$

Therefore, the vector direction cosine from the satellite to the receiver i is:

$$\frac{\partial \rho_j'}{\partial X^\rho} = \frac{X'' - X_{j0}}{\rho_{j0}'} = l_p' \tag{5}$$

$$\frac{\partial \rho_j''}{\partial Y^\rho} = \frac{Y'' - Y_{j0}}{\rho_{j0}^P} = m_\rho^j \tag{6}$$

$$\frac{\partial \rho_j^P}{\partial Z^\rho} = \frac{Z^p - Z_{j0}}{\rho_{j0}^P} = n_p^j \tag{7}$$

The carrier phase observation equation is linearly expressed as:

$$\varphi_j'\lambda = \rho_{j0}^P - cV_t + cV_{tp} - N_i^P\lambda - (V_{imA})_j^P - (V_{top})_i^P \tag{8}$$

Therefore, in the case of the same frequency carrier, the observation equations of the base station and the mobile station are:

$$\varphi_i^P\lambda = \rho_{i0}^P - cV_{t,} + cV_{tp} - N_i^P\lambda - (V_{ion})_i^P - (V_{trop})_i^P \tag{9}$$

$$\varphi_i'\lambda = \rho_{i0}^P - cV_t, +cV_t, -N_j^P\lambda - (V_{ian})_i^P - (V_{top})_j^m \tag{10}$$

The detection distances of the reference station and the mobile station relative to the buoy can be differentiated respectively, and the double difference observation equation of the two satellites can be obtained:

$$\Delta\varphi_{ij}^{Ai}\lambda = (l_i^P - l_i^o, m_i^P - m_i^o, n_i^P - n_i^o)(\partial x_j, \delta Y_j, j - \Delta N_i^{P0}\lambda) \tag{11}$$

The least squares method of ambiguity floating point solution, rewrite the formula as follows:

$$y = Aa + Bb = \begin{bmatrix} A & B \end{bmatrix}\begin{bmatrix} a \\ b \end{bmatrix} \tag{12}$$

According to the demand analysis and research goals of the system, the overall design of the system is carried out. The overall structure of the system is divided into a buoy unit and a shore station monitoring unit. The buoy unit includes a sensor group, data acquisition and data processing modules, and the shore station monitoring unit includes a shore-based monitoring center and a buoy remote monitoring system. According to the overall design analysis of the buoy unit, the overall composition of the buoy unit is divided into two parts: the data acquisition system and the buoy body. The data acquisition system is composed of acquisition main control service module, acquisition array element terminal, acquisition system cable array and sensor communication terminal. Among them, the acquisition main control service module mainly includes the main control CPU, RS232 interface circuit, CAN bus interface circuit, power supply module and so on. The acquisition array element terminal mainly includes cabin structure design and electronic system design. Secondly, the design index analysis and the structure design of the buoy body are carried out. According to the overall design analysis of the shore-based unit, the overall composition of the shore-based unit is divided into the shore-based unit hardware system and the shore-based unit software system. First of all, the hardware system is mainly composed of database server, data acquisition terminal and sensor communication terminal. Secondly, the shore-based unit software system mainly includes the shore-based unit software system architecture, function overview, functional module composition, database tables, and the design of various functional modules to ensure the quality of system operation.

3   Analysis of Results

In order to verify the practical application effect of the multi-sensor-based ocean observation buoy monitoring system, the following experimental tests were carried out.

The acquisition array terminal is placed in the sealed pressure tank, the sealed pressure tank is connected with a test port of the manual water pressure source through the connecting pipe, and the other test port is connected with the precision pressure gauge, and then the manual water pressure source is injected into the tap water. After pressurizing by manual pressure valve and micro pressure valve of water pressure source, the data of precision digital pressure gauge and the data collected by acquisition array terminal are read out and compared. This collection array terminal pressure collection and testing experiment started from SOKPa pressure points, with a collection point every SOKPa. A total of 40 pressure collection points were collected, and 200 sets of data were collected at each collection point.

A buoy is placed in a small scale wave tank, and the anchor chain link plate on the underside of the buoy is connected with the small scale tank. The purpose is to record the vertical displacement change of the water quality point at the fixed position measured by the buoy in the wave tank.

The frequency of the experimental sampled data is 5 Hz, and a buoy is placed 30 m away from the wave building plate. Comparable to test results, verify the accuracy of RTK technology buoy wave and the stability of the system, the selection model BG-1 high resistive wave sensor calibration, the measurement of the depth of the water for 3 m, the range of measurement of A resistive sensor resolution size was mainly affected by A/D converter circuit, the resistance sensor measurement accuracy theoretically to 1 mm.

RTK differential technique is used to measure the time series of buoy vertical displacement. The measured wave displacement signal series mainly includes wave signal and observation noise signal. A series of vertical displacement data of the wave surface obtained by the buoy is compared with the wave height and period of the resistive wave height sensor using the power period method. The comparison results are shown in Table 2 and Table 3.

Table 2. Resistive sensors and buoy regular waves

Wave measurement method	$H_{1/100}$/cm	H_{mean}/cm	$H_{1/3}$/cm	Tmean
RTK buoy wave measurement	83.4983	51.2202	83.4983	4.9673
Wave measurement with resistance sensor	82.6652	49.4253	82.6652	4.0227

Table 3. Irregular waves of resistive sensors and buoys

Wave measurement method	$H_{1/100}$/cm	H_{mean}/cm	$H_{1/3}$/cm	Tmean
RTK buoy wave measurement	110.7536	48.9234	78.6325	4.9683
Wave measurement with resistance sensor	108.5286	46.7288	76.4257	4.7864

Based on the results of the above table, a comparative analysis was performed, the standard deviation of the sensor terminal test was standardized, and the collected data of the collection element terminal was compared with the data calculated by the output curve of the collection element terminal, and the standard deviation was calculated as shown in Fig. 12.

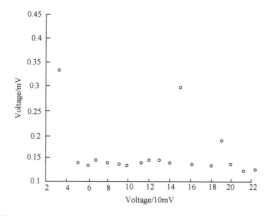

Fig. 12. The standard deviation of the sensor terminal test

Further compared with the traditional detection system, the detection data were adjusted down with an interval of 0.1 degree. After collecting the data, the data curve is obtained by fitting the data. On this basis, the collected data and quasi and curve contrast, with horizontal resistance, arrays of collection terminal to conduct stress tests, and contrast arrays arrays acquisition terminal and the data acquisition terminal output curve to calculate the data, calculate the standard deviation between them, to verify the correctness of the whole process from data collection to data warehousing and reliability, the specific test results are shown in Fig. 13.

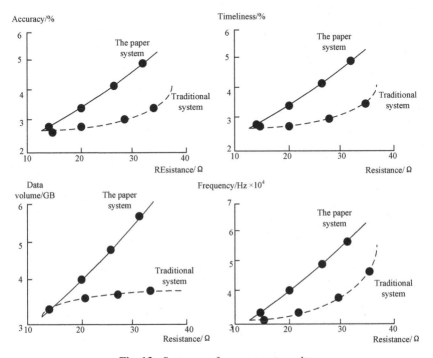

Fig. 13. System performance test results

Based on the comparative analysis of the detection results shown in Fig. 13, it can be seen that compared with the traditional sea breeze buoy detection system, the system performance in this paper is significantly better in the actual application process, which fully meets the research requirements.

4   Conclusion and Outlook

As a satellite navigation system with independent intellectual property rights in my country, multi-sensor has the characteristics of small investment, simple and cheap user equipment, high reliability, and strong confidentiality. Its appearance provides a way for the application of satellite communication and positioning systems in marine monitoring buoys. The new choice lays a technical foundation for the use of multiple sensors for ocean observation data transmission. Therefore, this research combines the characteristics of marine buoys and multi-sensors to design a multi-sensor-based marine monitoring buoy communication system.

Although the system has achieved some achievements, it still has some shortcomings, such as little communication data and data loss. Therefore, targeted measures should be taken to apply multi-sensor to Marine monitoring system in future research.

References

1. Shrestha, B., Saxena, N., Truong, H.T.T., Asokan, N.: Sensor-based proximity detection in the face of active adversaries. IEEE Trans. Mob. Comput. **18**(2), 444–457 (2019)
2. Sumorek, A., Jamińska-Gadomska, P., Lipecki, T.: Influence of ultrasonic wind sensor position on measurement accuracy under full-scale conditions. Sensors **20**(19), 1–17 (2020)
3. Maier, G., Pfaff, F., Wagner, M., et al.: Real-time multitarget tracking for sensor-based sorting: a new implementation of the auction algorithm for graphics processing units. J. Real-Time Image Proc. **16**(6), 2261–2272 (2019)
4. Zheng, Y., Li, D., Wang, L., et al.: Robustness of the planning algorithm for ocean observation tasks. Int. J. Performab. Eng. **16**(4), 629–635 (2020)
5. Purser, A., Marcon, Y., Dreutter, S., et al.: Ocean Floor Observation and Bathymetry System (OFOBS): a new towed camera/sonar system for deep-sea habitat surveys. IEEE J. Ocean. Eng. **44**(1), 87–99 (2019)
6. Venkatesan, R., Sannasiraj, S.A., Ramanamurthy, M.V., et al.: Development and performance validation of a cylindrical buoy for deep-ocean tsunami monitoring. IEEE J. Ocean. Eng. **44**(2), 415–423 (2019)
7. Zhang, H., Zhang, D., Zhang, A.: An innovative multifunctional buoy design for monitoring continuous environmental dynamics at Tianjin Port. IEEE Access **8**(11), 820–833 (2020)
8. Liu, S., Liu, X., Wang, S., Muhammad, K.: Fuzzy-aided solution for out-of-view challenge in visual tracking under IoT assisted complex environment. Neural Comput. Appl. **33**(4), 1055–1065 (2021)
9. Knight, P.J., Cai, O.B., Sinclair, A., et al.: A low-cost GNSS buoy platform for measuring coastal sea levels. Ocean Eng. **203**(9), 107–118 (2020)
10. Zhang, Y., Shang, H., Zhang, X., et al.: Development and sea test results of a deep-sea tsunami warning buoy system. Mar. Technol. Soc. J. **53**(3), 6–15 (2019)
11. Carlson, D.F., Pavalko, W.J., Petersen, D., et al.: Maker buoy variants for water level monitoring and tracking drifting objects in remote areas of Greenland. Sensors **20**(5), 1254–1255 (2020)
12. Johnson, J.E., Hooper, E., Welch, D.J.: Community marine monitoring toolkit: a tool developed in the pacific to inform community-based marine resource management. Mar. Pollut. Bull. **159**(9), 111–118 (2020)
13. Liu, S., Li, Z., Zhang, Y., Cheng, X.: Introduction of key problems in long-distance learning and training. Mob. Netw. Appl. **24**(1), 1–4 (2018)
14. Liu, S., Sun, G., Fu, W. (eds.): eLEOT 2020. LNICSSITE, vol. 339. Springer, Cham (2020). https://doi.org/10.1007/978-3-030-63952-5
15. Liu, G., Rui, G., Tian, W., et al.: Compressed sensing of 3D marine environment monitoring data based on spatiotemporal correlation. IEEE Access **12**(9), 1–10 (2021)
16. Zhang, S.W., Yang, W.C., Xin, Y.Z., et al.: Research progress of a mooring buoy system for sea surface and seafloor observation. Chin. Sci. Bull. **64**(28), 2963–2973 (2019)

Research on Context-Aware Recommendation Algorithm Based on RFID Application

Yan Zhao$^{(\boxtimes)}$

Ningbo City College of Vocational Technology, Ningbo 315199, China

Abstract. With the development of mobile Internet, more and more users get information through mobile terminals. However, the current personalized recommendation system is lack of the ability to perceive the user's situation and provide personalized information recommendation service for users. For this reason, this paper proposes a context aware recommendation algorithm based on RFID application, which collects user resource category preference learning features by combining with context features, and integrates different category preferences for collaborative filtering personalized information recommendation. The RFID method is used to learn the user's preference for each resource category in different contexts, and then the category preference is combined with the traditional RFID context-aware recommendation algorithm to generate personalized information recommendations that are in line with the user's current context. Experiments show that the context-aware recommendation algorithm proposed in this paper based on RFID applications can improve the accuracy of recommendation.

Keywords: RFID · Situational awareness · Information recommendation · Recommendation algorithm

1 Introduction

With the rapid development of Internet technology, the total amount of network resources is growing at an exponential rate, and it becomes more and more difficult for users to find the resources they need. One way to solve this problem is personalized information recommendation. In addition, due to the limited display size and processing capacity of mobile terminals such as smart phones, the above problems become more prominent. Traditional recommendation algorithms only consider two dimensions of users and resources in the recommendation process [1]. In different situations, the environment where the user is located is unstable, and these unstable factors affect the information demand of the user. For example, for information reading in a mobile environment, the network condition affects whether the user will choose multimedia resources. Therefore, it is necessary to expand the two dimensions of user and resource into multiple dimensions of user, resource and context to provide information services that conform to the user's current context [2]. In order to ensure the quality of personalized recommendation in a mobile environment, context is a factor that cannot be ignored. However,

© ICST Institute for Computer Sciences, Social Informatics and Telecommunications Engineering 2022
Published by Springer Nature Switzerland AG 2022. All Rights Reserved
S. Liu and X. Ma (Eds.): ADHIP 2021, LNICST 417, pp. 313–327, 2022.
https://doi.org/10.1007/978-3-030-94554-1_25

different contextual factors need to be considered for different applications, and there is currently a lack of unified models and algorithms that integrate contextual information into traditional recommendations. Therefore, this paper discusses the context awareness in the field of personalized reading, and proposes a context based user resource category preference learning based on RFID method and a collaborative filtering personalized information recommendation integrating the category preference, so that the traditional collaborative filtering has the ability of context awareness and can provide users with personalized resources in line with the current situation.

2   Context-Aware Recommendation Algorithm

2.1   Situational Awareness Information Category Planning

RFID context awareness recommendation algorithm can be divided into two directions: user based and project-based. The recommendation to target users based on user interest points is based on user collaborative filtering, among which neighbor users are users close to the target user scoring mode [3]. The collaborative filtering recommendation method based on the project first analyzes the scores of neighbor users, and then predicts the target score according to the score. The following figure describes the basic classification of RFID context awareness recommendation algorithm (Fig. 1).

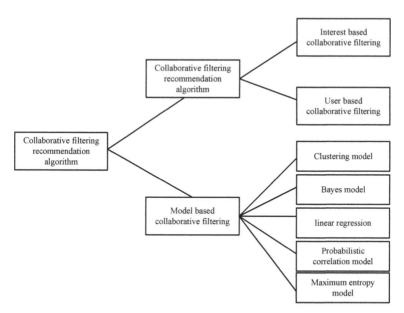

Fig. 1. Context-aware collaborative filtering information classification

The basic idea of user based collaborative filtering algorithm is to assume that users have similar behaviors, which means that some users have similar evaluation or operation on some items, while others still show similar evaluation [4]. This kind of people can be considered as neighbor users, and the similarity between non target users and target users can be calculated in the process of research.

Situational acquisition: This part mainly includes the content of acquiring situational information through a variety of ways.

Situation representation: Situation representation, as the name suggests, refers to the effective organization and storage of the collected situation information in various ways.

Situational use: Mainly the specific use of situational information.

In a word, situational perception is a new intelligent computing form, which senses the surrounding environment and makes functional response by simulating human cognition to achieve specific goals. Situational perception technology is originated from the research of pervasive computing, which is generally considered to collect and convert the current environment information of the equipment into a specific format through sensor technology or means, so as to analyze and process the system and better adapt to or monitor the environment. Generally speaking, situational perception system consists of the following steps (Fig. 2):

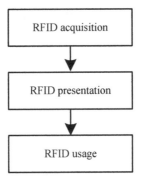

Fig. 2. Context-aware information processing flow

From the perspective of situational awareness processing, the subject can be divided into two parts: situational information utilization and organization. The purpose of the former is to specifically obtain information about the situation in order to put it into application. The purpose of the latter is to collect contextual information, then abstract it into a contextual model, give contextual explanations, and finally store contextual information for retrieval, etc. [6]. The current situational information collected is generally obtained through third parties such as mobile networks, the Internet of Things, digital maps, and weather. The situational abstraction first needs to clean up the collected situational information, including sampling, averaging, statistics and calibration, and then merge and classify different situational information for subsequent analysis and processing of the data [7]. Context interpretation is to obtain the semantic information hidden by context features and organize the context into knowledge, including the possible future behavior of users. Finally, the processed context information is stored and managed in a centralized or distributed way for future retrieval and utilization. As shown in the figure (Fig. 3):

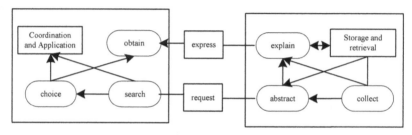

Fig. 3. User preference perception processing steps

It is generally believed that user preferences are used to describe the decision-maker's ranking of multiple projects. Users have different preferences in different situations. The introduction of context information into user preference analysis model has become one of the key technologies of user preference analysis based on context awareness. The main research methods are qualitative and quantitative analysis [8]. The former mainly considers the binary partial order relationship between users and items and attributes under the current situational constraints, and uses logical reasoning to mine the partial order relationship between users on any two items or attributes [9]. The latter quantifies user preferences into a numerical score table, and then calculates it according to the scoring rules. In terms of specific quantitative representation, generally multidimensional vector scoring models and hierarchical models are used details as follows (Table 1):

Table 1. Interest recommendation features of multidimensional vector scoring model and hierarchical model

Classification	Advantages and disadvantages	Typical technology
Quantitative analysis	Advantages: it is conducive to expand the quantitative representation and calculation, and is convenient to generate sorting	Cosine similarity, clustering, Bayesian probability model, decision tree
	Disadvantages: make the user preferences of each specific situation form a total order relationship	
Qualitative analysis	Advantages: it is helpful to describe which situational user preferences do not necessarily satisfy the full order relationship	Strict partial order model
	Disadvantages: it does not have the advantages of digital quantitative expression and large amount of calculation	

Relative contextual pre-filtering measures and selects user information in the data input stage. In the post-context filtering mode, the role of contextual information is to adjust the recommended result data set. The post-context filtering mode first directly uses the traditional recommendation algorithm to recommend in the user-item two-dimensional space, and then uses the user's current context information to filter the recommendation result data set or adjust the Top-N ranking [10]. For example, if a user wants to watch a movie on the weekend, and the user's habit is to watch comedy movies on the weekend, the recommendation system first generates a recommendation list for the user according to the traditional recommendation method, and then filters out movies that are not comedy types. More generally, the basic idea of context post filtering mode is to find specific items found by a given user in a specific situation based on context reference data. Since context is only used in the output stage of recommendation results, the relationship between context and users is loosely coupled.

2.2 User Interest Algorithm Optimization

At present, relatively mature scenario modeling schemes include key value pair model, markup language model, graph model, object-oriented model, logic model, vector model and ontology model [11]. In the distributed service architecture, the key value pair model is generally used. The key value pair is used to describe the attributes of the service, and then the matching degree between the described attributes and the service itself is calculated through the matching algorithm. The key value pair model is relatively simple, but there is a lack of mature framework, and there is a shortage in obtaining the usage context information. The markup language model uses markup languages such as logos and notes to represent contextual information. These markup languages can be interpreted by the device and are responsible for controlling the relationship between the various parts of the model structure, subject to the pre-defined hierarchical structure and rewrite level of the markup language Structure, markup language model is not widely used [12]. The graph model expresses the contextual information graphically, which is intuitive and vivid for sorting out the relationship between various contextual information. Universal UML can directly involve and query contextual data. Due to the lack of unified standards, the software tools obtained are not supported enough, so the application is the same Restricted. The object-oriented model encapsulates the context information in an object. Other system components access the object through the interface. Through reuse and inheritance, the minimum granularity attributes, functions and rules can be defined to simplify the knowledge expression. In the logic model, situation information is defined by facts, expressions and rules, which is highly normative, difficult to construct and difficult to manage [13]. The vector model quantifies the situation information and is easy to carry out the corresponding mathematical calculation. The ontology model embeds semantic information, has mature ontology language reference standards, has received strong support from related software, and is relatively widely used [14]. In specific research, the representation of multi-dimensional situations generally adopts two models: key-value pairs and vectors. Collaborative filtering algorithm can be described as follows: $U = \{u_1, u_2, \ldots, u_m\}$ represents a set of series of users; $I = \{i_1, i_2, \ldots, i_m\}$ is the set of a series of items, $R(t) = \{r_1(t), r_2(t), \ldots, r_m(t)\}$ is the user's interest in each item [15]. In general, the goods purchased by users only account for a small part of the total number of goods, which will inevitably lead to the sparsity of the item user matrix. The calculation process of collaborative filtering algorithm can be divided into three parts, as shown in the figure.

$$R(m, n) = \begin{bmatrix} R_{1,1} & R_{1,2} & R_{1,3} & \cdots & R_{1,n} \\ R_{2,1} & R_{2,2} & R_{2,3} & \cdots & R_{2,n} \\ R_{3,1} & R_{3,2} & R_{3,3} & \cdots & R_{3,n} \\ R_{m,1} & R_{m,2} & R_{m,3} & \cdots R_{m,n} \end{bmatrix} \tag{1}$$

Use user-item rating matrix R(m,n). Among them, m represents the number of users, in the row position, n represents the number of items, in the column position. User i is score for item J corresponds to the value of the element in the i row and j column of the matrix [16]. On this basis, a user-based collaborative filtering calculation model is established, which is established for the user's recommendation relationship, as follows:

$$E = \min_{\theta^{(1)},\dots,\theta^{(su)}} \frac{1}{2} \sum_{j=1}^{n_u} \sum_{ir(i,j)=1} \left(\left(\theta^{(j)} \right)^T x^{(i)} - y^{(i,j)} \right)^2 + \frac{\lambda}{2R(m, n)} \sum_{j=1}^{n_u} \sum_{k=1}^{n} \left(\theta_k^{(j)} \right)^2 \tag{2}$$

From the perspective of context aware processing, the subject can be divided into two parts: context information utilization and organization. The purpose of the former is to know the specific way of situational information for application [17]. The purpose of the latter is to collect situational information, then abstract it into a situational model, give situational explanation, and finally store the situational information for retrieval [18]. At present, the collection of situational information is generally obtained through mobile network, Internet of things, digital map, weather and other third parties. The situational abstraction first needs to clean up the collected situational information, including sampling, averaging, statistics, and calibration, and then integrate and classify different situational information for subsequent analysis and processing of the data [19]. Contextual interpretation is to obtain the hidden semantic information of contextual features and organize context into knowledge, including possible behaviors of users in the future (Fig. 4).

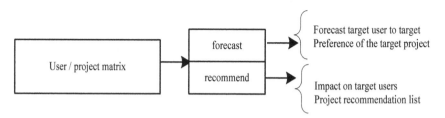

Fig. 4. Analysis principle of user interest behavior

In the recommendation process, the user's nearest neighbors can be effectively selected, and the user's neighbor selection method varies with different algorithms. The important calculation process of collaborative filtering is the nearest neighbor selection process [20]. In the algorithm input process, the purpose of the algorithm will directly affect the output result of the algorithm. The model is used to predict the user's preference for items, and a quantified preference value is obtained. When a recommendation list that the current user is interested in is generated, the user's item recommendation will be generated accordingly.

2.3 Implementation of Context Aware Recommendation

The collected data sample set is applied to the context system, and the reasoning operation is applied to the modeling and solution process of the information recommendation problem, so as to complete the solution process of the recommendation problem [21]. The following figure can summarize the entire integration process as a whole (Fig. 5):

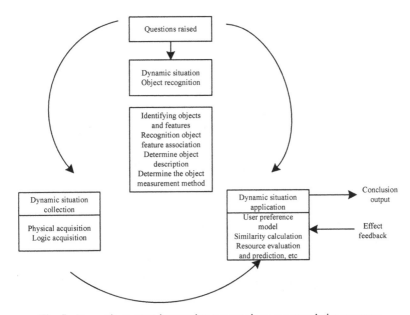

Fig. 5. Dynamic context integration process in recommendation process

Context also plays an important role in personalized recommendation. The user's contextual preference can be analyzed through the user's historical scoring data. This article uses contextual weights to measure user preferences [22]. Through the context weighting method, the weight is calculated for each user, and the different impact values of different context factors on the user are determined. Before calculating the context weight, we need to calculate the fluctuation of the context type, and the fluctuation coefficient represents the emotional fluctuation of the user' scoring in a certain context [23]. When the user's scoring in a certain context is very different, it indicates that the user's emotional fluctuation in this situation is very large, so we can think that the emotional fluctuation factors have a great impact on the user's decision-making. The formula for calculating the fluctuation coefficient of scenario type CT is as follows:

$$cof_t = \frac{\frac{1}{|C_t|}\sqrt{E\sum\limits_{i\in I, c_i \in C_t}\left(r_{ic_i} - \overline{r_i C_t}\right)^2}}{\overline{r_i C_t}}, \overline{r_i C_t} = \frac{1}{|C_t|}\sum\limits_{i\in I, c_i \in C_i} r_{ic_i} \qquad (3)$$

Ct is used to indicate the number of context intervals of the context type, and r is used to indicate the value in the context. The user's score for item Z is given by the specific algorithm.

$$\begin{cases} \omega C_t = \dfrac{cof_t}{\sum\limits_{t=1}^{m} cof_t} \\ cof_t \neq 0, \\ cof_t = 0 \\ \omega C_t = \dfrac{1}{m} \end{cases} \qquad (4)$$

The movie ticket recommendation model based on context aware recommendation is shown in the figure below. When the user logs on to the ticket platform, the system obtains the user's information according to the current user's context information. By analyzing the user's information, the system obtains the user's preference resources in different situations, and obtains the user's score on these resources. Then the formula is used to calculate the context weights of all the context attributes, so as to get the impact value of the context attributes on the recommended resources. Then use the score value of similar users in the same context to calculate the similarity value based on the target user's collaborative filtering, so as to find the neighbor users of the target user, and generate a recommendation list of information resources to complete the entire context-aware personalized recommendation process. In this process, the information resources and diary records that the system pushes to users are stored in the information database to provide queries for subsequent recommendations (Fig. 6).

Contextualized modeling is to integrate contextual information into the resource evaluation stage of the recommendation process. In the early stage of recommendation generation, the contextual data set is input into the personalized recommendation model, and then the multi-dimensional data generated by the fusion is processed through heuristic-based algorithms or predictive models. The final recommendation result. Different from the contextual pre-filtering and post-filtering methods, contextualized modeling takes into account the contextual factors to obtain user preference information in the resource evaluation stage, so the context and users are tightly coupled in this model. When using the personalized method of context aware recommendation to recommend movies to users, we must fully consider the user's context information and context factors, and list some context factors that are often considered in the process of recommending movie resources to users.

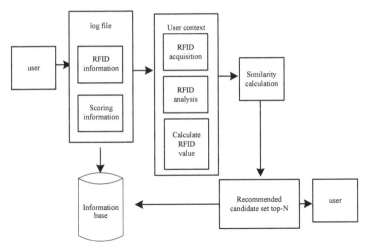

Fig. 6. Context aware personalized recommendation steps

3 Analysis of Experimental Results

The purpose of personalized recommendation is to help users find interesting resources faster and better, while Mae reflects the difference between the predicted score and the actual score, which can not fully reflect the quality of recommendation. Therefore, this paper adopts the accuracy rate as an improved algorithm, that is, the measurement standard of the quality of the context-aware collaborative filtering recommendation, and compares it with the traditional user-based collaborative filtering algorithm. The so-called accuracy rate refers to the proportion of items in the recommended list that are marked as favorites by the user to the total number of items in the list. The formula is as follows:

$$Precision = \frac{|A \cap B|}{|B|} \tag{5}$$

Among them, a and B represent the item set that users like and the item set in the recommendation list, respectively, $| a \cap B |$ represents the number of items that meet users' preferences in the recommendation list, and B represents the total number of items in the recommendation list. Recommendation system usually evaluates the recommendation quality of recommendation algorithm through offline experiment, user survey and network experiment. In this article, using the offline experiment method, the user data set is divided into a training set and a test set at a ratio of 5:1. There are many commonly used recommended evaluation indicators, and only three of them are used as metrics here, namely the average absolute error, the root mean square error and the recall rate. The definitions of the three evaluation indicators are as follows:

$$MAE = \frac{\sum_{i=1}^{N} |p_i - q_i|}{N} \tag{6}$$

The value of MAE is inversely proportional to the accuracy of recommendation, that is, the smaller the value of MAE, the higher the accuracy of recommendation.

$$RMSE = \frac{\sqrt{\sum_{i=1}^{N} |p_i - q_i|}}{N} \tag{7}$$

Similarly, the smaller the value of RMSE, the higher the accuracy of the recommendation. The larger the value of RMSE, the lower the recommended performance.

$$Recall = \frac{N_{\text{like}}}{N_{\text{test}}} \tag{8}$$

The higher the recall value, the better the recommended performance of the algorithm. The experiment in this paper is mainly to compare the performance of traditional recommendation algorithm, user based collaborative filtering algorithm and the improved algorithm. Each user is randomly selected as the target user, and the average of the test results of each user is the final test result. Meanwhile, the neighbor size of the experiment is selected from 15–90 and the step size is 15 for the test. The above three measures, MAE, RMSE and recall are evaluated respectively. The specific experimental results and analysis are as follows (Fig. 7):

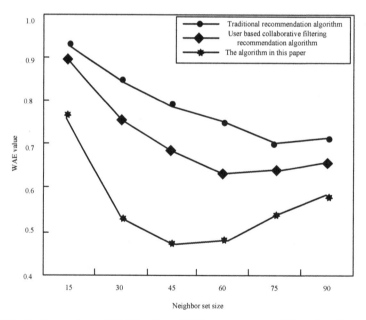

Fig. 7. Comparison of MAE values of different recommendation algorithms

As shown in the figure, the MAE value of our algorithm is lower than that of traditional recommendation algorithm and user collaborative filtering algorithm when the size of neighbor set is larger and larger, which indicates that our algorithm fully considers the factors of user context and product context on the basis of simple user commodity binary relationship, Therefore, we can get higher quality recommendation (Fig. 8).

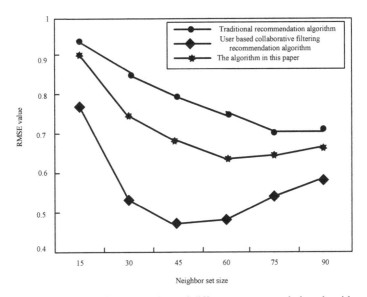

Fig. 8. Comparison of RMSE values of different recommendation algorithms

Compared with MAE, RMSE is more strict in the evaluation of recommendation performance, and the penalty for inaccurate prediction is increased on the basis of MAE. Therefore, the value of RMSE is relatively higher than that of MAE, but it still follows the principle that the smaller the value is, the more accurate the performance result is. There are differences in the application occasions of recommendation algorithms based on users and items. The former recommendation tends to be more socialized and will take into account neighbor users who have the same characteristics as the user, so as to reflect the popularity of items in the user's group, while the latter Recommendations emphasize the user's personal needs and are more personalized. The algorithm can reflect the user's own points of interest. In the current e-commerce industry, the use of project-based recommendation algorithm is more important, because in the e-commerce system, it is difficult to identify the user's neighbor groups, so the corresponding recommendation will be made according to the user's purchase history. The following table reflects the main advantages and disadvantages of user based and Project-based Collaborative filtering recommendation systems (Table 2).

Table 2. Comparison of user-based recommendation and item-based recommendation

	Based on user recommendation	Project based recommendation
Timeliness	Strong timeliness and quick update of recommendation list	Weak timeliness and slow update of recommendation list
Applicability	It is suitable for the situation where the number of users is less than the number of items	It is suitable for the situation where the number of items is less than the number of users
Expansibility	There are scalability problems, and the performance is limited by the number of users	There are scalability problems and performance is limited by the number of projects
Stability	Poor stability	Good stability
Novelty	The results are novel	The recommended results are similar to the projects users have contacted in the past

In order to verify the recommendation effect of the personalized recommendation system based on situational perception in mobile environment, 80% of movielens movie data set is used as training set and 20% as test set. The data in the training set is predicted by using the personalized recommendation algorithm based on situational perception and user based collaborative filtering algorithm, the mean absolute deviation of the two methods is calculated by comparing with the actual score data in the test set. The comparison of experimental results is shown in the table (Fig. 9):

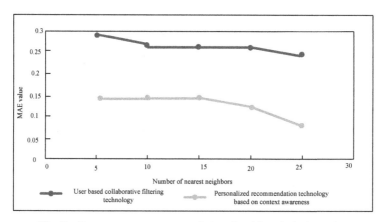

Fig. 9. Context-aware recommendation algorithm deviation value

It can be seen from the figure that when the number of neighbors increases in turn, the personalized algorithm based on context awareness proposed in this paper has smaller mean absolute deviation (MAE) compared with the traditional collaborative filtering algorithm. This result shows that the algorithm has more accurate recommendation under the condition of considering the situation factors, so it proves that the proposed recommendation algorithm has better recommendation effect.

The comparison between the improved collaborative filtering algorithm and the traditional collaborative filtering regarding the recommendation accuracy rate is shown in the figure (Fig. 10):

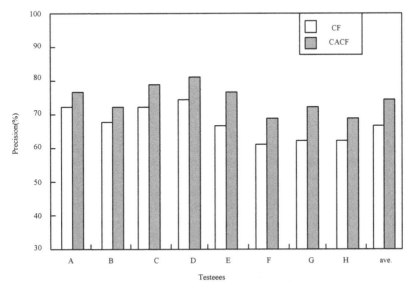

Fig. 10. Comparison of the practicability between the algorithm and the traditional collaborative filtering algorithm

Because each tester evaluates the recommendation list of multiple contexts, the recommendation accuracy rate in the figure is the mean value of the recommendation accuracy rates of multiple contexts. The last set of data is the average of the first eight sets of data. From this set of data, it is not difficult to see that the improved algorithm has improved the accuracy of the recommendation to a certain extent.

4   Conclusion and Outlook

It is a new direction of information recommendation to introduce context factors into personalized recommendation system and consider the ternary relationship among user resource context to recommend appropriate information services to meet the needs of mobile users in different situations and interests. Aiming at the problem that the current personalized recommendation system does not consider context factors, this paper proposes a context aware personalized recommendation algorithm based on RFID, which combines context information and traditional content filtering to push information suitable for the current situation for users. Experiments show the effectiveness of the algorithm.

Although there is a lot of research on context-aware recommendation technology, many improved algorithms have not ideal application effects, and in the era of big data there are higher requirements for the accuracy of recommendation and the performance of recommendation algorithms. Therefore, it is necessary to further improve the recommendation accuracy and algorithm through data mining, neural network, machine learning and other cross-fields.

References

1. Sakthivelan, R.G., Rjendran, P., Thangavel, M.: A video analysis on user feedback based recommendation using A-FP hybrid algorithm. Multim. Tools Appl. **79**(5), 3847–3859 (2020)
2. Zamil, N., Ansari, M.N.M., Jalil, Y.E., et al.: Design and simulation of henna-based composite substrate for UHF-RFID application on dielectric properties. IOP Conf. Ser. Mater. Sci. Eng. **1128**(1), 012031 (7pp) (2021)
3. Alsini, A., Datta, A., Du, Q.H.: On utilizing communities detected from social networks in hashtag recommendation. IEEE Trans. Comput. Soc. Syst. **7**(4), 971–982 (2020)
4. Gja, A., Rgmb, C., Smm, B., et al.: A meta-learning approach for selecting image segmentation algorithm - ScienceDirect. Pattern Recogn. Lett. **128**(11), 480–487 (2019)
5. Duan, S., Zhang, D., et al.: JointRec: a deep-learning-based joint cloud video recommendation framework for mobile IoT. IEEE Internet Things J. **7**(3), 1655–1666 (2020)
6. Bc, A., Hao, F.B., Sl, B., et al.: The recommendation service of the shareholding for fund companies based on improved collaborative filtering method. Procedia Comput. Sci. **162**, 68–75 (2019)
7. Bw, A., Wen, W.A., Zha, B., et al.: Multi-context aware user–item embedding for recommendation. Neural Netw. **124**(11), 86–94 (2020)
8. Amer, A.A., Abdalla, H.I., Nguyen, L.: Enhancing recommendation systems performance using highly-effective similarity measures. Knowl.-Based Syst. **34**(4), 106842 (2021)
9. Singh, V.P., Pandey, M.K., Singh, P.S., et al.: Neural net time series forecasting framework for time-aware web services recommendation. Procedia Comput. Sci. **171**(6), 1313–1322 (2020)
10. Huang, Z., Lin, X., Liu, H., et al.: Deep representation learning for location-based recommendation. IEEE Trans. Comput. Soc. Syst. **7**(3), 648–658 (2020)
11. Li, L., Chen, L., Dong, R.: CAESAR: context-aware explanation based on supervised attention for service recommendations. J. Intell. Inf. Syst. **3**, 1–24 (2020)
12. Musto, C., Lops, P., de Gemmis, M., Semeraro, G.: Context-aware graph-based recommendations exploiting Personalized PageRank. Knowl.-Based Syst. **216**, 106806 (2021)
13. Gomes, L., Almeida, C., Vale, Z.: Recommendation of workplaces in a coworking building: a cyber-physical approach supported by a context-aware multi-agent system. Sensors (Basel, Switzerland) **20**(12), 3597 (2020)
14. Chen, C., Wang, D., Ding, Y.: User actions and timestamp based personalize- -d recommendation for e-commerce system. IEEE Int. Conf. Comput. Inf. Technol. **35**, 183–189 (2016)
15. Zou, C., Zhang, D., Wan, J., Hassan, M.M., Lloret, J.: Using concept lattice for personalized recommendation system design. IEEE Syst. J. **11**(1), 305–314 (2017)
16. Tamhane, A., Arora, S., Warrier, D.: Modeling contextual changes in user behaviour in fashion e-commerce. In: Kim, J., Shim, K., Cao, L., Lee, J.-G., Lin, X., Moon, Y.-S. (eds.) Advances in Knowledge Discovery and Data Mining, pp. 539–550. Springer International Publishing, Cham (2017). https://doi.org/10.1007/978-3-319-57529-2_42
17. Das, J., Majumder, S., Malli, K.: Control aware scalable collaborative filtering. In: Internate Conference on Big Data, pp. 184–190 (2017)

18. Haim, M., Graefe, A., Brosius, H.-B.: Burst of the filter bubble?: Effects of personalization on the diversity of Google News. Digit. Journal. **6**(3), 330–343 (2017)

19. Zhang, Y., Chen, M., Huang, D., et al.: i Doctor: personalized and professionalized medical recommendations based on hybrid matrix factorization. Futur. Gener. Comput. Syst. **66**, 30–35 (2017)

20. Okura, S., Tagami, Y., Ono, S., et al.: Embedding-based news recommendation for millions of users. In: ACM SIGKDD International Conference on Knowledge Discovery and Data Mining, pp. 1933–1942. ACM, New York (2017)

21. Wei, J., He, J., Chen, K., et al.: Collaborative filtering and deep learning based recommendation system for cold start items. Exp. Syst. Appl. **69**, 29–39 (2017)

22. Guo, Q., Zhang, C., Zhang, Y., et al.: An efficient SVD-based method for image denoising. IEEE Trans. Circuits Syst. Video Technol. **26**(5), 868–880 (2016)

23. Zou, X., Gonzales, M., Saeedi, S.: A context-aware recommendation system using smartphone sensors. In: Information Technology, Electronics and Mobile Communication Conference, pp. 1–6. IEEE, New York (2016)

RFID Based Fast Tracking Algorithm for Moving Objects in Uncertain Networks

Yan Zhao[✉]

Ningbo City College of Vocational Technology, Ningbo 315199, China

Abstract. Mobility model is used to describe the mobility mode of nodes in the network and determine the mode of node movement. The selection of mobile model and its parameters is also of great significance to simulation results. Therefore, a fast tracking algorithm based on RFID is proposed. The performance evaluation based on RFID network related protocols and algorithms plays an important role in improving the network performance. The spatial distribution of nodes in various mobile models is studied and compared. The effects of velocity, velocity of different distribution, pause time of different distribution and simulation area on the spatial distribution of nodes are obtained. After summarizing the defects of fast tracking of network moving targets, based on the random direction motion model, the fast tracking algorithm of uncertain network moving targets is completed by setting the running time of nodes and changing the speed of nodes in different time periods. Through experimental analysis, it is shown that the RFID based fast tracking algorithm has uniform spatial distribution and good performance Low energy consumption.

Keywords: RFID · Uncertain network · Moving object tracking

1 Introduction

With the progress of information technology, mobile communication technology has also been a long-term development. New mobile communication technologies such as cellular communication, WiFi technology, Bluetooth communication are emerging. With the help of these technologies, people's life becomes very convenient, and the wireless communication technology is also promoted by them. Generally speaking, mobile communication technology adopts centralized control, that is, centralized control [1]. This kind of wireless network needs to deploy the infrastructure in advance to be able to use, for example, the cellular mobile communication system, which needs to rely on the support of the infrastructure such as base station and mobile switching center to operate; Access point and limited backbone network are the general working modes of WLAN. But for some special occasions, it is impossible to deploy fixed facilities in advance, so the above working mode can not be used. For example, rescue and disaster relief, field survey, marginal mine operation, temporary meeting, etc. [2]. In this case, we need a mobile communication technology that can quickly and automatically network temporarily. To sum up, the mobile model is to build a mobile network. Network

S. Liu and X. Ma (Eds.): ADHIP 2021, LNICST 417, pp. 328–341, 2022.
https://doi.org/10.1007/978-3-030-94554-1_26

simulation and test environment is one of the important parts of the environment [3]. The consistency between the mobile model and the real scene is an important factor affecting the reliability of network simulation results. Designing a mobile model in line with the real situation can provide a reasonable node mobile scene for mobile RFID network simulation. At the same time, studying the physical characteristics of the mobile model and its relationship with the network topology changes will help to better understand the simulation results, And on this basis, improve and perfect the relevant protocols and algorithms.

This method collects the characteristic data of uncertain network moving objects, and divides the structure according to the characteristic data; On this basis, the frequent activity pattern distance measurement based on sequence alignment is used to accurately track the moving objects; The tracking target is completed according to the time variation characteristics of regional activity heat.

2 Fast Tracking Algorithm for Moving Objects in Uncertain Networks

2.1 Feature Collection of Mobile Objects in Uncertain Network

Mobile RFID network is composed of a group of mobile terminals with wireless transceiver devices, and does not rely on fixed infrastructure, so it can not provide centralized control. Mobile RFID network is a kind of wireless network which is set up for fixed purpose. In mobile RFID network, nodes not only act as hosts, running user oriented applications, but also act as routers to find routes and forward data according to routing protocols [4]. At the same time, due to the limited communication range of nodes in RFID network, the source node and destination node of communication will not be within the scope of direct communication, so they need to have intermediate nodes to forward data for communication, that is to say, its routing is generally composed of multi hop, so adhc network is sometimes called wireless multi hop network [5]. Its multi hop characteristic is also the most fundamental difference between RFID network and other mobile communication networks. Mobile RFID network has flexibility. The network can be constructed at any time and anywhere. As long as there are two or more nodes connected with each other, they can communicate with each other if they are in the communication range of each other, and they can communicate with each other through intermediate mobile nodes when they are beyond each other's traffic range [6]. Because of the advantages of RFID network, such as flexibility, self-organization without infrastructure, easy maintenance and cost-effectiveness, it can be widely used in many fields, such as wireless sensor network, which has very important strategic significance. This also makes mobile RFID network become an important part of wireless network and research hotspot. Network nodes can be divided into three parts: host, router and wireless transceiver [7]. Among them, the host is used to run applications, complete data processing and other functions; the router is used to run routing protocols, complete routing, forwarding packets and so on. The wireless transceiver is used to complete the function of data transmission [8]. In terms of physical structure, nodes can be divided into single access, multiple access, single access and multiple access (Fig. 1).

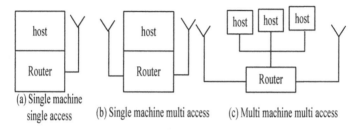

Fig. 1. Characteristic structure of uncertain network nodes

The advantage of hierarchical network is that cluster members do not need to maintain the information of road and mountain, and update the table of road and mountain. In this way, the number of road and mountain control information in the network will be greatly reduced, and the hierarchical form can have good scalability [9]. Its disadvantage is that maintaining hierarchical structure requires nodes to implement cluster head selection algorithm and maintenance mechanism among cluster members, which makes the calculation more complex; Cluster heads need to forward data centrally, which is a heavy task and may become the bottleneck of the network. Plane structure and hierarchical structure have their own advantages and disadvantages, which requires us to consider comprehensively when building the network and choose the appropriate network type. The following table lists the properties of the planar and hierarchical structures (Table 1).

Table 1. Network plane structure and hierarchical structure

Plane structure	Hierarchical structure
The status of all nodes is equal	Nodes are divided into cluster heads and cluster members
Fully distributed network	Network composed of multiple clusters
There is no network bottleneck, suitable for small and medium-sized network	Good scalability, suitable for large-scale network
Poor scalability	Cluster head nodes may become network bottlenecks

Mobile RFID network is a wireless communication network without any infrastructure, so it has different characteristics from traditional communication network. RFID network has the following main characteristics. In the RFID network, there is no control center, the nodes in the network have equal status, and it is a peer-to-peer network, so its invulnerability is also very strong. Network is a kind of communication network that can be constructed at any time and any place without any infrastructure. Due to the limited communication range of nodes in the RFID network, each node in the network may not be able to establish direct communication with other nodes. Therefore, when a node wants to transmit data with other nodes beyond its communication coverage, it needs to use the intermediate node to forward data and realize the required communication connection through the intermediate node. That is to say, the communication from the source node to the destination node is through multi hop connection [10] (Fig. 2).

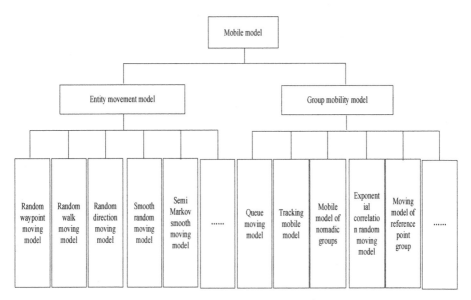

Fig. 2. Network mobility model classification

Because the nodes in the RFID network move all the time and are no longer fixed in the same location, the links between nodes can not be maintained all the time, which means that the network topology can not be stable and always changes dynamically with time[11]. At the same time, because the movement of nodes is random, it can not be predicted. Therefore, the change of network topology is unpredictable [12]. RFID network is a distributed network, its bandwidth resource is very limited, and its bandwidth is much lower than wired channel, and the quality of wireless channel is poor. At the same time, the communication process will be subject to various kinds of interference, such as electromagnetic wave, noise interference, collision, channel interference and other factors. The actual bandwidth that the mobile terminal can get will be far less than the theoretical maximum bandwidth, so it is necessary to save the network bandwidth resources as much as possible.

2.2 Moving Object Tracking Algorithm in Uncertain Networks

The uncertainty network mobile model can be used in RFID network to capture the correlation between time and speed of nodes. In this model, at the beginning, each node is given a speed, including the rate v and direction A. every time interval Δt is fixed, the node updates the speed and direction once, and the update is based on the previous time, for example, the speed and direction of time n are calculated on the basis of time n-1, The specific calculation method is as follows

$$v_n = \alpha v_{n-1} + (1 - \alpha)\overline{v} + \sqrt{1 - \alpha^2}\omega_{v_{n-1}} \tag{1}$$

$$\theta_n = \alpha\theta_{n-1} + (1-\alpha)\bar{\theta} + \sqrt{1-\alpha^2}\omega_{\theta_{n-1}} \tag{2}$$

By adjusting a to control the randomness of nodes, the model is widely used. When a = 0, the state update of nodes does not depend on the previous state, that is, no memory, so it is consistent with RW model; When a = 1, it is consistent with the previous state, and the node will run in a straight line; 0 < a < 1, the nodes will be in between two states, and the movement has a certain correlation. At the same time, the location of the node will be updated in the following way:

$$x_n = x_{n-1} + v_n \cos\theta_{n-1} \tag{3}$$

$$y_n = y_{n-1} + v_n \sin\theta_{n-1} \tag{4}$$

The frequent activity patterns of moving objects are identified from activity sequences by using sequence mining algorithm. These patterns represent the hidden behavior habits of moving objects. Then, the distance measure of frequent activity patterns based on sequence alignment is defined to calculate the similarity between two frequent activity patterns. Finally, all the frequent activity sequences of moving objects are integrated to calculate the similarity (Fig. 3).

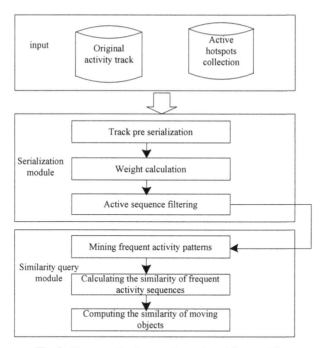

Fig. 3. Framework of network moving object tracking

Although mobile models have their own advantages and algorithms are relatively simple, they also have defects, and each node in the model is not connected and independent, which is inconsistent with the reality. Therefore, it is necessary to improve the model or propose a new model to make it more similar to the reality. The following table shows the performance comparison of entity mobility models (Table 2).

Table 2. Performance comparison of entity mobility models

Mobile model	RD model	RW model	RWP model	GM model	SMS model	SR model
Parameter	Speed, direction, time	Speed, direction	Speed, destination	Speed, direction	Speed, direction, time	Speed, direction
Mobile phase	Move, pause	Move, pause	Move, pause	Move	Accelerate, smooth, decelerate, pause	Move, pause
Smoothness	No	No	No	Yes	No	Yes
Velocity attenuation	No	-	Yes	No	No	No
Description angle	Macroscopic	Macroscopic	Macroscopic	Microcosmic	Microcosmic	Microcosmic
Is the node uniform	No	Near	No	Yes	Yes	Near
Controllability	Low	Low	Low	Middle	High	Middle

In the entity mobile model, there is no connection between nodes, while in the group mobile model, the movement of nodes has a certain connection, which is not completely independent and has a certain correlation. The common group mobility models include: queue mobility model, tracking mobility model, nomadic group mobility model, reference point group mobility model, etc. The queue movement model is that the nodes line up and move along a specific force, and each node can move in a small range. The model can well simulate the process of scanning and searching. In the queue mobility model, node location updating can be described as the following process:

$$R_m(t) = R_m(t - 1) + x_n r_m(t) \tag{5}$$

$$P_m(t) = R_m(t) + y_n p_m(t) \tag{6}$$

Where, Rm(t) is the position of the reference point, Rm (t−1) is the position of the reference point in the previous state, and Pm (t) is the displacement vector of the reference point. The following figure describes the movement mode of the node in the queue movement model, in which the black dot is the reference point, and the white dot is the node. From the figure, we can see that the reference points of the nodes are all in a straight line, and the nodes are all around the reference points (Fig. 4).

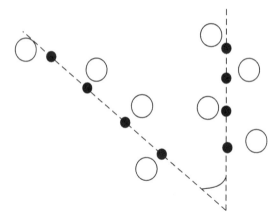

Fig. 4. Moving object tracking arrangement model

In the exponential correlation random mobility model, the mobility of a group is determined by creating the mobility function of the node. The node location is updated in the following way:

$$D(t) = D(t-1)P_m(t)e^{\frac{1}{\tau}} + \left(Q\sqrt{1 - \left(R_m(t)e^{-\frac{1}{\tau}}\right)^2}\right)r \tag{7}$$

Where e is the adjustment amount of the node's current time and the previous time position, and r is a Gaussian variable with variance Q. the motion mode of the model node is changed by adjusting the parameters.

2.3 Fast Tracking of Uncertain Mobile Network Objects

The location tracking of moving objects is based on the continuous location update to obtain location information. In urban road network, because of the continuous movement of moving objects, the position in the server database must be updated constantly to improve the real-time accuracy of location data. A large number of position update strategies and algorithms are proposed, but most of the algorithms are based on the basic update strategies first proposed. In addition, the basic strategies are improved and optimized by adding some other driving variables and parameters of moving objects. Forwarding mechanism is in the opportunity network, when meeting nodes, which strategy or standard will trust other nodes also called routing algorithm. At present, the forwarding mechanism of mobile opportunity network can be divided into two categories: information forwarding to zero information transfer assistant forwarding, and the classification is shown in the figure (Fig. 5):

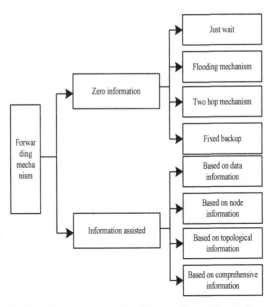

Fig. 5. Classification chart of mobile tracking object information

The original data is usually a two-dimensional coordinate with location information, or a three-dimensional coordinate with time information. For illustration, it is assumed that the moving object is a 2D coordinate point. To illustrate the distance measurements, the basic symbol definitions are given below. Define two moving objects as A and B, and the representation method is as follows:

$$A = a_1, a_2, \ldots, a_i, \ldots a_m (1 \leq i \leq m) \tag{8}$$

$$B = b_1, b_2, \ldots, b_j, \ldots b_n (1 \leq j \leq n) \tag{9}$$

Where m and n are the length of moving objects A and B, respectively. The moving object point is a two-dimensional position coordinate point without time information, as shown in the formula.

$$\Delta a_i = \left(A a_i^x - B a_i^y \right) / D(t) - 1 \tag{10}$$

Moving object data is spatiotemporal data, which is a series of sequence values with temporal and spatial information. Therefore, the method used to calculate the similarity of moving objects can not be used for static data in conventional algorithms. There are more general methods for distance measurement of static data, which can be used to calculate the similarity of moving objects. The Euclidean distance between two points $a(x1, y1)$ and $b(x2, y2)$ on the two-dimensional plane is calculated as follows.

$$d(a, b) = \Delta a_i \sqrt{(x_1 - x_2)^2 + (y_1 - y_2)^2} \tag{11}$$

The Euclidean distance between two points in three-dimensional space is calculated as follows.

$$\Delta d(a, b) = \sqrt{(x_1 - x_2)^2 + (y_1 - y_2)^2 + (z_1 - z_2)^2} \tag{12}$$

The calculation method of Euclidean distance between two n dimensional vectors is shown in the formula

$$S = \sqrt{d(a, b) \sum_{k=1}^{n} \left(x_k - x_k'\right)^2} \tag{13}$$

The regions with dense moving objects are potential active hot spots, which are called candidate hot spots. Firstly, the active space is meshed, and the general information of the moving objects is saved in the corresponding grid. The candidate hot spots with dense moving objects are identified by multi-density grid clustering. The clustering object is a grid with the profile information of moving objects, and its scale is much smaller than that of moving objects, so it can significantly improve the efficiency of candidate hot spots detection. The candidate hot spot regions find the regions with relatively dense moving objects, which do not consider the activity characteristics of moving objects. In the hot area filtering stage, this paper proposes a region heat calculation method based on the structure of moving objects to measure the activity heat of moving objects in candidate regions, and filters out the final active hot areas according to the time-varying characteristics of the activity heat (Fig. 6).

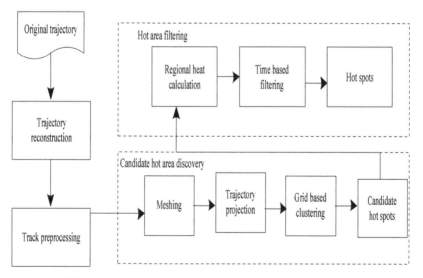

Fig. 6. Time variation characteristics of regional activity heat

Dense moving objects is a sufficient and unnecessary condition for dense activities. The basic idea of HSTS is to use the candidate region discovery algorithm to identify the dense regions of moving objects from the data of moving objects as candidate regions, and then further filter them according to the activity characteristics of moving objects, so as to improve the tracking efficiency and accuracy.

3 Analysis of Experimental Results

This experiment is carried out on the real Geolifev1.30 data set. The Geolife data set contains 182 moving objects in five years, a total of 17621 moving objects, with a total length of more than 1.29 million km. Further verify the effectiveness of TSPM. Speed is a very important parameter in the mobile model. The change of speed will affect the spatial distribution of nodes, and then affect the performance of the mobile model and the performance of the whole RFID network. The final simulation results will also change. Therefore, it is necessary to study the influence of the change of speed on the spatial distribution of nodes in the mobile model of RFID network (Table 3).

Table 3. Comparison of activity pattern mining results

	Time	PatternNwn	AvgPattemNwn	AvgLength
Filter ($\delta w = 0.015$)	411	227	4.54	3.87
No filtering ($\delta w = 0$)	1368	243	4.86	4.11

Because the verification of the effect based on WiFi P2P location tracking experiment needs a lot of energy and time to achieve, it needs users to install apps, to achieve in the real road network, and need to negotiate to participate in the forwarding of opportunistic network, which can not be achieved under the current realistic conditions, but some performance data can be obtained through the measurement between several mobile phones, These performance data can provide reference for further research in the future. The performance data are as follows: the time of device connecting to WiFi P2P is about 3–8 s; the distance between devices and WiFi P2P can reach 150 m; Average transmission rate: 1.1 m/s. Next, we will study this in two simulation regions, so that we can not only get the spatial distribution of nodes in different speed mobile models, but also compare the impact of different simulation regions on the spatial distribution of nodes. Since most of the mobile models used in RFID network are entity mobile models, this paper also uses entity mobile models RW model, RD model, RWP model, GM model and SMS model for simulation to explore the spatial distribution of nodes under different speeds of mobile models and the influence of different simulation areas on the spatial distribution of RFID network nodes. In order to calculate the spatial distribution of nodes, it is necessary to divide the simulation area. The following will introduce the partition method of rectangular simulation area and circular simulation area used in the experiment and the calculation method of probability density of mobile model nodes (Fig. 7).

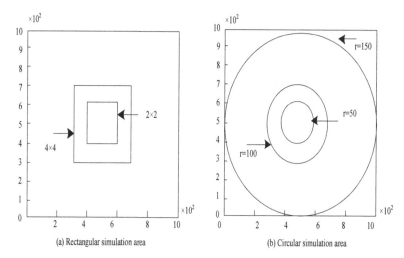

(a) Rectangular simulation area   (b) Circular simulation area

Fig. 7. Simulation area of uncertain network tracking object

As shown in the figure, the statistical information of moving objects in the two candidate regions numbered # 56 and # 76 is normalized for each attribute (Fig. 8).

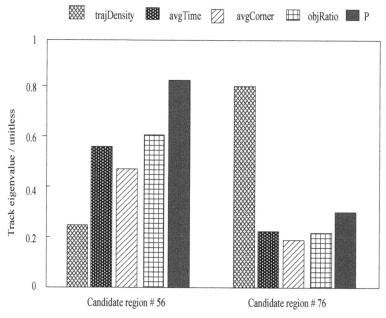

Fig. 8. Comparison of trajectory structure characteristics of moving objects in candidate regions

It can be seen from the figure that although the density of moving objects in the # 56 area is relatively low, it is shown that the area is more suitable for the characteristics of the active area in terms of the time, rotation times and object coverage of the moving object. Although the density of moving objects is high, the activity characteristics of moving objects are not obvious, which indicates that # 76 may only be the necessary place for traffic intersections. The region heat calculated based on the structure of moving objects reflects the difference of moving object activity between the two candidate regions. The regional heat of candidate region is calculated in the total time. In order to further verify the necessity of time filtering for candidate regions, this paper selects two candidate regions with close overall regional heat degree, namely, 1.09 and 1.13 respectively. The work history heat of these two candidate regions is calculated by using the month as the time interval. The active moving objects are serialized by AWTS, and the average length of the activity sequence is calculated under different activity thresholds (Fig. 9).

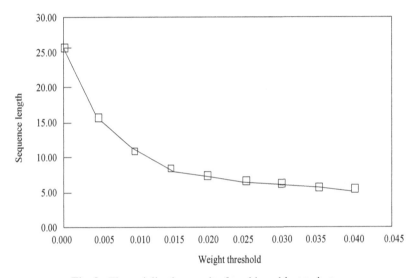

Fig. 9. The serialization result of tracking object trajectory

The following figure shows the comparison of frequent activity pattern recognition effects under different scale of moving objects (Fig. 10).

From the moving object graph and location graph of nodes, we can see that the change of node direction in the improved random direction moving model does not only occur at the edge of the region, but also within the region. And it can be seen that the distribution of nodes is small, and then there are more edges and less questions, but they are more evenly distributed in the whole simulation area (Fig. 11).

When the weight threshold aw is higher and higher, more and more hot spots are filtered out in the moving object sequence, so the average length of the sequence will be shorter and shorter. When the threshold value of activity weight is less than <0.015, the activity weight of hot spots is mainly concentrated in the range of [0,0.015), so a large number of low activity weight location points will be filtered out, which leads to

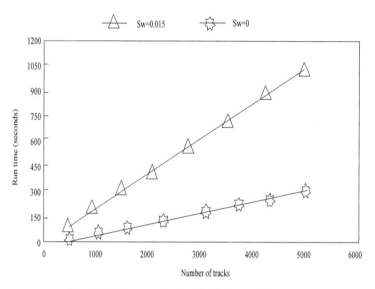

Fig. 10. Time complexity of mining activity mode

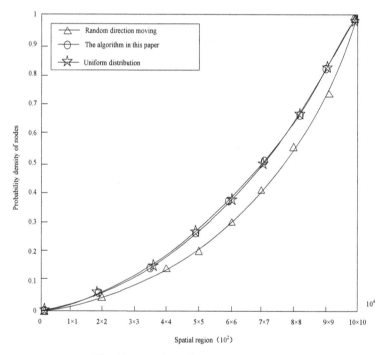

Fig. 11. Tracking effect contrast detection

the rapid shortening of the average length of activity sequence. The weight threshold is greater than 0.015, and fewer active hotspots meet the threshold conditions. Therefore, the less the filtered location points are, and the average length of the activity sequence decreases slowly.

4 Conclusion and Prospect

With the rapid development of intelligent devices in recent years, GPS, GSM network and WiFi Positioning Technology has been widely used, it is easy to track mobile objects, resulting in a large number of spatio-temporal data. How to effectively discover valuable information from these massive spatio-temporal data has become an important research topic. Similarity query of moving objects has broad application prospects in the fields of spatiotemporal data mining and location-based services. This paper focuses on the phase tracking effect of moving objects.

References

1. An, Z., Lin, Q., Yang, L., et al.: Acquiring bloom filters across commercial RFIDs in physical layer. IEEE/ACM Trans. Netw. **15**(99), 1–14 (2020)
2. Yu, J., Lin, C., Zhang, R., et al.: Finding needles in a haystack: missing tag detection in large RFID systems. IEEE Trans. Commun. **65**(5), 2036–2047 (2019)
3. Duncombe, T.A., Dittrich, P.S.: Droplet barcoding: tracking mobile micro-reactors for high-throughput biology. Curr. Opin. Biotechnol. **60**(1), 205–212 (2019)
4. Liu, S., Liu, D., Muhammad, K., Ding, W.: Effective template update mechanism in visual tracking with background clutter. Neurocomputing **458**, 615–625 (2020). https://doi.org/10.1016/j.neucom.2019.12.143
5. Vera-Amaro, R., Rivero-Angeles, M.E., Luviano-Juarez, A.: Design and analysis of wireless sensor networks for animal tracking in large monitoring polar regions using phase-type distributions and single sensor model. IEEE Access **11**(8), 1 (2019)
6. Curtis, C., Mckenna, M., Pontes, H., et al.: Predicting in situ nanoparticle behavior using multiple particle tracking and artificial neural networks. Nanoscale **11**(46), 22515–22530 (2019)
7. Wang, E., Wang, D., Huang, Y., et al.: Siamese attentional cascade keypoints network for visual object tracking. IEEE Access **78**(19), 1–5 (2020)
8. Liu, S., Liu, X., Wang, S., Muhammad, K.: Fuzzy-aided solution for out-of-view challenge in visual tracking under IoT assisted complex environment. Neural Comput. Appl. **33**(4), 1055–1065 (2021)
9. Spilger, R., Imle, A., Lee, J.Y., et al.: A recurrent neural network for particle tracking in microscopy images using future information, track hypotheses, and multiple detections. IEEE Trans. Image Process. **29**, 3681–3694 (2020)
10. Liu, S., Li, Z., Zhang, Y., Cheng, X.: Introduction of key problems in long-distance learning and training. Mob. Netw. Appl. **24**(1), 1–4 (2018)
11. Shi, T.T., Han, W.-J., Tao, N.: Mining aggregation moving pattern of moving object from spatio-temporal trajectories. J. Chin. Comput. Syst. **40**(5), 1099–1106 (2019)
12. Zhou, Q., Huang, S., Cheng, H.-L.: Research on multiple access protocol of Internet of Things nodes based on probability detection. Comput. Simulat. **37**(12), 148–152 (2020)

Position Sensorless Control of Brushless DC Motor Based on Sliding Mode Observer

Li-jun Qiu[1]([⊠]), Ming-fei Qu[1], and Xin-ye Liu[2]

[1] College of Mechanical and Electrical Engineering, Beijing Polytechnic, Beijing 100176, China
[2] College of Economics and Management, China Agriculture University, Beijing 100083, China

Abstract. In order to realize the position sensorless control of brushless DC motor, a position sensorless control method of brushless DC motor based on sliding mode observer is proposed. In order to reduce the chattering of the system, a smooth hyperbolic tangent function is introduced. Therefore, the control system can obtain a smooth linear back-EMF estimate without adding a low-pass filter and a phase compensation module, thereby avoiding the phase lag of the back-EMF estimate. The estimated BLDC motor position sensorless back EMF signal corresponds to 3 virtual Hall signals, and 6 discrete commutation signals are directly obtained, thus eliminating the need for fixed phase shift circuit and phase shift angle calculation. Simulations and experiments show that the proposed method can accurately estimate the line back EMF of the position sensorless brushless DC motor, and achieve the research goal of precise control of the position sensorless brushless DC motor.

Keywords: Sliding mode observer · DC motor · Sensor

1 Introduction

Brushless DC motors have the advantages of simple structure, high reliability, and high power density, and have been widely used in many fields. Brushless DC motors usually use the rotor position sensor to detect the rotor position in real time to achieve commutation. However, the disadvantages of the position sensor limit the application of brushless DC motors. Therefore, in recent years, the position sensorless control of brushless DC motors has become researched. Among the hot spots, the back EMF method is currently the most widely used brushless DC motor sensorless control method. The back EMF method usually detects the opposite EMF or the zero-crossing point of the line back EMF and calculates the ideal commutation point [1].

In this paper, a sensorless control method of Brushless DC motor based on sliding mode observer is proposed. The specific research ideas are as follows:

Firstly, by improving the sign function of the sliding mode observer, the continuous and smooth function of the sensor is obtained, and the data investigation and correction control are carried out to effectively solve the lag caused by the weakening of the observer, which leads to the oscillation of the estimated value up and down.

S. Liu and X. Ma (Eds.): ADHIP 2021, LNICST 417, pp. 342–355, 2022.
https://doi.org/10.1007/978-3-030-94554-1_27

Secondly, based on the line voltage equation of Brushless DC motor, taking the line back EMF as the expanded state quantity, a new state equation is established.

Then, the sliding mode observer is improved to realize the accurate estimation of line back EMF of Brushless DC motor on the premise of reducing chattering

Finally, the estimated line back EMF signal is corresponding to three virtual Hall signals, so as to establish the commutation logic of Brushless DC motor, realize the double closed-loop control of Brushless DC motor without position sensor, and draw a conclusion through experiments.

2 Brushless DC Motor Sensorless Control

2.1 Brushless DC Motor Without Position Sensing Information Collection

With the increasing application of high-performance permanent magnet motors, the requirements for its control performance are getting higher and higher. In order to meet the high performance needs of the brushless DC motor speed control system, the closed-loop control link is essential [4]. The traditional method is to install mechanical sensors on the motor shaft, such as resolvers, photoelectric encoders, tachogenerators, etc. These sensors generally can accurately obtain the motor speed and rotor position information in real time, and realize the closed-loop control of the motor. However, the use of mechanical sensors will increase system cost and reduce system reliability, and mechanical sensors are susceptible to environmental factors such as temperature and humidity, which limit the application of motors. Therefore, position sensorless control technology has received more and more attention and research [5]. Therefore, a position sensorless control method of brushless DC motor based on sliding mode observer is proposed. Mechanical sensors are not installed on the shaft of the motor. Instead, the voltage, current and other signals measured in the motor are used to estimate the motor by a certain method. The rotor position and speed information of the rotor, and then realize the closed-loop control of the motor. During the rotation of the brushless DC motor, the permanent magnet excitation field generates induced electromotive force in the stator winding [6]. The method of estimating the spatial position off by means of the induced electromotive force generated in the stator winding by the permanent magnet excitation flux linkage is an estimation method based on the electromagnetic relationship. Among them, the direct calculation method and the estimation method based on the back electromotive force or the stator flux linkage belong to the estimation method based on the basic electromagnetic relationship of the brushless DC motor. For the surface type brushless DC motor, in the three-phase static ABC coordinate system, there are:

$$u_s = R_s i_s + L_s \frac{di_s}{dt} + j\omega_r \psi_f \tag{1}$$

Among them, $j\omega_r \psi_f$ is the induced electromotive force, and the position d of the rotor in the shaft system is the axis component of the induced electromotive force L_s. Using the strategy of separately controlled starting and automatic control operation, we can get:

$$e_0 = j\omega_r \psi_f = j\omega_r \psi_f(u_s \cos\theta_r + j\sin\theta_r)$$
$$= -\omega_r \psi_f u_s \sin\theta_r + j\omega_r \psi_f \cos\theta_r \qquad (2)$$
$$= e_\alpha + ju_s e_p$$

In the formula, θ_r is the electrical angle between the rotor flux vector R_s and the stator i_s axis. After the rotor is positioned, a rotating magnetic field is given to the motor stator to start the motor. After starting, the sliding mode observer continuously observes the current and speed of the motor. When the motor starts to its current and voltage meet the predicted calculation, then switch to automatic control mode operation [7]. Using the current following characteristics of the sliding mode observer, the observed current is close to the actual current value. The block diagram of the position sensorless control system is as follows (Fig. 1):

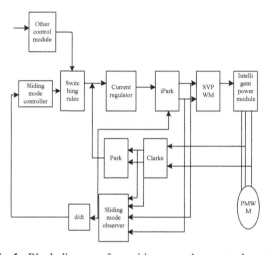

Fig. 1. Block diagram of a position sensorless control system

The control system is mainly composed of sliding mode controller, sliding mode observer, other control modules, switching rules, current regulators, Park converters, Clarke converters, Park inverters, SVPWM modules, smart power modules and brushless DC motors, etc. Link composition [8]. Among them, SCM is used to realize the zero-speed start of the motor. SMO collects the stator voltage and current signals and estimates the rotor position and speed; SMC realizes the smooth transition of the system from the other control mode to the automatic control mode when the output current error of the two control modes approaches zero [9]. The stator voltage equation in the two-phase stationary coordinate system is

$$\begin{bmatrix} u_a \\ u_\beta \end{bmatrix} = \begin{bmatrix} R_s & 0 \\ 0 & R_s \end{bmatrix}\begin{bmatrix} i_\alpha \\ i_\beta \end{bmatrix} + p\begin{bmatrix} L & 0 \\ 0 & L \end{bmatrix}\begin{bmatrix} i_\alpha \\ i_\beta \end{bmatrix} + \begin{bmatrix} E_\alpha \\ E_\beta \end{bmatrix} \qquad (3)$$

Where u_α, u_β is the stator voltage in the $\alpha\beta$ coordinate system; i_α, i_β is the stator current in the $\alpha\beta$ coordinate system, R_s is the stator resistance, L is the stator inductance, and p is the differential operator; E_α, E_β is the back electromotive force in the $\alpha\beta$ coordinate system, And is further described as

$$\begin{bmatrix} E_\alpha \\ E_\beta \end{bmatrix} = \begin{bmatrix} \cos\theta \\ \sin\theta \end{bmatrix} E = 4.44N_s k_{Ns}\phi_m \begin{bmatrix} \cos\theta \\ \sin\theta \end{bmatrix} f_1 \tag{4}$$

In the formula, N_s is the number of stator windings in series per phase; k_{Ns} is the stator fundamental winding coefficient, ϕ_m is the air gap magnetic flux per pole, f_1 is the stator frequency; θ is the angle between the back electromotive force and its α axis component, then:

$$T_{em} = J\dot{\omega}u_\alpha + B_m\omega u_\beta + T_L(E_\alpha - E_\beta) \tag{5}$$

In the formula, $\dot{\omega}$ and J are the motor speed and its first derivative respectively; ω is the moment of inertia, B_m is the friction coefficient; T_L is the load torque, and T_{em} is the electromagnetic torque. The open-loop estimation method based on the mathematical model is simple in calculation and intuitive, and because it does not use integral links and regulators, it has a fast dynamic response [10]. This open-loop estimation method relies on the mathematical model of the motor. The motor parameters selected in the model are usually the parameters when the motor is working in a steady state, and the motor parameters will be in a dynamic state due to temperature changes, magnetic circuit saturation effects and other factors during operation. Changing. Changes in motor parameters will affect the accuracy of rotor position and speed estimation.

2.2 Position Sensorless Adaptive Algorithm for DC Motor

Model reference adaptive system is an effective method of rotor position and speed estimation, which has developed into a mature positionless control. The accuracy of the reference model in the model reference adaptive method directly affects the accuracy of identification. The position accuracy estimated by this method is related to the selection of the model, and it is necessary to ensure that the position deviation can be accurately estimated. After obtaining the position deviation, there must be a reasonable adaptive law to ensure the stability of the system and the robustness to parameter changes while ensuring the convergence speed. This method is computationally intensive and requires a digital signal processor with high-speed computing capabilities. The principle is shown in the figure (Fig. 2).

The essence of the state observer is to reconstruct the state of the control system, use the directly measurable variables in the original system as the input signal of the newly constructed system, and make the output signal of the constructed new system equal to the state of the original system under certain conditions. The reconstructed system is called an observer. The observer-based estimation method has the advantages of good dynamic performance, high stability, and strong parameter robustness. The disadvantage is that the speed regulation effect is not ideal at low speed, and the algorithm is complicated and the calculation amount is large. In recent years, the rapid development of some

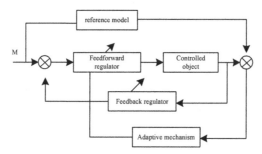

Fig. 2. The principle of position sensorless adaptive DC motor

high-speed signal processors DSP has greatly promoted the application of the observer method in position sensorless control. At present, the commonly used observers include Kalman filter, sliding mode observer and so on. Sliding mode variable structure control means that according to the current state of the system, the structure of the feedback controller changes purposefully and continuously in accordance with the predetermined control law. Through the continuous change of the control system structure, the structure of the system is finally switched back and forth at a very high frequency, and the state point of the system performs a high-frequency up and down movement, that is, sliding mode operation. The advantages of the sliding mode variable structure control system are that it is insensitive to parameter changes, robust to external disturbances, and has fast dynamic response performance. The sliding mode observer method is to design the control loop in the general state observer into a sliding mode variable. The form of the structure makes the sliding mode observer method have parameter robustness that the general state observer does not have. Because in the actual variable structure control system, the switching device always exhibits certain time and space relay characteristics and system inertia, etc., these factors will make the variable structure system exhibit inherent chattering. The existence of jitter makes the steady-state accuracy of the control system poor, so debounce is the key problem to be solved by the sliding mode observer method. The mathematical model of the brushless DC motor in the three-phase static ABC coordinate system is a set of variable coefficient differential equations, which are more complicated. In order to simplify the mathematical model of the permanent magnet motor, a two-phase stationary a coordinate system and a two-phase rotating L coordinate system are introduced. Transform each physical quantity from the three-phase coordinate system to the two-phase static voltage equation:

$$\begin{cases} \frac{di_\alpha}{dt} = -R_i \frac{i_\alpha}{L_c} - \frac{e_\alpha}{L_c} + \frac{u_\alpha}{L_c} \\ \frac{di_p}{dt} = -R_s \frac{i_s}{L_s} - \frac{e_\beta}{L_s} + \frac{u_\beta}{L_s} \end{cases} \tag{6}$$

The standard current of the dynamic equivalent circuit of the brushless DC motor in the two-phase rotating coordinate system is Z. When the motor is at a standstill, the voltage equation can be simplified to:

$$\begin{bmatrix} i_d \\ i_q \end{bmatrix} = \begin{bmatrix} \frac{1}{Z_i} & 0 \\ 0 & \frac{1}{Z_q} \end{bmatrix} \begin{bmatrix} u_d \\ u_q \end{bmatrix} \tag{7}$$

The back EMF equation is:

$$\begin{cases} e_\alpha = -\psi_f \omega_x i_d \sin\theta \\ e_\beta = \psi_f \omega_r i_q \cos\theta \end{cases} \qquad (8)$$

The voltage equation of the brushless DC motor in the coordinate system is:

$$\begin{cases} u_d = R_s i_d + \frac{d\psi_d}{dt} - \omega_r \psi_q \\ u_q = R_s i_q + \frac{d\psi_q}{dt} + \omega_r \psi_d \end{cases} \qquad (9)$$

As the brushless DC motor speed control system is widely used in some high-precision fields, the requirements for the performance of the control system and the speed control system are getting higher and higher. By detecting the three-phase current and obtaining the alternating and direct axis currents after coordinate transformation, the current inner loop is formed; the feedback speed n is detected by the position sensor, and the outer speed loop is formed. The parameters of the surface-mount brushless DC motor are shown in Table 1.

Table 1. Self-adaptive parameters of brushless DC motor

Parameter	Numerical value	Company	Parameter	Numerical value	Company
Rated voltage U_N	310	V	Resistance per phase R_S	1.5	Ω
Rated power P_N	0.8	kW	d-axis inductance L_d	1.48	mH
Rated speed n_N	3000	rpm	q-axis inductance L_q	1.48	mH
Rated torque T_N	2.6	N.m	Line number of optical encoder	2500	p/r
Rated current I_N	4.2	A	Flux per pole Φ	1.779×10^{-3}	Wb
Mechanical moment of inertia J	1.03×10^{-4}	Kg.m^2	Polar logarithm P_n	2	

The basic requirements of the brushless DC motor control system can be summarized as the system's wide speed range, simple control, and high reliability. Torque control has fast response, high precision, small waveform, etc. From the motion equation of the motor, it can be seen that the control of the brushless DC motor is the control of the electromagnetic torque, and the control of the speed and position can be achieved through the control of the electromagnetic torque of the motor. The torque equation of the brushless DC motor can be transformed into:

$$\begin{aligned} T_e &= p_n(\psi_d i_q - \psi_q i_d) \\ &= p_n \psi_f i_q + p_n(L_d - L_q)i_d i_q \end{aligned} \qquad (10)$$

The speed and current double closed-loop control structure and p_n current control strategy are adopted. The control system mainly includes: speed setting i_d,i_q speed loop PI regulator, current loop PI regulator, Park/Clarke inverse transformation, SPWM modulation, three-phase inverter, surface-mount brushless DC motor, current sensor, Clarke transformation, Park transformation and a position sensor for detecting the position and speed of the rotor.

2.3 Realization of Sensor Control of Brushless DC Motor

The brushless DC motor system adopts a vector control scheme based on physical sensors, and the rotor information fed back in the system is directly measured by the position sensor. At the same time, a new sliding mode observation algorithm is attached to the brushless DC motor, and the adaptive detection method of the rotor position of the brushless DC motor system has different effects at low speed and high speed. The rotor position estimation method suitable for low speed and the rotor position estimation method suitable for high speed have different mechanisms for estimating the rotor position. The rotor position estimation method suitable for low speed is to obtain the rotor position information by detecting the salient poles of the motor, applying different excitation methods and different signal detection and separation methods. The rotor position estimation method suitable for high-speed uses the back EMF of the motor to obtain the rotor position information. The sliding mode observers and full-dimensional and reduced-dimensional observers mentioned above are all suitable for high-speed rotor information detection methods. Therefore, in order to improve the performance of the brushless DC motor system at startup, an open-loop/closed-loop switching module is designed. The principle of open-loop/closed-loop switching module is introduced below. The figure below is the block diagram of the open-loop/closed-loop switching module (Fig. 3).

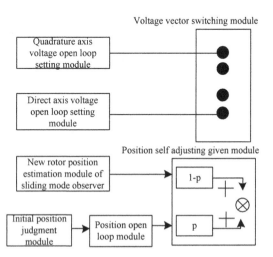

Fig. 3. Block diagram of the open-loop and closed-loop switching control module

Combining the above analysis, in a complete chopping excitation interval, the winding has gone through three stages: current establishment, current chopping, and free-wheeling shutdown. When the winding chopper is turned on, the power supply voltage Udc. For forward access, the winding voltage balance equation is shown in the formula.

$$+U_{dc} = Ri + T_e L(\theta)\frac{di}{dt} + T_e i\omega\frac{dL(\theta)}{d\theta} \tag{11}$$

There are two types of winding chopper shut-off: zero voltage freewheeling and negative pressure freewheeling. The input voltages are 0 and Udc respectively. The winding voltage balance equations in the zero voltage and negative pressure freewheeling states are as shown in formulas (12) and (13). Shown:

$$0 = Rj + L(\theta)\frac{di}{dt} + i\omega\frac{dL(\theta)}{d\theta} \tag{12}$$

$$-U_{dc} = Ri + L(\theta)\frac{di}{dt} + i\omega\frac{dL(\theta)}{d\theta} \tag{13}$$

From the theoretical analysis of sliding mode variable structure control, the sliding mode can be designed according to the actual expected control performance. In addition, due to the characteristics of the continuous change of the control system structure during the control process, the variable structure control system can maintain the original control performance under the influence of the internal parameter changes of the system and the disturbance of the external environment, and has good robustness. However, from the perspective of the entire control process, the control quantity u of sliding mode variable structure control is constantly switched, which is a discontinuous function of the state quantity. This control method is a discontinuous control method. In addition, because the actual switching device always has time relay characteristics and space relay characteristics, as well as the influence of factors such as system inertia, state measurement errors, etc., in the actual sliding mode variable structure control, the sliding mode motion is not like theoretical The above is generated on the pre-designed sliding mold surface, but a kind of back and forth traversal near the sliding mold surface, resulting in high-frequency vibration. This phenomenon is the chattering of the sliding mode control system.

The existence of chattering will have an adverse effect on the control system: on the one hand, it will affect the control accuracy of the control system and increase the energy consumption of the system. On the other hand, it can also cause unmodeled high-frequency motion components in the system and cause high frequency in the system oscillation. In the actual current chopping control, in order to make the torque output more smooth and stable, the width of the chopping loop is usually set to be small, and the current change between each chopping on and chopping off cell is small, and the chopping state The duration is short and the rotor displacement angle is not significant. Therefore, it can be approximately considered that the phase current and phase inductance remain unchanged in each chopping on and off interval, which is reflected in the formula, that is, the adjacent chopping on and off on the right side of the equation The voltage drop of the winding of the first term and the electromotive force of the third term are equal in the process of chopping off.

The current position sensorless control system of brushless DC motors is a vector control system. When the motor speed is within the low speed range of zero speed or nearby, it is impossible to avoid the control dead zone and observing the dead zone. At this time, the speed loop in the vector control system is in an open loop state. Therefore, it is necessary to adopt other control methods to realize the zero-speed start or stop of the motor without changing the structure of the vector control system. In order to ensure the smooth start or stop of the motor, it is necessary to ensure that the amplitude of the stator circular rotating magnetic field generated by the system and the The phase can increase or decrease steadily. The change law of the voltage space vector during the starting process is given (Fig. 4):

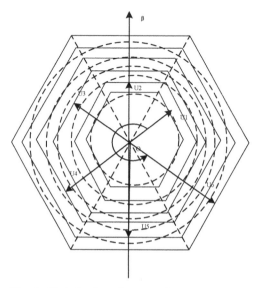

Fig. 4. Variation law of stator voltage space vector

In the figure, the α axis and β axis respectively represent the components of the voltage space vector in the stationary two-phase α, β coordinate system; the dotted circles nested in each other are the state trajectories produced by the voltage space vector rotating one circle when the amplitude is constant; U is For the six voltage space vectors that increase in the order of counterclockwise rotation, with the increase of the vector amplitude, the motor speed will increase accordingly, so as to realize the smooth start of the motor. In the separate control mode, the electromagnetic torque is easily affected by various disturbances of the power grid, load and the motor itself, which causes the rotor speed and the synchronous speed of the stator rotating magnetic field to be inconsistent, that is, the motor is out of step. Torque is easily affected by various disturbances of the power grid, load and the motor itself, which causes the rotor speed to be inconsistent with the synchronous speed of the stator rotating magnetic field, that is, the motor loses synchronization.

$$
\begin{cases}
T_{cm} = \frac{uE}{|Z|\omega} \sin(\delta + \alpha_Z) - \frac{E^2 R_s}{|Z|^2 \omega} \\
Z = -U_{dc} R_s + j\omega L \\
\alpha_Z = 90^\circ - \phi_Z = 90^\circ - \arctan\left(\frac{\omega L}{R_s}\right) \\
\omega = 2\pi f + U_{dcl}
\end{cases}
\tag{14}
$$

In the above algorithm, E is the control interference coefficient, in order to shorten or even eliminate the approaching motion process in sliding mode control, make the system speed as low as possible when switching from the other control mode to the automatic control mode, and strengthen the robustness of the entire speed regulation process. Awesome.

3   Analysis of Results

In order to verify the effectiveness of the method proposed in this paper, a simulation model is established using Matlab/Simulink, and the traditional sliding mode observer is compared with the improved sliding mode observer proposed in this paper. The motor parameters in the simulation are shown in the table (Table 2):

Table 2. Brushless DC motor parameters

Parameter	Numerical value	Parameter	Numerical value
Rated power P/W	70	Stator resistance R/Ω	0.3
Rated voltage U_N/V	24	Stator inductance L_I/mH	0.45
Rated torque T_N/(N · m)	0.23	Rated speed n_t/(r/min)	3000
Rated current I_N/A	4	Polar logarithm p	5

On the basis of applying the variable voltage and frequency conversion control strategy to the brushless DC motor control start, the SMO is added to predict the rotor position and the motor speed, and the position sensorless control system of the brushless DC motor is completed in real time according to the prediction results Switching of automatic control mode. In order to verify the system response characteristics of the algorithm proposed in this paper when the motor is located near zero speed and the control mode is switched, the system simulation results with the superposition of ramp and sine wave signals as input are given. The simulation uses Matlab as a tool, and the speed output response and following error with the superposition of ramp and sine wave signal as input are shown in the figure (Fig. 5):

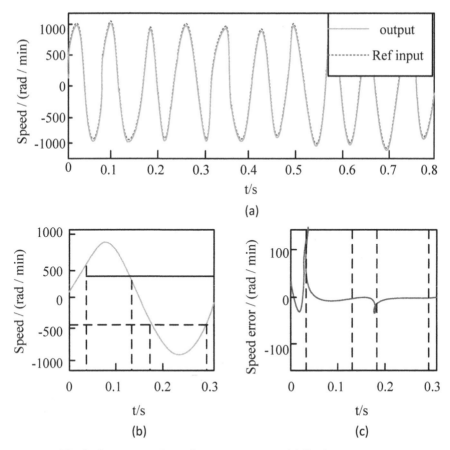

Fig. 5. Sensor control speed output response and following error curve

nges step by step is given. In the simulation, the brushless DC motor set speed is switched from 30 r/min to 50 r/min at 0.4 s, and the load torque is 100 Nm. Among them, Figure a is the actual rotor position waveform, Figure b is the estimated rotor position, Figure c is a partial enlarged view of the actual and estimated rotor position in 0.4 s (Fig. 6).

The above simulation results show that the two main problems of traditional sensor control are: when the motor starts, the rotor position estimation error is large; the rotor position estimation value and the back EMF estimation value have large chattering and have a certain hysteresis. Chattering is an inherent problem of sliding mode observers. The discontinuity of sliding mode variable structure control makes the estimated value chatter larger. The hysteresis of position estimation is caused by the phase delay caused by the first-order low-pass filtering. The control method in this article has solved the above problems to a great extent, can better guarantee the safety of countless DC motors, solve the control dead zone problem under the position sensorless control mode, and better realize the control mode from the other control mode to the automatic control mode. The smooth transition of the system fully meets the research requirements.

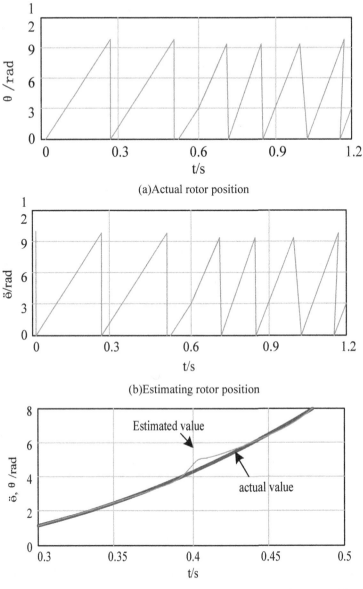

(a)Actual rotor position

(b)Estimating rotor position

(c) Compared with the estimated rotor position, the actual rotor position is partially enlarged at 0.4s

Fig. 6. Waveform when the sensor control speed step changes

In order to verify the effectiveness of this method, the sensorless control time of reference [3] method and this method is adopted, and the results are shown in Table 3.

Table 3. Control time without position sensor under different methods

Number of iterations	Sensorless control time/s	
	Literature [3] methods	Paper method
10	15	0.1
20	18	0.2
30	22	0.5
40	24	0.3

It can be seen from Table 3 that the control time without position sensor is different under different methods. When the number of iterations is 10, the sensorless control time of the method in document [3] is 15 s, and the sensorless control time of the text method is 0.1 s. When the number of iterations is 30, the sensorless control time of the method in document [3] is 22 s, The sensorless control time of the text method is 0.5 s. The sensorless control time of this method is much lower than that of other methods, which shows that the control efficiency of this method is high.

4   Concluding Remarks

A sensorless control method of Brushless DC motor based on sliding mode observer is proposed in this paper. The variable voltage variable frequency control strategy is used for other control starting. Combined with the prediction function of sliding mode observer for rotor angle and speed, according to the principle of current following and speed following, the time-varying sliding mode controller is used to switch the control system from other control mode to automatic control mode. Simulation and experiments show that the control strategy proposed in this paper has good application value and achieves high comprehensive control performance.

In the future, we will study how to obtain more accurate sensor control accuracy.

Fund Projects. Research on Sensorless Control of Brushless DC motor based on sliding mode observer (YZK2015045).

References

1. Liu, C., Liang, H., Ueda, N., Li, P., Fujimoto, Y., Zhu, C.: Functional evaluation of a force sensor-controlled upper-limb power-assisted exoskeleton with high backdrivability. Sensors **20**(21), 63–79 (2020)

2. Aaa, B., Ak, A., Bm, A., et al.: Feedback linearization based sensorless direct torque control using stator flux MRAS-sliding mode observer for induction motor drive. ISA Trans. **98**(10), 382–392 (2020)

3. Rastegari, A., Arefi, M.M., Asemani, M.H.: Robust H ∞ sliding mode observer-based fault-tolerant control for One-sided Lipschitz nonlinear systems. Asian J. Control **21**(1), 114–129 (2019)

4. Liu, S., Li, Z., Zhang, Y., Cheng, X.: Introduction of key problems in long-distance learning and training. Mob. Netw. Appl. **24**(1), 1–4 (2018). https://doi.org/10.1007/s11036-018-1136-6

5. Narzary, D., Veluvolu, K.C.: Higher order sliding mode observer-based sensor fault detection in DC microgrid's buck converter. Energies **14**(6), 1586–1599 (2021)

6. Adjissi, N., Sait, B.: Monitoring of multicell converter and DC motor. Prog. Electromag. Res. B **86**, 159–176 (2020)

7. Litak, G., Syta, A., Wasilewski, G., et al.: Dynamical response of a pendulum driven horizontally by a DC motor with a slider–crank mechanism. Nonlinear Dyn. **99**(3), 1923–1935 (2020)

8. Liu, S., Liu, X., Wang, S., Muhammad, K.: Fuzzy-aided solution for out-of-view challenge in visual tracking under IoT assisted complex environment. Neural Comput. Appl. **33**(4), 1055–1065 (2021)

9. Qian, P.Q., et al.: Compressive measurement identification of linear time-invariant system application in DC motor. J. Beijing Inst. Technol. **28**(03), 21–31 (2019)

10. Saxena, S., Hote, Y.V.: Design and validation of fractional-order control scheme for DC servomotor via internal model control approach. IETE Tech. Rev. **36**(1), 49–60 (2019)

Research and Analysis with Intelligent Education

The Evaluation Method of Online and Offline Blended Teaching Quality of College Mathematics Based on Bayes Theory

Xing-jun Li$^{(\boxtimes)}$, Pan Xue, and Zhi-wei Shao

Eurasia University, Xi'an 710065, China

Abstract. Due to the current online and offline mixed teaching quality evaluation methods of mathematics, the index weight evaluation criteria have not been determined, resulting in low accuracy of evaluation index weight calculation results and small discrimination of evaluation results. For this reason, a Bayesian online and offline college mathematics online and offline based on Bayesian theory is proposed. Mixed teaching quality evaluation method. Based on the domestic and foreign research results of teaching quality evaluation and the principle of determining the price index, the teaching quality evaluation index is set, and the online and offline mixed teaching quality evaluation index of mathematics is divided into three levels; the index weight evaluation standard is determined, and the teaching quality evaluation index weight is calculated. Carry out consistency test; use the expected value and revised probability of Bayesian theory to construct the teaching quality evaluation model, design the teaching quality evaluation set, and complete the online and offline mixed teaching quality evaluation of mathematics. The experimental results show that the method for evaluating the quality of online and offline mixed teaching of mathematics in this study, when evaluating the quality of online and offline mixed teaching of mathematics in colleges and universities, the calculation results of the evaluation index weights are accurate and the evaluation results are highly differentiated.

Keywords: Bayesian theory · College mathematics · Online and offline mixing · Teaching quality evaluation

1 Introduction

The current rapid development of computer technology has driven the change of teaching mode, and the online and offline mixed teaching mode has become the main teaching mode of current teaching [1]. However, in this teaching mode, the quality of teaching varies, and the teaching effect has not reached the desired effect [2]. Therefore, China attaches great importance to teaching quality evaluation research. In China, new perspectives and directions have been found in the evaluation of the quality of mathematics teaching. Frontline workers and researchers have realized the shortcomings of traditional classroom education quality, and constantly strive to correct the shortcomings of traditional classroom evaluation and give new evaluations. The way is active

S. Liu and X. Ma (Eds.): ADHIP 2021, LNICST 417, pp. 359–370, 2022.
https://doi.org/10.1007/978-3-030-94554-1_28

and more comprehensive objective exploration [3]. Some scholars believe that the current classroom evaluation should be carried out from six angles: "people-oriented and objective", "qualitative and quantitative evaluation", "summative evaluation and formative evaluation", "other evaluation, self-evaluation, mutual evaluation", "diversification of evaluation indicators and generalization of evaluation field", "guiding evaluation and artistic evaluation"; in addition, some scholars put forward fuzzy evaluation On the basis of price method, it constructs the three-level index of College English evaluation; uses matter-element evaluation method and fuzzy comprehensive evaluation to construct the secondary index of evaluation and sets weight for it; divides learning process into three dimensions, and takes process evaluation and result evaluation as teaching evaluation standard [4]. Literature [5] thinks that only by adopting four ways of evaluation, paying attention to process evaluation, combining students' self-evaluation with peer, teacher and parent evaluation, and establishing students' Growth Portfolio, can the integrity and dynamics of teaching evaluation be reflected.

However, when the above-mentioned teaching quality evaluation method evaluates the quality of online and offline mathematics teaching in colleges and universities, the calculation result of the evaluation index weight is inaccurate, resulting in a large error in the evaluation result. The Bayesian theory method is a basic method in statistical model decision-making and an important part of subjective Bayesian induction theory. Bayesian decision-making is to estimate part of the unknown state with subjective probability under incomplete intelligence, then use Bayesian formula to modify the probability of occurrence, and finally use the expected value and modified probability to make the optimal decision. This paper uses Bayesian theory to study the online and offline hybrid teaching quality evaluation method of college mathematics, and through simulation experiments, it verifies the effectiveness of this method and solves the problems of traditional methods.

2 The Evaluation Method of Online and Offline Blended Teaching Quality of College mathematics Based on Bayes Theory

2.1 Setting up Teaching Quality Evaluation Index

Design the online and offline mixed teaching quality evaluation index of college mathematics, draw on the domestic and foreign research on the quality evaluation of online and offline mixed teaching of college mathematics, put forward the principles for determining the quality evaluation index of online and offline mixed teaching of college mathematics, and determine the quality of online and offline mixed teaching of college mathematics Evaluation indicators, as shown in Table 1.

In the online and offline mixed teaching quality evaluation indicators of college mathematics as shown in Table 1, four first-level indicators, 8 s-level indicators, and 18 third-level indicators are selected, covering classroom culture, structure, content, effects, etc. There are four aspects of mathematics online and offline teaching. Among them, the first level index establishes the method framework, corresponding to the result elements of the project, and the first level index increases the teaching effect, which serves as the basis for the second and third level decomposition and refinement. Level indicators.

Table 1. Quality evaluation index of online and offline mathematics teaching in Colleges and Universities

First level indicator	Secondary indicators	Level three indicators
Classroom Culture A	Teacher-student relationship A1	Listen to A11
		Support A12
	Learning atmosphere A2	Organization A21
		Interactive A22
Course structure B	Expression B1	Evaluation reflection B11
		Communication display B12
	Plan B2	Design plan B21
		Evaluation plan B22
	Question B3	Ask question B31
		Defining problem B32
Course content C	Teaching feature C1	Authenticity C11
		Result clarity C12
		New mosaic technology C13
		Task ambiguity C14
Teaching effect D	Organizational Association and Application D1	Application D12
		Knowledge understanding D13
	Learning Orientation D2	Cooperation D21

The first level indicator of classroom culture a mainly evaluates the classroom situation, atmosphere and the relationship between teachers and students. It can be regarded as the integration of norms, values, ideas and behavior patterns that occur in the process of classroom teaching and shared by all members. Classroom culture is to limit the culture in the specific classroom situation, with its own characteristics. Classroom teaching reform is not only the change of behavior, concept, mode and other teaching activities, but also the reconstruction of classroom culture. Therefore, in the first dimension of classroom culture A, teacher-student relationship A1 and learning atmosphere A2 are taken as the second dimension.

The first-level indicator curriculum structure B is mainly reflected through teaching design, curriculum implementation and other links, focusing on whether teachers effectively organize classroom activities and promote student learning. Taking into account the confusion of evaluation caused by the order of the classroom teaching links and the intersection of processes, following the principle of designing evaluation indicators, design expression B1, plan B2, question B3 and three secondary indicators.

The implementation includes task driven, theme activities, experiential learning, game learning, inquiry learning, experimental learning and other specific ways. Online and offline teaching is the most important implementation method. therefore, for the first level index course content C, the teaching feature C1 is selected as a secondary index.

The first-level index teaching effect D is the output of classroom teaching, which evaluates the learning achievements of students through the performance of students' homework or achievement works. The effectiveness of teaching is measured by how well the teaching goals are achieved. Developmental evaluation requires that evaluation be integrated into classroom teaching and become an integral part of the teaching process. It focuses on evaluating students' performance and the development of emotional attitudes in real situations. Finally, meaningful learning can be realized. through classroom learning orientation and students' learning evaluation, organizational association and application D1 and learning orientation D2 are regarded as the secondary indicators of teaching effect D.

2.2   Calculate the Weight of Teaching Quality Evaluation Index

Based on the evaluation index of online and offline mixed teaching quality of College mathematics, the data is normalized according to the data calculated from the selected indicators. Therefore, the range transformation method is used, and the dimensionless processing method is adopted. If the original data information is y and the dimensionless data is x, then:

$$y_{ki} = \frac{x_{ki} - x_{\min(i)}}{x_{\max(i)} - x_{\min(i)}} \times 100 \tag{1}$$

In the formula, y_{ki} represents the original data of the i second-level index in the k first-level index, x_{ki} represents the dimensionless data of the i second-level index in the k first-level index, and $x_{\min(i)}$ represents the i second-level index. The data with the smallest dimensionless index, $x_{\max(i)}$ represents the data with the largest dimensionless i secondary index, and in Table 1, the value range of each index is concentrated between [0, 100] [7, 8]. At this time, after normalizing each index in Table 1 according to formula (1), the weight of the evaluation index of the online and offline mixed teaching quality of college mathematics can be calculated. Therefore, suppose that the set of the online and offline mixed teaching quality evaluation index U of college mathematics is $U = \{u_1, u_2, u_3, \cdots, u_n\}$, where u represents the first and second index in the online and offline mixed teaching quality evaluation index U of college mathematics, and n represents the online and offline hybrid of college mathematics. The number of teaching quality evaluation indicators. In this study, rough set theory is used to calculate the weight of online and offline mixed teaching quality evaluation index of College mathematics. It is necessary to use 1–9 scales to compare the evaluation indexes of online and offline mixed teaching quality of College mathematics in Table 1 to form the evaluation index of online and offline mixed teaching quality of College mathematics. The evaluation criteria of weight are shown in the Table 2.

Table 2. Evaluation criteria of index weight

Serial number	Pairwise compare the importance of the first index and the first index	Scale P_{mj} assignment
1	The first indicator is as important as the first indicator	1
2	The first indicator is slightly more important than the first	3
3	The first indicator is more important than the first	5
4	The first indicator is much more important than the first indicator	7
5	The first indicator and the first indicator are extremely important	9
6	Between the above two adjacent judgments	2,4,6,8
7	The comparison between the importance of the first index and the first index is the reciprocal of each other	$P_{mj} = \frac{1}{P_{mj}}$

At this point, you can get the evaluation index set $U = \{u_1, u_2, u_3, \cdots, u_n\}$ of the online and offline mixed teaching quality of n major colleges and universities. Through pairwise comparison, the obtained judgment matrix U is:

$$U = \begin{bmatrix} u_{11} & u_{12} & \cdots & u_{1n} \\ u_{21} & u_{22} & \cdots & u_{2n} \\ \vdots & \vdots & \ddots & \vdots \\ u_{n1} & u_{n2} & \cdots & u_{nn} \end{bmatrix} = (u_{mj})_{n \times n} \tag{2}$$

At this time, u_{mj} in the formula (2) has properties such as $u_{mj} > 0$, $u_{mj} = \frac{1}{u_{mj}} (i \neq j)$, $m = j$ $(m = j; m, j = 1, 2, \cdots, n)$ [9]. According to the calculation result of formula (2), assuming that the relative weight of the online and offline mixed teaching quality evaluation index of college mathematics is $W_{U(i)}$, then:

$$W_{U(i)} = \frac{2}{n(n-1)} \sum_{i=1}^{n} u_{ki} \tag{3}$$

At this time, according to the value of $W_{U(i)}$ obtained by formula (3), the composite weight W_{U_k} of each index to the target layer is calculated, then:

$$W_{U_k} = \left\{ W_{U_{k_1}}, W_{U_{k_2}}, W_{U_{k_3}}, \cdots, W_{U_{k_n}} \right\} \times W_{U(i)} \tag{4}$$

The relative weight $W_{U(i)}$ and composite weight W_{U_k} calculated based on Eqs. (3) and (4) need to be tested for consistency to verify whether the index weight matrix established according to the online and offline mixed teaching quality evaluation indicators of colleges and universities meets Consistency test results. When the consistency test results are not met, the index weight matrix needs to be rebuilt. Therefore, in this study, the consistency index is set to C, and the maximum characteristic root is $\lambda_{\max}$, then:

$$C = \frac{\lambda_{\max} - n}{n - 1} \tag{5}$$

In the formula, when $\lambda_{\max} = 0$ is the consistency index $C = 0$, the index weight matrix constructed at this time is completely consistent, and the larger the consistency index C value, the worse the consistency of the index weight matrix. In addition, when $\frac{C}{R} \leq 0.1$, the index weight matrix can pass the consistency test. where R is the average random consistency index, and according to the matrix order, there is a fixed constant value.

At this time, in the calculated consistency index C, there is an average random consistency index R, that is, an average random consistency index R. According to the data ratio $C.R.$, the consistency of the weight calculation results can be determined, then:

$$C.R. = \frac{C}{R} \tag{6}$$

According to the calculation result of formula (6), when $C.R. < 0.1$, matrix $C.R. \geq 0.1$ meets the consistency requirement; when U, matrix D must re-modify the matrix value, and change the online and offline mixed teaching quality evaluation index of college mathematics determined in Table 1.

After the above calculation process, the online and offline mixed teaching quality of college mathematics is evaluated this time, the online and offline mixed teaching quality evaluation index of college mathematics is determined, and the calculated weight vector meets the consistency index evaluation result.

On this basis, the quality evaluation model of online and offline mixed teaching of college mathematics is constructed. According to the quality evaluation of online and offline mixed teaching of college mathematics, this paper designs the quality evaluation set of online and offline mixed teaching of college mathematics, and evaluates the quality of online and offline mixed teaching of college mathematics.

2.3 Evaluation of Online and Offline Mixed Teaching Quality of Mathematics in Colleges and Universities

Constructing Teaching Quality Evaluation Model Based on Bayesian Theory
According to the formula (3) and formula (4), the weight of the online and offline mixed teaching quality evaluation index for college mathematics can be known. The evaluation factors are related. The more the evaluation factors are calculated, the more important the evaluation index of the online and offline mixed teaching quality of the college mathematics teaching. Therefore, suppose the number of factors of the online and offline mixed teaching quality evaluation index of college mathematics is x, and

x ≥ 0, the linear relationship of this factor can be expressed by B, then:

$$B(x) = \begin{cases} 0 & x \leq 3 \\ 2\left(\frac{x-3}{15-3}\right)^2 & 3 < x \leq 9 \\ 1 - 2\left(\frac{x-3}{15-3}\right)^2 & 3 < x \leq 15 \\ 1 & x > 15 \end{cases} \tag{7}$$

Through formula (7), the linear relationship of each evaluation factor in the online and offline mixed teaching quality evaluation index of college mathematics can be obtained. At this time, it is also necessary to determine the linear order of each factor. For group $U = \{u_1, u_2, u_3, \cdots, u_n\}$ of the online and offline mixed teaching quality evaluation index U of college mathematics, the fitness value of each factor in the online and offline mixed teaching quality evaluation index of college mathematics is arranged in descending order, namely $f(U_1) \geq f(U_m) > f(U_{m1}) \geq f(U_{mj})$, then the probability P of each factor being selected is:

$$P = \frac{1}{x}\left[\eta^+ - \frac{\eta^+ - \eta^-}{x - 1}(x - 1)\right], x = 1, 2, \cdots, x \tag{8}$$

In the formula, η^+ and η^- are the most important and worst factors of each evaluation factor in the online and offline mixed teaching quality evaluation indicators of college mathematics, and $1 \leq \eta^+ \leq 2$, $\eta^- \geq 0$ [10–13]. At this time, substituting the online and offline mixed teaching quality evaluation index, index weight calculation, and the weight of each factor in the evaluation index in college mathematics, you can establish an English autonomous learning ability evaluation model, as shown in Fig. 1.

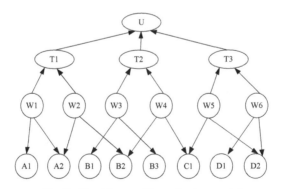

Fig. 1. Teaching quality evaluation model

In Fig. 1, U represents the online and offline mixed teaching quality of college mathematics, T represents the evaluation data, and W represents the weight of the evaluation index of the online and offline mixed teaching quality of college mathematics. At this point, according to the teaching quality evaluation model shown in Fig. 1, the online and offline mixed teaching quality evaluation set of college mathematics can be designed to evaluate the quality of online and offline mixed teaching of college mathematics.

Design Teaching Quality Evaluation Set

According to the index weights obtained from formulas (3) and (4), using the teaching quality evaluation model shown in Fig. 1, the online and offline mixed teaching quality evaluation indicators of college mathematics determined in Table 1 are ranked according to the index weight levels. At this point, you can determine the comment level of the online and offline mixed teaching quality evaluation index of college mathematics, and establish the evaluation matrix V_i:

$$V_i = \begin{bmatrix} v_{11} & v_{12} & \cdots & v_{1\sigma} \\ \vdots & \cdots & \ddots & \vdots \\ v_{\sigma 1} & v_{\sigma 2} & \cdots & v_{\sigma\sigma} \end{bmatrix} \tag{9}$$

In the formula, v represents element level and σ represents comment. At this time, suppose the online and offline mixed teaching quality evaluation index set of college mathematics is $u = \{u1, u2, \cdots, u_i, \cdots, u_j, \cdots, u_n\}$, according to the evaluation matrix V_i of formula (9), the determined online and offline mixed teaching quality evaluation index comment E of college mathematics is:

$$\sigma = \{\text{Good, good, average, poor, poor}\}$$
$$= \{v_1, v_2, v_3, v_4, v_5\} \tag{10}$$

According to formula (10), the quality of online and offline mixed teaching of mathematics in colleges and universities can be evaluated.

To sum up, it is determined that the online and offline mixed teaching mode is adopted for mathematics teaching. The existing online and offline mixed teaching quality evaluation indicators of mathematics in colleges and universities are used to calculate the performance evaluation index weights and conduct consistency tests to ensure the performance evaluation results Objectivity, establish a teaching quality evaluation model, set performance evaluation comment sets, and evaluate the quality of online and offline mixed teaching of mathematics in colleges and universities.

3 Analysis of Experimental Demonstration

This paper verifies the quality evaluation method of online and offline mixed teaching of mathematics in Colleges and universities. It selects a school in a certain region that adopts the online and offline mixed teaching mode to evaluate the quality of online and offline mixed teaching of mathematics in the school. This paper takes the online and offline mixed teaching quality evaluation method of College mathematics as experimental group A, and takes the two teaching quality evaluation methods mentioned in the introduction as experimental group B and experimental group C respectively. Three teaching quality evaluation methods are used to evaluate the mathematics teaching quality of the selected schools. The weight results of the evaluation indexes calculated by the three evaluation methods, the consistency with the weight results of the expert evaluation, and the differences of the evaluation methods are compared.

3.1 Experiment Preparation

To verify the quality evaluation method of online and offline mixed mathematics teaching in colleges and universities designed this time, the school that adopts the online and offline mixed teaching mode to conduct mathematics teaching has already conducted one semester of online and offline mathematics teaching for newly enrolled students. Students have adapted to the online and offline mixed teaching mode of mathematics. Therefore, for the school's mathematics teaching, the quality evaluation index of the school's online and offline mixed teaching is unified, as shown in Table 3.

Table 3. Evaluation indicators of the school's online and offline mixed teaching quality

First level indicator	Secondary indicators	Level three indicators
Course structure A	Question A1	Ask question A11
	Model A2	Form an opinion A21
	Plan A3	Design plan A31
	Express A4	Communication display A41
Course content B	Teaching feature B1	Authenticity B11
		Teaching Technology B12
	Content meaning B2	Accuracy B21
		Bracket and material B22
Classroom Culture C	Learning atmosphere C1	Interactive C11
		Organization C12
	Teacher-student relationship C2	Listen to C21
		Support C22
Teaching effect D	Learning Orientation D1	Creativity D11
		Cooperation D12
	Organizational Association and Application D2	Knowledge understanding D21
		Application D22

Based on the characteristics of mathematics online and offline mixed teaching quality evaluation methods, three teaching quality evaluation methods are used to calculate the design of evaluation indicators, and the three evaluation methods are analyzed from two aspects of evaluation index weight and differentiation. Check the three evaluation methods, calculate the weight results of the evaluation indicators shown in Table 3, and whether the results are consistent with the weight results after expert evaluation; compare the differences of the three evaluation methods.

3.2 The First Group of Comparative Experiments

Based on the experimental parameters set in this experiment, the school's online and offline mixed teaching methods, teaching process, and teaching survey results were distributed to 20 mathematics education experts. These 20 experts calculated the evaluation index weights of the experimental design, And the obtained evaluation index weight result is used as the experimental control group. For the three evaluation methods, the calculated evaluation index weights were compared with the control group, and the comparison results are shown in Fig. 2.

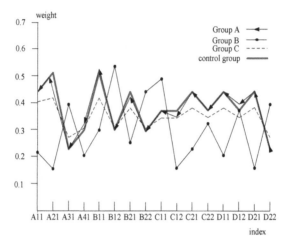

Fig. 2. Comparison Chart of evaluation index weight

It can be seen from Fig. 2 that the experimental group B calculated the teaching quality evaluation weight result of the school's online and offline hybrid teaching mode, which is quite different from the control group's calculation result. There is only D12 index. The calculated weight value is compared with the control Groups are the same; experimental group C calculates the school, mathematics online and offline mixed teaching mode, teaching quality evaluation weight results, and the control group calculation results, although the difference is smaller than the experimental group B, the index weight fluctuation range is similar to the control group, but The index weight value differs greatly from the control group. In experiment A, not only the fluctuation range of the index weight is very similar to that of the control group, but also the index weight value is less different from the control group, and the indexes overlap in many places. It can be seen that the online and offline mixed teaching quality evaluation method designed this time calculates the evaluation index weights and the results are more accurate, which can accurately evaluate the quality of online and offline mixed mathematics teaching.

3.3   The Second Set of Comparative Experiments

Based on the evaluation index designed in this experiment, on the basis of the first group of comparative experiments, the second group of comparative experiments was carried out. Three evaluation methods were used to calculate the discrimination of evaluation indexes. At this time, the discrimination degree of evaluation index is defined as:

$$\eta = \sum_{i=1}^{n} \left(K_i - \overline{K}\right)^2, \overline{K} = \sum_{i=1}^{n} \frac{K_i}{n} \qquad (11)$$

In the formula, η represents the degree of discrimination, n represents the number of evaluation indexes, i represents the number of rating index levels, K represents the weight of the evaluation index, K_i represents the weight of the i-th grade evaluation index, and $\overline{K}$ represents the average value of the weight of the evaluation index. At this time, the discrimination degree of the evaluation index can be calculated according to formula (11), and the comparison results are shown in Table 4.

Table 4. Comparison results of evaluation indexes

Score	A	B	C	D	η
Evaluation value of experiment a group	90.9	70.6	81.7	88.3	857.6
Sorting of evaluation values of experiment a	1	4	3	2	
Evaluation value of experiment group b	72.5	80.0	70.3	76.9	737.2
Sorting of evaluation value of experiment b group	3	1	4	2	
Evaluation value of experiment group c	73.6	72.5	69.7	76.3	700.7
Sorting of evaluation values of experiment group c	2	3	4	1	

It can be seen from Table 4 that the discrimination between group B and group C is significantly smaller than that of group A. It can be seen that the evaluation results of experiment a are more distinguishable and have certain advantages in evaluating the quality of online and offline mathematics teaching.

Based on the above two sets of experiments, it can be seen that the online and offline hybrid teaching quality evaluation method designed this time has accurate calculation results of evaluation index weights, and the evaluation results are highly distinguished. The obtained online and offline hybrid teaching quality evaluation results are clear and clear.

4   Conclusion and Outlook

In summary, the online and offline hybrid teaching quality evaluation method used in this study makes full use of Bayesian theory. Under incomplete information, the characteristics of subjective probability estimation for some unknown states are used to evaluate the online and offline hybrid language. Teaching Quality. However, the mathematics online and offline mixed teaching quality evaluation method in this study still lacks a certain theoretical basis for the determination of various indicators, which is only based on the current common practice of colleges and universities, and has not been verified by scientific methods. Therefore, in the future research, it is necessary to further study the evaluation index of mathematics online and offline mixed teaching quality, so as to improve the breadth and depth of the evaluation of mathematics online and offline mixed teaching quality through the evaluation index.

Fund Projects. School level key courses of Engineering Mathematics (2019KC030); Construction of mathematics curriculum system for applied talents training of new engineering – Taking Xi'an Eurasia University as an example (2020GKYB006); Existence of positive solutions for a Gause type predator prey model with cross diffusion term (2020JK0815).

References

1. Peng, Y., Ren, W., Qian, R.: Millimeter wave MIMO channel estimation based on bayesian compressive sensing. Comput. Simul. **37**(4), 151–154,334 (2020)
2. Tao, Q.: Research on the teaching evaluation methods of senior chemistry and the application of real-time evaluation. Sci. Educ. Art. Cult. **19**, 136–137 (2019)
3. Wang, T.: College English teaching evaluation and classroom teaching behavior. Jiaoyu jiaoxue luntan **2**, 241–242 (2019)
4. Zhang, X.: The improvement of teaching evaluation methods and analysis of influenced degree of evaluation index to evaluation results. J. Ili Norm. Univ. (Natural Science Edition) **12**(2), 61–67 (2018)
5. Li, H., Wang, H.: The Influence of aestheticized teaching on the teaching evaluation of " novice" teachers. J. Aesthet. Educ. **11**(1), 19–22 (2020)
6. Chen, X.L.: Analysis of Ideological and Political Classroom Teaching Evaluation. Teach. Forest. Reg. **3**, 65–67 (2018)
7. Chen, Y., Chen, G.: A method of police intelligence prediction based on Bayes theory. J. Shandong Inst. Light Ind. (Natural Science Edition) **33**(2), 69–77 (2019)
8. Haiyan, W., Li, Y., Li, W.: Pedestrian detection for surveillance video based on gradient histogram Bayesian theory. Comput. Eng. Des. **39**(6), 1679–1684 (2018)
9. Zhou, Y., Jia, F., Xi, S.: Experiment research on multi-model structural identification based on Bayesian theory. J. Hunan Univ. (Natural Sciences) **45**(5), 36–45 (2018)
10. Zheng, P., Shuai, L., Arun, S., Khan, M.: Visual attention feature (VAF): a novel strategy for visual tracking based on cloud platform in intelligent surveillance systems. J. Parall. Distrib. Comput. **120**, 182–194 (2018)
11. Zhang, M., Yu, X.: The construction of teaching quality evaluation system of modern apprenticeship based on big data. J. Phys. Conf. Ser. **1578**, 012124 (2020)
12. Guo, M., Xiong, X., Feng, G., et al.: A research on behavioral deviation and reform program of teaching quality evaluation in experiment and training of colleges. Lab. Res. Explor. **038**(005), 229–232,274 (2019)
13. Zhao, H.: Research on the evaluation method of teaching quality of Chinese as a foreign language based on immersion teaching. Manag. Sci. Res. Chin. English Vers. **1**, 8–10 (2019)

Design of Public Course Teaching Evaluation System in Colleges and Universities Under Mobile Education Platform

Xiao-yu Wang[✉], Jian-rong Li, and Chang-zhu Li

College of General Education, Heihe University, Heihe 164300, China

Abstract. In the construction of the public curriculum system in colleges and universities, it is necessary to evaluate the teaching of public curriculum in colleges and universities in order to improve the control of the key links of the curriculum system. Therefore, the design of a mobile education platform under the university public course teaching evaluation system. The module design of the system is as follows: the client module uses C/S architecture, the client is mainly responsible for data processing, the processed data is sent to the server, and the server stores it. The main task of server module is data preparation and data storage. The hardware configuration module configures the hardware of client and server respectively. Teaching evaluation module is mainly used for teaching evaluation of public courses in colleges and universities. The main function of the evaluation information query module is to query the evaluation information that has been made. The main function of the evaluation result management module is to view the evaluation result information. Evaluation summary module is the continuation of teaching evaluation module. The database module is responsible for storing the evaluation data. Through the combination of hardware and software, the teaching evaluation of public courses in colleges and universities can be realized. And the system is tested. The test results show that the function and performance of the system are good. It has great promotion value.

Keywords: Mobile education platform · University public curriculum · Teaching evaluation system · C/S architecture

1 Introduction

Public courses in Colleges and universities refer to the courses that students of any department or major in colleges and universities need to study, as well as the compulsory courses of basic theory, basic knowledge and basic skills for students of a certain discipline or major. Courses are offered in the form of compulsory or elective courses. The purpose of this kind of curriculum is to form the basic literacy of students and the basic knowledge and ability needed by a major, that is to form the public, general ability and common professional ability of students [1]. Public courses in colleges and universities should dare to make bold attempts, actively explore the goal, direction and

S. Liu and X. Ma (Eds.): ADHIP 2021, LNICST 417, pp. 371–382, 2022.
https://doi.org/10.1007/978-3-030-94554-1_29

way of reform, make overall planning for the current basic courses, and further optimize the curriculum model.

According to the characteristics and actual needs of college students' future career, the content of professional courses is constantly increased, break the single course public curriculum system in colleges and universities, and highlight the basic, comprehensive and professional functions of new public courses. For example, reference [2] studies the evaluation of structured teaching effect of hand hygiene knowledge for medical staff, and reference [3] proposed a study on users' attitude towards image archiving and communication system in Kum teaching hospital of Iran based on technology acceptance model. In short, the reform of the public curriculum system must be clear about the curriculum objectives, that is, to build a comprehensive public curriculum system from the three directions of curriculum quality, efficiency and development, to carry out the basic curriculum reform according to the special professional curriculum content, and to emphasize the "necessity" of future development, such as real-time reduction of curriculum, not only to reform a single basic curriculum, but also to improve the quality of curriculum, pay more attention to the overall optimization and construction of multiple courses.

The construction of the public curriculum system in colleges and universities should highlight the principle of people-oriented and ability-oriented, comprehensively and seriously select the curriculum content and knowledge, pay attention to the development direction of professional development, improve the application effect of theoretical knowledge in the future real life, and realize the teaching efficiency with the smallest curriculum structure as far as possible. Therefore, this paper designs a mobile education platform under the teaching evaluation system of public courses in colleges and universities. In order to evaluate the teaching of the public courses in colleges and universities in the construction of public course system in order to strengthen the control of the key links of the course system.

2   Design of University Public Course Teaching Evaluation System Based on Mobile Education Platform

2.1   Design Client Module

The client module is designed based on mobile education platform. The mobile client of the system adopts C/S architecture, and the client is mainly responsible for processing the data, sending the processed data to the server and storing it by the server [4, 5]. The advantage of C/S architecture is that it can separate data processing from data storage. The client is developed based on the mobile education platform. Finally, the client is installed on the user's Android mobile phone or other mobile devices. On the premise of accessing the network, the client logs in to the system and sends the teaching evaluation request to the web page. After receiving the user's request, the web page accesses the database to obtain the information of the user's evaluation, and finally returns the request result to the user in the end. The functions developed by the client include: user login, teacher information modification, user password modification and teaching evaluation. The main users are teachers and students. Mobile education platform client provides a good interface for teachers and students to operate the corresponding functions conveniently [6].

The development tools used to develop the client module include:

① Java development kit - Java SE Development Kit JDK, the version used is JDK 8 UPDATE 101. After downloading, complete the installation according to the prompts.

② Eclipse, the IDE tool of Java, uses Eclipse 4.2.1 and installs ADT, a general plug-in supporting Android projects and tools. Eclipse is installation free. After downloading it, unzip it to the directory F:\Android\Eclipse.

③ Mobile education platform Android development kit, the version used is Android SDK_ r17-windows. After downloading, unzip it to the hard disk. Run the SDK manager, and the program will automatically detect whether it is updated. After selecting Updates/New and Installed, the file will be automatically downloaded to the F:\Android\SDK directory. Then add the tools path in the SDK to the system variable tools. The steps are as follows: right click "computer" on the desktop, select "properties" option, select "advanced system settings" on the properties page, and then click "advanced", "environment variables" and "system variables". Add "F:\Android\SDK\tools" to the system variable Path and click "OK" to save Save.

④ Android simulator can be created by using the Android SDK. The program can be tested directly through the virtual device AVD. After starting eclipse, select "Android virtual device manager" under the "window" menu to open the Android Virtual Device Manager window. Select new to create a new Android simulator. After the creation is successful, click the start button to start the virtual device. The virtual device interface consists of two parts, the left part is the display part of the simulated mobile phone, and the right part is the input part of the simulated mobile phone [7, 8].

2.2 Design the Server Module

The main task of server module is data preparation and data storage. The system server is written in Java and realizes the Web page through MyEclipse. The database mainly contains the information data of user, class, department, course and so on. On the server side, the system administrator has the highest level of authority, responsible for maintaining and managing the data, and can query, add or delete the data. The functions developed by the server include: user login, password modification, teaching evaluation and evaluation result query. The function diagram is shown in Fig. 1 [9].

2.3 Design Hardware Configuration Module

Under the mobile education platform, the main users of university public course teaching evaluation system are concentrated in the campus network, with a large number of people, large amount of business data and high requirements for system processing speed, which requires that the hardware environment of the system deployment must have advanced performance and strong processing ability. The hardware configuration of the server is determined by the system design, the amount of data, the number of users and the amount of concurrent access. Hardware configuration requirements are mainly considered from storage capacity, stability, running speed and other aspects. For this system, according

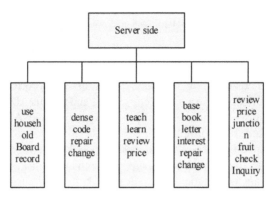

Fig. 1. Function diagram of server module

to the demand and the equipment that the school can provide, the IBM × 3850 server is used as the system server and Gigabit network card.

The hardware configuration of the client is CPU dual core 2.6 GHz and above, memory 2 G and above, hard disk 160 g and above, using more than 100 m network card.

2.4 Design Teaching Evaluation Module

Teaching evaluation module is mainly used for teaching evaluation of public courses in Colleges and universities. The student user can select the evaluation student to log in on the login interface, and input the student's account and password. If it is consistent with the database, the user can enter the student function information page, where the user can select the evaluation subsystem and query subsystem for corresponding operation. If the account password does not match, enter the prompt page. After successful login, students can modify and edit personal data in the query subsystem, such as personal basic information, account password modification and other operations.

In addition, students can view the courses that need to be evaluated, as well as the detailed introduction of these courses, such as the corresponding books and class hours, so as to have an intuitive understanding of the courses that need to be evaluated. In the teaching evaluation part, the system will judge the type of users, distinguish the users and students, teachers, school management, expert group members, and then issue different types of questionnaires for different types of users. Student users can fill in the questionnaire according to their own basic situation in class and submit it [10].

In the teaching evaluation module, using fuzzy mathematics theory to establish the teaching evaluation fuzzy evaluation model is the basis of the design. Taking teacher evaluation as an example, based on the theory of fuzzy mathematics, the evaluation factor set and comment set of teaching evaluation in the module are given in detail.

Evaluation factor set: for teacher evaluation, evaluation users are divided into four categories: students, teachers at the same level, expert group and leaders in charge. According to the different attention characteristics of four types of users, four different evaluation factor sets are designed to meet the personalized needs, as shown in Table 1, 2, 3 and 4.

Table 1. Student evaluation factor set

First level indicators	Secondary indicators	Score
Lesson preparation	Teaching preparation, content and attitude (10)	10
Lecture	Key and difficult points (10), language (10), inspiration (10), teaching aids (10), blackboard writing (10)	50
Task	Attitude (10), timeliness (10)	20
Educating people	Work attitude (10), requirements (5), care (5)	20

Table 2. Set of evaluation factors for teachers at the same level

First level indicators	Secondary indicators	Score
Teaching objectives and content	Teaching objectives (10), teaching contents (10), ideological education (10)	30
Teaching process and methods	Teaching methods (5), teaching media (5), the role of teachers and students (5), feedback control (5)	20
Instructional design	Organization ability (5), language expression (5), education mechanism (5), demonstration blackboard writing (5)	20
Instant effect	Student feedback (10), goals achievement (10), time standard (10)	30

Table 3. Evaluation factor set of expert group

First level indicators	Secondary indicators	Score
Teaching objectives	Thinking training (10), teamwork (10), value cultivation (10)	30
Instructional design	Objective design (5), student foundation (5), resource utilization (5), teaching students in accordance with their aptitude (5)	20
Teaching organization	Teaching ability (5), situation creation (5), teaching effect (5), classroom organization (5)	20
Teaching art	Classroom atmosphere (10), comprehensive measurement (10), teaching mechanism (10)	30

Table 4. Evaluation factor set of leaders in charge

First level indicators	Secondary indicators	Score
Routine assessment	Professional ethics (10), subject knowledge (10), teaching ability (10), cultural literacy (10)	40
Teamwork	Sense of participation (10), good at working together (10), unity and cooperation (10)	30
Continuing Education	Lifelong learning (10), changing ideas (10), research and innovation (10)	30

The evaluation factor set is stored in tables according to the first level factor set and the second level factor set.

The weight of comment set and evaluation factors is as follows:

There are four grades in the comment set: excellent, good, medium and poor. The corresponding scores of each two-level evaluation factor are: excellent: 9–10 points, good: 7–9 points, medium: 5–7 points, poor: less than 5 points; if it is a five point system, the corresponding scores are: excellent: 4–5 points, good: 4–3 points, medium: 3–4 points, poor: less than 2 points.

According to the analysis of the evaluation factors and considering the opinions of all parties, the weight values of the first level evaluation factors and the second level evaluation factors are set (see the table above respectively), and the weight values of the first level and the second level evaluation factors are given according to the percentage system in the table. According to the above weight value and the actual scoring situation, and calculate the total score of different teachers to be evaluated. In addition, the weights of students, teachers at the same level, experts and leaders in charge are set as 0.2, 0.3, 0.3 and 0.2 respectively, so that the scores of different evaluation users can be comprehensively considered and the final results can be given [11].

The fuzzy operators of evaluation factors are as follows:

According to the primary and secondary evaluation factors in the evaluation factor set and the collected student evaluation data, its membership can be determined. The fuzzy operator is weighted average operator. This method can take into account the weight of all evaluation factors, so that the evaluation results reflect the effect of all evaluation factors.

Finally, the fuzzy evaluation model is established: according to the set of evaluation factors, the set of comments, the weight value, and the user evaluation recorded after the system running, the collected evaluation records can be calculated according to the fuzzy evaluation method to get the final evaluation results. The main function of the module is to record the evaluation contents of different users in the database. In the client page design, the evaluation factors are presented in the web page design. After user evaluation, the data submitted to the background are processed according to different users, and the evaluation scores corresponding to the evaluation factors are obtained and stored in the database.

2.5   Design Evaluation Information Query Module

The main function of the evaluation information query module is to query the evaluation information that has been made, so as to facilitate students, teachers and management personnel to understand the evaluation information.

In the evaluation information query module, the personnel who need to query the evaluation information enter their own account name and password to log in to the module. For student users, you need to input the information to query, such as the course name, to query their own evaluation information of the course; students can view the basic information of the course and their own evaluation content, and can modify their own evaluation; as student users, they have no authority to view the evaluation information of other students. Teacher users and expert users also have the right to view their own evaluation contents of a course and the summary information of their own teaching courses, understand the students' evaluation of their own teaching, and understand and correct the students' feedback information; teacher users can only query their own teaching evaluation. School management users have the right to view their own evaluation information of some courses, and can query the course evaluation information of all teachers and the summary of evaluation results.

2.6   Design Evaluation Result Management Module

The main function of the evaluation result management module is to view the evaluation result information. This module contains permission control. Different users can only query the evaluation result information within their own permissions. The evaluation result management module can collect all the evaluation information in a certain way and analyze the data. The evaluation result management module is the most important module in the system, which classifies and summarizes the specific index results of student evaluation, peer teacher evaluation, expert evaluation and leadership evaluation, so as to obtain the overall evaluation results of a teacher. The evaluation results represent the opinions and suggestions of students, teachers, experts and leaders on the teaching level of the teacher. Teacher users can view the evaluation results of their own courses, so as to adjust the way of teaching. The users of the leadership can view the evaluation results of all teachers, so that the school leadership can have a comprehensive grasp of the teaching ability of the whole teaching team, and reasonably formulate the corresponding incentive measures.

The evaluation results management module can also generate evaluation reports according to needs for users with permission to query and download. In addition, the specific method of classification and summary of the evaluation result management module can be changed. If the administrator gets a better evaluation and statistical method, the new algorithm can be implemented in the module.

2.7   Design Evaluation Summary Module

Evaluation summary module is the continuation of teaching evaluation module, the core content and goal of the system, which reflects the existence value of teaching evaluation system. The summary function of evaluation results reflects the level of a teacher's teaching ability, is the feedback of the teacher's teaching level, and also provides the school leadership with the opportunity to fully understand the teaching level of teachers. The evaluation summary module needs to summarize all the evaluation contents in the database, which involves the operation of a large number of data. Therefore, the evaluation summary function can only be implemented after all users or the vast majority of users have completed the evaluation. The system manager will submit the evaluation summary request, and the system will start the evaluation summary and get the evaluation summary results, and save the evaluation summary results to avoid errors. The summary without evaluation is carried out many times, resulting in unnecessary waste of resources.

The evaluation summary module is started by the system administrator, which will call Java background class to access the database, classify and summarize according to the first level evaluation factors of different users, and store the summary results in the database table. After the summary results of the four types of users are completed, the total evaluation results are obtained, which are also saved in the database.

2.8   Design Database Module

In order to achieve the functional requirements of the teaching evaluation system, it is necessary to store the evaluation data. There are a large number of users in the teaching evaluation system, and the main users are teachers and students. The personal information storage of a large number of users can not simply use the file mode, but should use the efficient data storage scheme database. Database was born to solve the problem of large amounts of data storage. Database table is the core infrastructure of the whole teaching evaluation system. The design of database table is generally directly related to the efficiency of data extraction and preservation, especially for big data.

Database conceptual structure design is the core part of database design. Its purpose is to find out the relationship between different entities in the real world, and get the entity relationship diagram (ER diagram). ER diagram reflects the relationship between objects and their attributes in the database [12, 13]. The system database file is to be deployed and managed on the specific database management system, so the database management system used in the system is selected, and the corresponding restrictions should be followed when designing the conceptual structure of the database. E-R relational model is the most commonly used tool for database conceptual structure design. E-R relational model is an abstract concept in reality, which is not restricted by the specific implementation environment of the system. E-R diagram consists of entity, attribute and relation. Entity: entity is the real existence in the objective world, which can distinguish each other. Entity can represent a kind of concrete things, can also represent a kind of abstract things. For example: customers, orders, contracts, etc. are entities. Attribute: the characteristics of things in the objective world are called attributes. The order number, signing date, goods, quantity, etc. are attributes of the order entity. Any entity must have one or a group of primary attributes. The primary attribute can distinguish the entity from other entities. The primary attribute of an order is the order number. Connection: the relationship between entities is called connection. Connections include "many to many", "one to one" and "one to many". For example, a demand plan can only belong to one department of the company, while the relationship between demand plan and department of the company is many to one. There is no definite boundary between entity and attribute. At present, the commonly used method of dividing entity and attribute is to follow two principles: one is that the attribute does not need to be described any more, and it is already the smallest data item, such as the order number, which is a string of symbols without any other characteristics; the other is that the attribute cannot be associated with the entity. The E-R diagram is designed according to the main function modules, the main user information and the information to be stored of the teaching evaluation system obtained in the demand analysis stage.

The logical structure design of database is to transform the conceptual structure model of database into the relational model supported by database management system. The design of the logical structure of the database is restricted by the selected database management system. The implementation of the database of the university public course teaching evaluation system under the mobile education platform is realized on the MySQL database management system.

As the E-R relational model of database is an abstract expression of the real world, it is not limited by the specific database system, but also can not be directly applied by the database system. When designing database conceptual model, it is necessary to transform per model into data relation logic supported by database system. The basic requirement of database logical structure design standardization is to meet the paradigm of 3NF. The logical model of teaching evaluation system is processed according to the standard of 3NF. After analyzing the concept model, the specific database table is designed.

3 Simulation Experiment Test

3.1 Experimental Design

From two aspects of function test and performance test, this paper tests the teaching evaluation system of public courses in Colleges and Universities under the mobile education platform.

System testing is to deploy the prepared software and hardware, network equipment and other equipment to form a whole system and run it, and then carry out various assembly tests and confirmation tests of the system by professional testers. System testing is to test the whole application system. The objective is to find out the difference between the realized system and the customer's requirements, find out the dissatisfaction with the customer, and then respond. In this test of teaching evaluation system, two different test methods are used: black box test and white box test. Black box testing refers to the ability to complete the given output according to the given input, which is generally completed by the program testing engineer. In this test process, the black box test is used to test whether the function modules of the software system are completed, and check whether the software has completed all the required functions of the requirements specification. White box testing means that developers can export the correct results through the analysis of program code and structure, which is generally completed by developers.

3.2 Test Results

Function Test Results

In the function test of the university public course teaching evaluation system under the mobile education platform, observe whether the input and output functions of the system can operate normally; whether the functions of student evaluation, user query, evaluation factor management are consistent with expectation.

Take user login, student evaluation function and evaluation factor modification function as examples, as shown in Table 5.

Table 5. Test examples of user login function, student evaluation function and evaluation factor modification function

Serial number	Test content	Test method	Test results	Remarks
1	User login function	Enter the correct student account and password, click the login button, you can enter the system, enter the wrong student account and password, and jump to the account and password error page	(1) The correct account can enter the system, and student users can not enter the unauthorized module (2) Wrong account and password cannot enter the system, jump to the error page	Nothing
2	Student evaluation function	Click the evaluation button to enter the evaluation page. After filling in the evaluation content, you can submit the evaluation normally	(1) Be able to enter the evaluation page (2) Be able to fill in the evaluation content (3) Be able to submit evaluation content normally	Nothing
3	Modification function of evaluation factors	Administrator users can log in with their own account, enter the maintenance subsystem, enter the evaluation factor modification page, change and save the evaluation factors, and view the changed evaluation factors on the evaluation page	(1) The administrator can modify the evaluation factors after entering the evaluation factors modification page (2) The administrator checks the evaluation page to confirm that the evaluation factor has changed	Nothing

Performance Test Results

In the performance test of the public course teaching evaluation system in colleges and universities under the mobile education platform, the running performance of the system is tested, such as the system response time, and further determine the timeliness and stability of the system response.

The performance test results show that the system has good timeliness. After a certain period of operation, the system shows high reliability.

After the function test and performance test, the teaching evaluation system solves the problems found in the test. At present, the system has been running, and the customers reflect that the system is running well.

4   Conclusion and Prospect

Under the mobile education platform, the use of university public course teaching evaluation system has changed several aspects of teaching

(1) The management mode of the teaching management department has been changed: the teaching management department is no longer limited to the traditional management mode. It can quickly and accurately grasp the evaluation data of students and teachers through the real-time statistics of the system, which makes the teaching method reform faster and more efficient.

(2) Promote interaction between teachers and students: students can truly and objectively evaluate a teacher's teaching level, so that teachers can quickly know the result. In fact, it is also interaction between teachers and students.

(3) Improve the teaching level and quality of teachers: through the system, teachers can quickly know the evaluation of students and teachers, and also enable teachers to quickly adjust teaching methods to adapt to students' learning and improve the quality of teaching.

(4) Greatly increase the enthusiasm and interest of students to participate in teaching activities: through real-time teaching evaluation, let students feel that they have a way to express the views of teachers' teaching level, greatly increasing the enthusiasm of students to participate in teaching activities.

(5) It evaluates the labor value of teachers more scientifically, systematically and efficiently.

(6) However, this study only tests a small number of teaching content and samples, so the next research will expand the teaching sample data to test and optimize the effectiveness of the design system.

Fund Projects. The Higher Education Teaching Reform Research Project in 2019, Title: Research on Construction and Practice of Teaching Quality Evaluation System for Public Basic Courses in Local Universities under the Background of Deepening Comprehensive Reform (SJGY20190452).

References

1. Whitlock, A.M., et al.: Teaching elementary social studies during snack time and other unstructured spaces. J. Soc. Stud. Res. SciVerse ScienceDirect **43**(3), 229–239 (2019)
2. Ahuja, S., Anita, P., et al.: Assessing the effectiveness of structured teaching on knowledge of hand hygiene among healthcare workers - ScienceDirect. Clin. Epidemiol. Glob. Health SciVerse ScienceDirect 7(3), 396–398 (2019)
3. Hamta, A., Mohammadzadeh, M., Hemati, M., et al.: Study of the attitude of users towards picture archiving and communication system based on the technology acceptance model in teaching hospitals of Qom, Iran. Qom Univ. Med. Sci. J. **14**(6), 1–8 (2020)
4. Shreedhara, A.K., Shanbhag, S., Joseph, R.C., et al.: A study of maternal breast feeding issues during early postnatal days. SciMed. J. **2**(4), 219–224 (2020)
5. Liu, S., Li, Z., Zhang, Y., et al.: Introduction of key problems in long-distance learning and training. Mob. Netw. Appl. **24**(1), 1–4 (2019)
6. Fitzpatrick, C., van Hoover, S., et al.: A: DBQ in a multiple-choice world: a tale of two assessments in a unit on the Byzantine Empire. J. Soc. Stud. Res. SciVerse ScienceDirect **43**(3), 199–214 (2019)
7. Lay-Khim, G., Bit-Lian, Y.: Simulated Patients' Experience towards simulated patient-based simulation session: a qualitative study. SciMedicine J. **1**(2), 55–63 (2019)
8. Liu, S., Glowatz, M., Zappatore, M., et al. (eds.): e-Learning, e-Education, and Online Training. Springer International Publishing, pp. 1–374 (2018). https://doi.org/10.1007/978-3-319-49625-2
9. Kbrs, T., Sisteminde, E., Retimi, T., et al.: Teaching history in the turkish cypriot education system. Gazi niversitesi Gazi Eitim Fakültesi Dergisi **40**(3), 1193–1217 (2020)
10. Lazi, A., Mili, S., Fier, Z.: BLS training: standard vs virtual reality BLS training. J. Resuscitatio Balcanica **6**(15), 227–231 (2020)
11. Erba, K., Hüseyin, N.: Prediction validity of teaching efficacy on task-centered anxiety: a study on physical education teacher candidates. Kuramsal Eitimbilim, **13**(4), 701–715 (2020)
12. Cesur, K.: Opinions on the use of magic card tricks in teaching English to young learners. Trakya Eitim Dergisi **10**(1), 285–297 (2020)
13. Fu, W., Liu, S., Srivastava, G.: Optimization of big data scheduling in social networks. Entropy **21**(9), 902 (2019)

Quality Evaluation Model of Mobile Internet Innovative Education Personnel Training Based on Fuzzy Analytic Hierarchy Process

Cong-gang Lyu[✉]

Jiangxi Tourism and Commerce Vocational College, Nanchang 243000, China

Abstract. Aiming at the problems of low accuracy and long evaluation time of traditional evaluation methods of mobile Internet innovative education talent cultivation quality, this paper proposes a fuzzy AHP based evaluation model of mobile Internet innovative education talent cultivation quality. According to the quality elements of mobile Internet innovative education personnel training, the system of mobile Internet innovative education personnel training is established. This paper uses the fuzzy analytic hierarchy process to construct the evaluation model of mobile Internet innovative education talent training quality, sets up the evaluation factor set and comment set, establishes the fuzzy evaluation matrix, determines the weight of each evaluation factor, calculates the fuzzy comprehensive evaluation vector and comprehensive evaluation value, and evaluates the mobile Internet innovative education talent training quality according to the principle of maximum membership. The experimental results show that the evaluation time of the method based on the fuzzy analytic hierarchy process is short, and it can effectively improve the quality evaluation accuracy of mobile Internet innovative education.

Keywords: Fuzzy analytic hierarchy process · Mobile social network · Mobile education · Personnel training · Quality evaluation

1 Introduction

At present, with the popularity of intelligent mobile devices and the improvement of functional scalability, people are beginning to use intelligent mobile devices to realize mobile office, mobile entertainment and mobile shopping. Various mobile applications are increasingly used in people's studies, work and life [1]. With the popularity of mobile devices, the rapid development of computer hardware and software technology, Internet technology, teaching mode and content need to be constantly updated and enriched, it is more and more difficult to achieve a higher level in the original teaching hours. However, traditional classroom teaching methods are limited by time and space, which can not change the current situation of teaching methods limited by time and space. The mobile internet teaching mode can better solve this problem. With the emergence, development and application of high and new technology, mobile Internet innovation education is an education mode adapting to the development of the knowledge economy

© ICST Institute for Computer Sciences, Social Informatics and Telecommunications Engineering 2022
Published by Springer Nature Switzerland AG 2022. All Rights Reserved
S. Liu and X. Ma (Eds.): ADHIP 2021, LNICST 417, pp. 383–393, 2022.
https://doi.org/10.1007/978-3-030-94554-1_30

era [2]. Mobile Internet innovation education is a new education concept and mode, which aims at cultivating high-quality talents with innovative consciousness, innovative thinking, innovative ability and innovative personality suitable for the development of the new era, and helps college students establish innovative consciousness, stimulate innovative spirit, master innovative knowledge and improve innovative ability through multiple channels such as schools, governments and enterprises.

Teaching quality refers to the degree of students' development under the condition of education, and also the degree of education results. Students are the only carrier and evaluation object of education quality [3]. The quality standard of mobile Internet innovative education personnel training is the basic premise of quality evaluation. Knowledge, ability and quality are the basic concepts in the current quality education theory research, and also the basic indicators of mobile Internet innovative education personnel training quality evaluation. At present, some scholars have done some research on the evaluation of teaching quality, and have made some achievements. Reference [4] proposed a learning quality evaluation model based on neural network RBF algorithm. Based on the analysis of the main problems and complexity of the evaluation system of learning quality, this paper studies the evaluation system of College English teaching quality by using the RBF regularized network method and RBF neural network decision algorithm. The model can monitor the learning quality in real time, but the accuracy of educational quality evaluation is low. Based on the above background, the quality evaluation of mobile Internet innovative education personnel training is increasingly concerning, and the evaluation of training quality has become a problem that all educational institutions must face. In order to improve the quality evaluation accuracy of mobile Internet innovative education and shorten the quality evaluation time of mobile Internet innovative education, this paper proposes a quality evaluation model for mobile Internet innovative education based on the fuzzy analytic hierarchy process. According to the quality elements of education and training, the education personnel training system is established. Using the fuzzy analytic hierarchy process, this paper constructs the training quality evaluation model, determines the evaluation weight through the evaluation factor set and matrix, and evaluates the training quality of educational talents. This method can shorten the evaluation time and improve the accuracy of quality evaluation.

2   Construction of Talent Training System for Mobile Internet Innovation Education

2.1   Building the Functional Structure of Mobile Internet Teaching

Mobile Internet technology consists of computer programming, mobile network communication and computer network. In order to improve the quality of innovative talent training by using mobile internet teaching, the implementers should have the ability to use mobile applications and products, acquire information, transmit information, process information by using mobile devices, and use mobile Internet resources. The architecture of mobile internet teaching network is as Fig. 1.

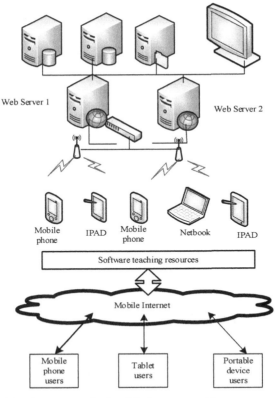

Fig. 1. Framework of mobile internet teaching structure

Teachers, students and administrators can connect to the web server through mobile phones, ipads and netbooks via wireless network, and the web server will apply for business processing to the DB server according to the user's request [5]. Further put forward the implementation scheme of mobile internet teaching mode transformation for colleges and universities. As a mobile internet teaching, the management of teaching resources and teaching behavior is its most important function, and its specific functions are as Fig. 2.

Mobile internet teaching mode is a new teaching mode, which is a teaching activity carried out in the mobile Internet environment with computer software technology, mobile Internet technology and multimedia technology [6]. Through the mobile Internet and teaching resources, and through the role of load balancer, the user requests are distributed to different servers, so as to improve the responsiveness and performance of the system.

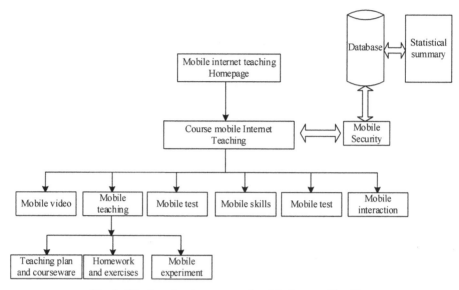

Fig. 2. The functional structure of mobile Internet Teaching

2.2 Construction of Mobile Internet Innovative Education Personnel Training System

In order to improve the quality of innovative personnel training in colleges and universities, it is necessary to clarify the elements of the quality system of personnel training. Clearly defined the training objectives, training specifications, curriculum system, teaching staff, teaching conditions, quality management and guarantee and other elements [7]. From a macro point of view, any factors that affect the process of personnel training can be regarded as elements in the model; from the perspective of colleges and universities, combined with the characteristics of innovative education focusing on practice, the

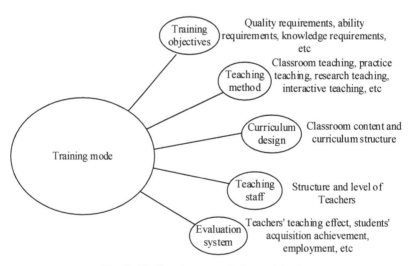

Fig. 3. Quality elements of talent cultivation

elements of innovative education quality mainly focus on training objectives, teaching methods, curriculum, teaching staff and evaluation system [8]. To sum up, the quality elements of talent training are summarized as Fig. 3.

Furthermore, the research contents of "theoretical curriculum system", "practical teaching system", "school enterprise cooperation" and "teacher team construction" are unified as a whole, and the training mode of mobile Internet innovative talents is established under the unified framework. The overall research scheme of innovative education personnel training system is as Fig. 4.

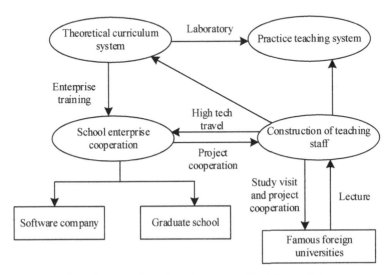

Fig. 4. Innovative education personnel training system

In the new theoretical curriculum system, the courses are classified into theory series, tool series, engineering series, other courses and special practice. With the continuous consolidation of students' foundation, the theoretical courses are gradually refined according to different application directions [9]. So that students not only have a solid theoretical foundation, but also have strong professional basic knowledge, to lay a good theoretical foundation for practical training and enterprise training.

3   Quality Evaluation Model of Mobile Internet Innovative Education Personnel Training

3.1   Fuzzy Analytic Hierarchy Process

Fuzzy analytic hierarchy process is the application of analytic hierarchy process in fuzzy environment [10]. It decomposes decision-making problems and related factors into objectives, criteria and schemes. On this basis, it reasonably combines qualitative analysis and quantitative analysis, and carries out decision-making process according to human thinking and judgment rules [11]. The fuzzy comprehensive evaluation method is based on the determination of evaluation factors, evaluation grade standard and the weight of each factor, using the principle of fuzzy set transformation, using membership degree to describe the fuzzy boundary of each index factor, constructing fuzzy evaluation matrix, and finally determining the object grade through multi-layer compound operation. The main steps are to set up the set of evaluation factors and comments, establish the fuzzy evaluation matrix, determine the weight of each evaluation factor, calculate the fuzzy comprehensive evaluation vector and comprehensive evaluation value, and analyze the results to get the evaluation conclusion. The quality evaluation process of mobile Internet innovative education personnel training based on fuzzy analytic hierarchy process is as Fig. 5.

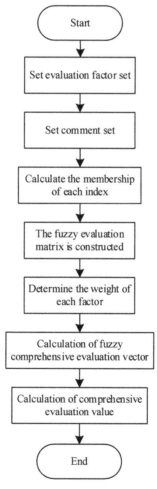

Fig. 5. Quality evaluation process of mobile Internet innovative education personnel training

3.2 Establishment of Fuzzy Evaluation Matrix

Applying the aforementioned combination weighting technology to the fuzzy comprehensive evaluation method, you will get a higher vocational education talent training quality evaluation model, set the mobile internet innovation education talent training quality evaluation factor set as: $Q = \{Q_1, Q_2, ..., Q_m\}$, $Q_i = \{Q_{i1}, Q_{i2}, ..., Q_{im}\}$, $Q_{ij} = \{Q_{ij1}, Q_{ij2}, ..., Q_{ijm}\}$, adopt the popular five-level standard, set comments set, that is, $W = \{W_1, W_2, W_3, W_4, W_5\}$, the corresponding assignment is $E = \{100, 80, 60, 40, 20\}$.

The membership degree of each index is calculated. The membership degree of each three-level index is expressed by the proportion of the number of evaluators who choose each evaluation level to the total number of evaluators. The membership degree of the second level index and the third level index to each evaluation grade constitutes the fuzzy evaluation matrix:

$$A_{ij} = \begin{bmatrix} a_{ij11} & \cdots & a_{ij15} \\ \cdots & \cdots & \cdots \\ a_{ijk1} & \cdots & a_{ijk5} \end{bmatrix} \tag{1}$$

In formula (1), $a_{ij11}, a_{ij15}, a_{ijk1}, a_{ijk5}$ is the membership degree of the k three-level index to each evaluation level in the factor set.

3.3 Determine the Weight of Each Factor

According to the aforementioned method for determining the subjective and objective combination weights [12], the weight vector $S = [s_1, s_2, ..., s_m]^T$ of the first-level factor set, the weight vector $S_i = [s_{i1}, s_{i2}, ..., s_{im}]^T$ of each second-level factor set, and the weight vector $S = [s_1, s_2, ..., s_m]^T$ of each third-level factor set are respectively obtained.

3.4 Calculation of Comprehensive Evaluation Value

The weight vector of each factor set and the fuzzy evaluation matrix are subjected to the weighted average fuzzy synthesis operation to obtain the corresponding fuzzy comprehensive evaluation vector. This model needs to perform first-level fuzzy comprehensive evaluation first, and let D_{ij} be the fuzzy comprehensive evaluation vector of the factor set, then the formula is [13]:

$$D_{ij} = S_{ij}A_{ij} \tag{2}$$

Then use the fuzzy evaluation matrix A_i of the constituent factor set to perform the fuzzy operation with the corresponding weight vector S_i to obtain the secondary fuzzy comprehensive evaluation vector D_i as shown in the following formula:

$$D_i = S_iA_i \tag{3}$$

Then the fuzzy evaluation matrix A, which constitutes the factor set, is subjected to fuzzy operation with the weight vector S to obtain the first-level fuzzy comprehensive evaluation vector D as shown in the following formula:

$$D = SA \tag{4}$$

Finally, the comprehensive evaluation value Z is calculated as shown in the following formula:

$$Z = D\,A = (d_1, d_2, ..., d_5)^T (100, 80, ..., 20)^T \tag{5}$$

The fuzzy comprehensive evaluation vector D is normalized, and the result is analyzed according to the principle of maximum membership degree, and the evaluation conclusion is obtained.

4   Experimental Analysis

In order to verify the effectiveness of the quality evaluation model of mobile internet innovation education personnel training based on the fuzzy analytic hierarchy process, the experiment uses a computer with the configuration: Inter E 1400 2.0 GHz processor, 4.00 G memory, 400 G hard disk, and 32-bit Windows 7 operating system. Perform simulation experiment analysis in MATLAB R2013B environment. In this study, the fuzzy analytic hierarchy process is used to build the evaluation model of mobile Internet innovative education talent training quality, set the evaluation factor set and comment set, establish the fuzzy evaluation matrix, determine the weight of each evaluation factor, calculate the fuzzy comprehensive evaluation vector and comprehensive evaluation value, and evaluate the mobile Internet innovative education talent training quality according to the maximum membership principle. In order to improve the persuasion of the experimental results, 100 teaching samples were selected, and the control group and the experimental group were set up in the same environment, different testers evaluated the quality of mobile Internet innovative education personnel training through the evaluation model of mobile Internet innovative education personnel training quality based on fuzzy analytic hierarchy process. The basic information of the tester is as Table 1.

Table 1. Basic information statistics of testers

Item	Option	Frequency	Effective percentage
Gender	Male	51	34.5%
	Female	97	65.5%
Academic degree	Bachelor	7	4.7%
	Master	79	53.4%
	Doctor	62	41.9%
Title	Assistant	5	3.4%
	Lecturer	89	60.1%
	Associate professor	36	24.3%

(continued)

Table 1. (*continued*)

Item	Option	Frequency	Effective percentage
	Professor	8	5.4%
	Other	10	6.8%
Major	Communication (Journalism)	66	44.6%
	Advertising	26	24.3%
	Management	8	5.4%
	Art design	10	6.7%
	Other	28	18.9%

Based on the above information, further comparative analysis of the traditional method and this method of mobile Internet innovation education personnel training quality evaluation accuracy, the specific comparison results are as Fig. 6.

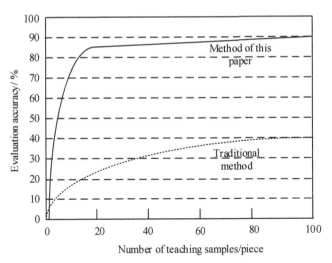

Fig. 6. Comparison results of evaluation accuracy of talent cultivation quality by different methods

According to Fig. 6, when the number of teaching samples is 100, the evaluation accuracy of the traditional method is 40%, while the evaluation accuracy of this method is 90%. Compared with traditional methods, the evaluation model of mobile Internet innovative education talent cultivation quality based on fuzzy analytic hierarchy process proposed in this paper has higher evaluation accuracy and better talent cultivation quality evaluation effect.

On this basis, in order to further verify the evaluation time of the proposed mobile Internet innovative education talent training quality evaluation model based on the fuzzy analytic hierarchy process, the traditional method is compared with the mobile Internet innovative education talent training quality evaluation time of this method, and the comparison results are as Fig. 7.

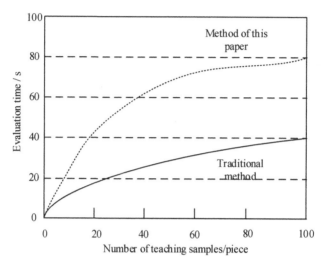

Fig. 7. Comparison results of talent cultivation quality evaluation time of different methods

According to Fig. 7, with the increase of the number of teaching samples, the evaluation time of talent cultivation quality with different methods increases. When the number of teaching samples is 100, the evaluation time of traditional method is 80 s, while the evaluation time of this method is 40 s. Therefore, the evaluation time of the model based on the fuzzy analytic hierarchy process is short.

5 Conclusion

In order to improve the quality evaluation accuracy of mobile Internet innovative education personnel training and reduce the evaluation time of mobile Internet innovative education personnel training, an evaluation model of mobile Internet innovative education personnel training quality based on Fuzzy Analytic Hierarchy Process is proposed. Through the mobile Internet innovative education personnel training system, the fuzzy analytic hierarchy process is used to establish the fuzzy evaluation matrix, determine the weight of each evaluation factor, calculate the fuzzy comprehensive evaluation vector, and realize the quality evaluation of mobile Internet innovative education personnel training. The model can accurately evaluate the quality of mobile Internet innovative education personnel training and shorten the evaluation time of mobile Internet innovative education personnel training quality.

References

1. Ai, F., Wang, N.: Integration of urban-rural planning and human geography for online education under the impact of COVID-19. J. Intell. Fuzzy Syst. **39**(6), 8847–8855 (2020)
2. Lenihan, A.S., Foley, A.R., Cary, B.W.A., et al.: Developing engineering competencies in industry for chemical engineering undergraduates through the integration of professional work placement and engineering research project - ScienceDirect. Educ. Chem. Eng. **32**(5), 82–94 (2020)
3. Babu, M., Suman, K., Rao, P.S.: Drafting software as a practicing tool for engineering drawing-based courses: content planning to its evaluation in client-server environment. Int. J. Mech. Eng. Educ. **47**(2), 118–134 (2019)
4. Chen, Y.: College English teaching quality evaluation system based on information fusion and optimized RBF neural network decision algorithm. J. Sens. **2021**(5), 1–9 (2021)
5. Temperature-induced phase transition of two-dimensional semiconductor GaTeProject supported by the National Natural Science Foundation of China (Grant No. 62004080), Postdoctoral Innovative Talents Supporting Program (Grant No. BX20190143), China Postdoctoral Science Foundation (2020M670834), and Jilin Province Science and Technology Development Program, China (Grant No. 20190201016JC). Chinese Physics B, 30(1), 016402 (2021). (6pp)
6. Padri, M., Boontian, N., Piasai, C., et al.: Cultivation process of microalgae using wastewater for biodiesel production and wastewater treatment: a review. IOP Conf. Ser. Earth Environ. Sci. **623**(1), 012025 (2021). (6pp)
7. Ma, Y., et al.: Characteristics of groundwater pollution in a vegetable cultivation area of typical facility agriculture in a developed city. Ecol. Indicat. **105**, 709–716 (2019)
8. Huo, S., et al.: Filamentous microalgae Tribonema sp. cultivation in the anaerobic/oxic effluents of petrochemical wastewater for evaluating the efficiency of recycling and treatment. Biochem. Eng. J. **145**, 27–32 (2019)
9. Matsuo, S., Fujita, T., Iwasaki, Y.: Evaluation of yield and energy of greenhouse tomato production in each asian monsoon area using cultivation simulator considering growth condition. MATEC Web Conf. **333**(4), 12004 (2021)
10. Tung, T.V., Thao, N.T.P., Vi, L.Q., et al.: Waste treatment and soil cultivation in a zero emission integrated system for catfish farming in Mekong delta. Vietnam. J. Clean. Prod. **288**(10), 125553 (2020)
11. Liu, S., Liu, D., Srivastava, G., Połap, D., Woźniak, M.: Overview and methods of correlation filter algorithms in object tracking. Comp. Intell. Syst. **7**(4), 1895–1917 (2020). https://doi.org/10.1007/s40747-020-00161-4
12. Liu, S., Bai, W., Zeng, N., et al.: A fast fractal based compression for MRI images. IEEE Access **7**, 62412–62420 (2019)
13. Liu, S., Li, Z., Zhang, Y., et al.: Introduction of key problems in long-distance learning and training. Mob. Netw. Appl. **24**(1), 1–4 (2019)

Design of Ideological and Political Online Education System Based on Neural Network

Gui-yan Quan[✉]

Guangzhou College of Technology and Business, Guangzhou 528138, China

Abstract. In order to improve the performance of the traditional online ideological and political education system, the design of the online ideological and political education system based on neural network is proposed. According to the background of the online ideological and political education project and the analysis of the system demand, the neural network is applied to the structure design of the online ideological and political education system, and the online ideological and political education system module is designed by using the information acquisition module, the standardization module, the self-training module, the self-defining module and the management module of the information model of the online ideological and political education. The online ideological and political education system module is designed by using the language model education module, the optimization module of educational results, the text input management module and the output management module of educational results. The test results show that the online ideological and political education system based on neural network has better performance.

Keywords: Neural network · Ideological politics · Online education system · Feature extraction

1 Introduction

Ideological and political education is the basic way for the Party to exercise ideological and political leadership in colleges and universities, and the fundamental guarantee for effectively fulfilling the historical mission of ideological and political education in the new era. Promoting the quality and benefit of ideological education by means of informationization is of great significance to cultivating college students in the new era, an inevitable choice of the development of the times, and the meaning of constructing powerful and modern applied talents [1]. Ideological and political education, as an important part of the construction of spiritual civilization in China, is also one of the important pushing forces to solve social contradictions and problems. With the development of the times and the continuous progress and deepening of Ideological and political education, the increasing number of Ideological and political education-related documents, how to better access to information and further analyze data on a large number of Ideological and political education documents is also the top priority of Ideological and political

© ICST Institute for Computer Sciences, Social Informatics and Telecommunications Engineering 2022
Published by Springer Nature Switzerland AG 2022. All Rights Reserved
S. Liu and X. Ma (Eds.): ADHIP 2021, LNICST 417, pp. 394–407, 2022.
https://doi.org/10.1007/978-3-030-94554-1_31

education research. It is a feasible and efficient program to summarize and deeply analyze a large number of documents with the help of computer related knowledge and natural language processing tools.

Ideological and political education system aims at providing intelligent support for ideological and political education in institutions of higher learning by using Internet, Internet of Things, multimedia and other technologies. At present, a variety of educational models emerge one after another, ideological education needs to be constantly improved and optimized [2]. The ideological and political education system pays attention to the use of new media. The new media has the characteristics of multimedia, which includes not only the visual symbol system, but also the sound symbol system. Using the new media, the content of ideological and political education can be transformed into a "cultural feast" with both visual enjoyment and auditory enjoyment, which makes the content of ideological and political education more vivid. Alice's Rational Emotive Therapy lists 11 irrational beliefs and characteristics existing in people's minds, and develops and establishes new platforms and positions of ideological education. The ideological and political education system is intelligent, integrated, interactive, real-time and other functional characteristics, which highly meets the multi-dimensional needs of colleges and universities to carry out ideological and political education in the new era. It can greatly improve the quality and efficiency of ideological and political education, and has strong vitality and high research and development value [3]. First, it can reduce the burden of teaching preparation. Based on ubiquitous and broadband network, the ideological and political education system can realize the interconnection of massive information, and the sharing of high-quality lesson plans, courseware and learning materials can greatly improve the efficiency of preparing lessons for political teachers; secondly, it can improve the accuracy of teaching. The system can provide functions such as thought investigation and analysis, online communication, forum and so on, and can accurately grasp students' realistic thoughts and different demands. Third, it can enhance the objectivity of teaching evaluation. The system has the function of real-time recording, which can be registered and archived, no matter whether the teaching plan is prepared for class preparation, induction video recording is made during class teaching, or the discussion and cooperation activities are organized after class, and the implementation of personnel, time, content and system is clear at a glance, and the assessment results are more objective and more reliable.

Different from the ideological and political education in our country, there are few such education abroad as "ideological and political education" or "moral education", and there are no teachers specialized in ideological and political education. But this does not mean that there is no ideological and political education abroad, but ideological and political education abroad using relatively hidden education strategies, skillfully avoid the frequency of these political and cultural terms. At present, there is not much research and development of online education platform for ideological and political education abroad, and some domestic scholars have put forward research results. Lee et al. [4] build an educational system based on Moodle by providing a friendly interface to adapt to most students' online learning. The implementation of the website is studied by taking the course of "Multimedia Implementation by Using JAVA" as an example. From the modified Moodle -based educational system, the time for students to browse each web page can be obtained. By analyzing the recorded information, teachers can find out the factors that affect students' learning performance. Therefore, teachers can evaluate students' learning performance by using the proposed learning performance evaluation mechanism to provide sufficient supporting learning materials for individual students. Zhang J et al. [5] In order to solve the problem that information island and organization are lack of effective management in the sharing of traditional base education resources, a model of education information system based on geographic information system is put forward, including system organization model and system architecture. The principles for the construction of the platform are summarized. Problems that need to be considered are pointed out. The system provides centralized management, expression and scientific analysis of relevant geographic and comprehensive information, describes the spatial distribution of education resources, combines spatial attributes with statistical analysis of education information, and realizes the distribution of education institutions, nearby security schools, micro-natural disaster prevention planning, macro-cutting and regional planning, as well as the implementation and optimization of information query and maintenance tools for the implementation of education resources.

The application of big data and artificial intelligence technology in ideological education is the inevitable choice of the times. Carrying out rich and colorful ideological education is the need of the times and the development of socialist countries. Advanced tools and means of ideological education are the indispensable part of carrying out ideological education in colleges and universities. Combined with the exploration of strengthening and improving college education, this paper puts forward the demand of developing college ideological and political education system, and determines the key technologies to be adopted through detailed investigation, on the basis of which an online ideological and political education system based on neural network is designed and implemented.

Considering the needs of ideological and political education, this paper can extract the texts in different formats and convert them into document formats for subsequent use. The language module is added, and different language modes can be exchanged. It enhances the applicability of the system, provides an operable platform for online education of ideological and political education, and enhances the convenience of education.

2   Structure Design of Online Ideological and Political Education System

According to the background of the online ideological and political education project and the analysis of the system needs, the online ideological and political education system is divided into four layers: presentation layer, business layer, support layer and data layer [6]. The overall structure of the ideological and political online education system is shown in Fig. 1.

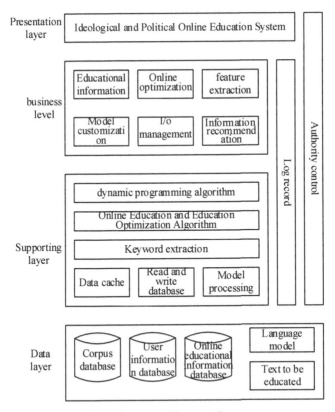

Fig.1. System architecture diagram

Presentation layer

The presentation layer provides the user with the system operation interface by the various options and icons in the system interface, which is convenient for the user to operate the various functional modules.

Business layer

The business layer is the main part of transaction processing, which includes model self-definition, input-output management, online education, online optimization, feature extraction, information recommendation and so on.

Supporting layer

The basic layer provides back-end support for the main functions of the business layer, including algorithm support, data caching, corpus reading and writing, and language model processing.

Data layer

The data layer provides the data needed for the function of the system. The database mainly includes system corpus, user corpus, user information database and so on.

3 Module Design of Online Ideological and Political Education System

The online ideological and political education system is mainly divided into three functional modules: the ideological and political education information training module, the online education module and the feature extraction module. The main functional modules are divided into several sub-modules according to different functions. The overall functional diagram of the system is shown in Fig. 2.

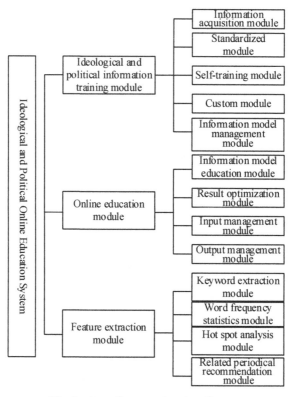

Fig. 2. Overall system function diagram

3.1 Ideological and Political Education Information Training Module

The information training module of ideological and political education includes the information acquisition module, the information standardization module, the self-training module, the self-definition module and the management module [8]. The main function is to deal with ideological and political education information and model training and management.

Ideological and Political Education Information Acquisition Module
In order to meet the needs of users for the individuation of ideological and political education information model, it is responsible for receiving different types of text documents provided by users and extracting the text from them, so as to prepare for the subsequent training and use.

Information Standardization Module of Ideological and Political Education
After obtaining the information of ideological and political education provided by users, the information can not be used in direct training, so the information of ideological and political education can be standardized and processed to meet the needs of follow-up training.

Model Self-training Module
Based on the standardized ideological and political education information and the corresponding algorithm, some parameters of the language model can be modified to meet the needs of the users.

Model Customization Module
Considering that some segments of online educational results may not meet the needs of users, further optimize the trained language model or system-owned language model through the model customization module, and add some high-weight words to the model to meet the needs of users to the greatest extent [9].

Ideological and Political Education Information Model Management Module
Considering the personalized components of different users and the privacy protection of the model, the customized language model of different users is managed and protected to ensure the privacy and information security [10].

3.2 Online Education Module

Online education module includes language model education module, education result optimization module, text input management module and education result output management module. Users can use the system language model to educate the educational texts online [11, 12], optimize the results and get the final educational results, and output different types of files according to the needs of users.

Language Model Word Segmentation Module
The language model education module is mainly based on the language model and uses the Viterbi algorithm to segment the text provided by the user.

Result Optimization Module
According to the preliminary results, the result optimization module optimizes the preliminary results by using the optimization algorithm, and can check the difference of the results according to the user's needs.

Text Input Management Module
According to the user's requirement, the text management module is mainly responsible for reading the text in different types of text documents provided by the user, and preserving the original text format.

Result Output Management Module
The result output module is mainly responsible for the output of the final results, and the results can be exported to a number of types of text documents according to the needs.

3.3   Feature Extraction Module

The feature extraction module includes the keyword extraction module, the word frequency statistics module, the research hotspot analysis module and the related journal recommendation module. After getting the results, users can get the key words and word frequency statistics of the text, and analyze the research hotspots according to the key words, and recommend the ideological and political education periodicals related to the hotspots as reference.

Keyword Extraction Module
Keyword extraction module is mainly responsible for the final results have been obtained, through keyword extraction algorithm, access to the corresponding ideological and political education keywords, and through keyword extraction results to prepare for the follow-up analysis.

Word Frequency Statistics Module
Word frequency statistics module is mainly responsible for the final results of the word frequency statistics, and ranking, according to the word frequency statistics results can be more comprehensive understanding of the corresponding text of the main research direction and written expression. At the same time, considering that it can reflect the text information more intuitively, this module provides the function of drawing the word cloud. By analyzing the text information, we can show the primary and secondary information of the text directly by drawing the word cloud, and provide reference for the text analysis.

Research Hotspot Analysis Module
The research hotspot analysis module can analyze the research hotspot of the ideological

and political text according to the keyword extraction module, and can count the research hotspot of the user's history result.

Recommendation Module for Related Journals

The relevant literature recommendation module is mainly responsible for referring to the main research directions of the text to be taught, recommending the periodicals with high relevance in the relevant ideological and political education periodicals that have been included according to the research directions, and recommending the relevant literature on the home page and in the literature recommendation column for the convenience of users to browse and reference.

3.4 System Flow Design

Through the analysis of user needs and in combination with the reality, this paper designs a set of standardized processes for the online ideological and political education system. The complete processes include logging in the user account password, recording the corresponding account education history, customized language model and other data,

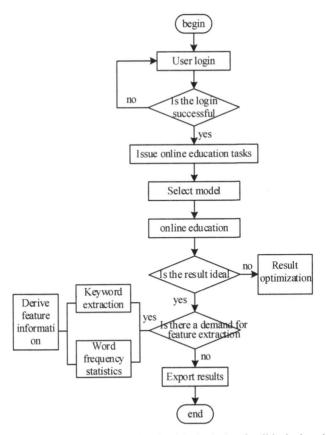

Fig. 3. Flow chart of main functions of online ideological and political education system

and providing online education and follow-up data analysis services in light of user needs [13]. Through the above analysis of the basic process, get the ideological and political online education system main function flow chart, as shown in Fig. 3.

4 System Testing

The system testing mainly includes three parts: system security testing, system function testing and system performance testing. The three parts include different testing contents and emphases.

4.1 System Security Testing

he system security test mainly verifies whether a user can use the system or check the user information of others under the circumstances that the logon information is inconsistent with that of the user, whether the system can operate normally or feed back in time after the user misoperates, whether the corresponding functions of user authority control are improved, and other items. The test results are shown in Table 1.

Table 1. System Security Test Results

Test project	Operation steps	Expected results	Test results
Users access the system without logging in	Try accessing the system without logging in	Unable to access the system, display login interface	Consistent with expected results
User session timeout exit	Multiple user input errors in the login system interface	Prompt login warning and close system	Consistent with expected results
User ultra vires access	Log in the system to enter the person's role account and access the personal information of the non-user	Unable to obtain other user information, access failed	Consistent with expected results
User information encryption	Viewing other user information through database operations and online education information	Database encryption, inaccessible to other users	Consistent with expected results

4.2 System Functional Testing

The results of the system function test on the corresponding functions of the ideological and political education information training module, online education module and feature extraction module are shown in Table 2.

Table 2. Functional Test of Ideological and Political Education Information Training Module

Test project	Operation steps	Expected results	Test results
Viewing Information on Ideological and Political Education	Access to the training interface to view ideological and political education information	Basic information on ideological and political education can be viewed in the file bar of standardized information	Consistent with expected results
Standardization of ideological and political education information	Enter the training interface to standardize the development of ideological and political education information	The information of ideological and political education is standardized and processed into standardized data	Consistent with expected results
View standardized ideological and political education information	Enter the training interface to view the uploaded standardized ideological and political education information	You can view standardized ideological and political education information in the model training file bar	Consistent with expected results
Training on standardized ideological and political education information	Enter the information training interface of ideological and political education model training operation	According to the standardized information of ideological and political education training and get the corresponding information model	Consistent with expected results
Standardized ideological and political education information derived	Entering the Information Training Field of Ideological and Political Education to Export the Standardized Data	The standardized ideological and political education information is exported to the designated location according to the user's needs	Consistent with expected results
Training Information Model Export	To enter into the ideological and political education information training field to face the training information model derived	Export the standardized information model to the specified location according to user requirements	Consistent with expected results

(*continued*)

Table 2. (*continued*)

Test project	Operation steps	Expected results	Test results
Viewing the Optimized Information Model Information	Viewing Optimization Information Model Information into Model Optimization Interface	The model information to be optimized is displayed centrally in the file box consistent with the expected results	Consistent with expected results
Add new information to existing information models	Enter Model Optimization to add new information to the specified model	Optimize the model by adding specified new information to the existing information model	Consistent with expected results

The function test of online education module mainly includes online education, optimization education, uploading documents and exporting results. The test results are shown in Table 3.

Table 3. Online education module functional test

Test project	Operation steps	Expected results	Test results
Upload to be participle document	Enter the online education interface and click the upload document button to upload the education document	The educational document is uploaded successfully, and the relevant information of the educational document is displayed in the file box	Consistent with expected results
Online education	Enter the online education interface and click the online education button to operate online education	The online education process is successful and the result is output to the result output box	Consistent with expected results
Optimizing education	Enter the online education interface and click the optimization button to optimize the results	Optimized education processing successfully, output the result to the result output box	Consistent with expected results

(*continued*)

Table 3. (*continued*)

Test project	Operation steps	Expected results	Test results
Model selection	Enter the online education interface and click the Select Model button to select the existing model for subsequent selection	Show the existing model and provide the basis for subsequent operations by selecting the corresponding model	Consistent with expected results
Output results	Click the Output Document button to output the	Outputs the corresponding file type result document to the specified address based on the final result	Consistent with expected results

The function test of feature extraction module mainly includes keyword extraction and hotspot analysis, and the test results are shown in Table 4.

Table 4. Functional test of feature extraction module

Test project	Operation steps	Expected results	Test results
Upload educational documents	Enter the feature extraction interface and click the upload document button to upload the educational document	The educational document is uploaded successfully, and the relevant information of the educational document is displayed in the file box	Consistent with expected results
Keywords extraction	Enter the feature extraction boundary and click the keyword extraction button to educate the educated text online	Keyword extraction is successful, output keyword extraction results in the left result output box	Consistent with expected results
Hot spot analysis	Enter the feature extraction interface and click the hot spot analysis button to analyze the key words	The results of hot spot analysis show that it is successful, and the hot spot analysis of the corresponding keywords is carried out according to the different results of keyword extraction	Consistent with expected results

4.3 System Performance Testing

The system performance test mainly tests the response time of the system to different operations of the users. In order to guarantee the reliability of the test as much as possible, in the process of testing various purposes, 100 tests are conducted for different test items, and the final results are summarized and the average value is taken as the final test results. In order to ensure the consistency of the test and reduce the deviation caused by different text lengths, 100 texts of similar length are extracted as the test set in the process of treating the ideological and political texts of education, and in the process of testing, a number of testers are found to use the system and recorded to obtain a more appropriate expected result of system performance, which is compared with the final test results as an indicator. Table 5 is the test results of system performance. By analyzing Table 5, it can be found that the online ideological and political education system designed in this paper meets the expected results in terms of user login time, page loading time, response time, optimized response time, keyword extraction time, word frequency statistics time, word cloud mapping time, hot spot analysis time, model self-training time, model optimization time, document uploading time and document output time, and the results are far superior to the expected values, with superior application performance.

Table 5. System performance testing

Test project	Expected results	Actual results	Result
User login time	<3 s	1.01 s	Pass
Page load time	<3 s	1.11 s	Pass
Response time	<3 s	2.88 s	Pass
Optimized response time	<4 s	1.71 s	Pass
Keywords extraction time	<1 s	0.23 s	Pass
Word Frequency Statistics Time	<1 s	0.12 s	Pass
Word cloud mapping time	<2 s	0.89 s	Pass
Hot spot analysis time	<2 s	0.87 s	Pass
Model self-training time	<30 s	21 s	Pass
Model optimization time	<3 s	2.28 s	Pass
Document upload time	<2 s	0.82 s	Pass
Document output time	<2 s	1.12 s	Pass

5 Conclusion and Outlook

This paper puts forward the design of online ideological and political education system based on neural network, and designs the structure of online ideological and political education system according to the background of online ideological and political education project and the analysis of system demand. The results show that the online ideological and political education system based on neural network has better performance. However, the online ideological and political education system designed by this method does not give much consideration to the design of interactivity. In the future research, the system will be further improved on interactivity to enhance the application performance of the system.

Fund Projects. This paper is a university-level scientific research project of Humanities and Social Sciences in Guangzhou College of Technology and Business "Research on the Innovative Practice of College Students' Formative Education under the Platform of "Five Entry"(Item Number:KA202026).

References

1. Deng, H.: The design and application system of educational administration management based on Web2.0 technologies. Int. J. Technol. Manag. **3**, 20–22 (2013)
2. Long, Y., Aleven, V.: Educational game and intelligent tutoring system: a classroom study and comparative design analysis. ACM Trans. Comput. Hum. Interact. **24**(3), 1–27 (2017)
3. Tan, J.: Design and evaluation of a web classification system for educational resources. Int. J. Contin. Eng. Educ. Life-Long Learn. **23**(3/4), 345–356 (2013)
4. Lee, Y.-C., Terashima, N.: A Distance instructional system with learning performance evaluation mechanism: moodle-based educational system design. Int. J. Dist. Educ. Technol. **10**(2), 57–64 (2012)
5. Zhang, J., Sun, Y., Zhang, Y.: The design and implementation of the information management system of the elementary educational resources based on GIS. Int. Conf. Geoinformat. 1–4 (2011)
6. Seidmann, A., Jiang, Y., Zhang, J.: Introduction to information issues in supply chain and in service system design minitrack. Decis. Support Syst. **51**(4), 831–832 (2011)
7. Starcic, A.I.: Competence management system design in international multicultural environment: registration, transfer, recognition and transparency. Br. J. Edu. Technol. **43**(4), E108–E112 (2012)
8. Hui, S.Y.R.: An extended photoelectrothermal theory for LED systems: a tutorial from device characteristic to system design for general lighting. IEEE Trans. Power Electron. **27**(11), 4571–4583 (2012)
9. Sotnikova, M., Zhabko, N., Lepikhin, T.: Control systems analysis and design labs with educational plants. IFAC Proc. Vol. **45**(11), 212–217 (2012)
10. Zhang, Y.: Exploring the learning mechanism of web-based question-answering systems and their design. Br. J. Edu. Technol. **41**(4), 624–631 (2010)
11. Liu, S., Li, Z., Zhang, Y., et al.: Introduction of key problems in long-distance learning and training. Mob. Netw. Appl. **24**(1), 1–4 (2019)
12. Liu, S., Glowatz, M., Zappatore, M., et al. (eds.): e-Learning, e-Education, and Online Training. Springer International Publishing, pp. 1–374 (2018). https://doi.org/10.1007/978-3-319-49625-2
13. Fu, W., Liu, S., Srivastava, G.: Optimization of big data scheduling in social networks. Entropy **21**(9), 902 (2019)

Construction of Mobile Internet Distance Education Teaching Platform from the Perspective of Industry Education Integration

Cong-gang Lyu[✉]

Jiangxi Tourism and Commerce Vocational College, Nanchang 243000, China

Abstract. Aiming at the problems of traditional education and teaching platform, such as low use satisfaction, low platform security and poor operation stability, this paper constructs a mobile internet distance education and teaching platform from the perspective of industry education integration. Through the mobile terminal, wireless network and learning resource server, the structure of distance education teaching platform is constructed. The framework of distance education data processing platform is built by using mobile learning course module, mobile course test module, mobile course question answering module and download zone module. Optimize the function of mobile Internet distance education teaching platform, according to the construction goal of Internet distance education platform, improve the data management function structure of mobile distance education platform, and realize the operation of mobile Internet distance education teaching platform. The experimental results show that the stability of the platform is good, which can effectively improve the user satisfaction and platform security.

Keywords: Industry education integration · Mobile internet · Distance education · Teaching platform

1 Introduction

Network teaching refers to the use of computer hardware and network environment to achieve remote teaching. Network teaching is an important form of information-based teaching and a powerful supplement to traditional on-site teaching. It breaks the restrictions of traditional teaching time, place and personnel, and allows more people to arrange their own time, study, discuss and share knowledge in distance [1]. To better carry out network teaching, we must rely on the excellent network teaching platform. As a new stage in the development of modern distance higher education, mobile learning is transforming from theoretical research to practical application. The mobility of learning form, the customization of learning content, the portability of learning equipment and the relevance of learning situation fit the essential characteristics of the separation of time and space in distance education and the value concept of promoting learners'

S. Liu and X. Ma (Eds.): ADHIP 2021, LNICST 417, pp. 408–418, 2022.
https://doi.org/10.1007/978-3-030-94554-1_32

development and meeting learners' personalized learning needs. In the process of transformation from radio and TV University to open university, how to build a network teaching platform and how to effectively implement teaching management are directly related to whether the open university can better meet the challenges of the times and expand new development space in the new historical period [2]. At present, some scholars have done some research on the construction of teaching platform in related fields, and have achieved some research results. Reference [3] proposed the construction of distance learning platform based on mobile communication technology. By analyzing the application types of mobile communication technology in distance education, this paper uses mobile communication technology to build the framework and functions of distance learning platform, and draws the specific platform system framework and function module diagram to complete the construction of distance learning platform based on mobile communication technology. The system can meet the design requirements, can achieve distance learning and mobile learning, but the stability of the system is poor. Therefore, the mobile Internet distance education teaching platform is constructed from the perspective of industry education integration. Using mobile terminal, wireless network and learning resource server, combined with mobile learning course testing, question answering and download module, the distance education teaching platform is constructed. Optimize the function of teaching platform, according to the construction goal of education platform, improve the data management function of education platform, and realize the establishment of mobile Internet distance education teaching platform. Network teaching platform includes hardware facilities and software supporting network teaching. It is an important tool for network teaching, and its advantages and disadvantages directly affect the effect of network teaching. Through the analysis of the existing network teaching platform, thinking about its construction mode, summarizing its structure and function, analyzing its common problems, to lay the foundation for building a more mature network teaching platform.

2   Construction of Mobile Internet Distance Education Teaching Platform

2.1   Structure Construction of Distance Education Teaching Platform

Distance education can make full use of the characteristics of network technology for high-quality teaching. Specifically, network technology can realize the following forms of education in distance education. According to their own situation, learners can selectively study the course. As long as the network is covered, learners can learn all day long through smart phones or tablets [4]. At the same time, when constructing course learning resources, distance education institutions should consider allowing students to download course content and learning resources in real time, so as to meet the needs of students' offline learning. In the learning process of students, if they encounter difficult problems, the distance education platform still needs to provide learning assistance. Distance education institutions can publish various teaching information in a timely manner on the learning platform for students to view. Dynamic information includes various learning tasks, expert lectures, real-time Q&A information, online test information, etc. The distance education learning platform is mainly composed of three parts: mobile terminal, wireless network and learning resource server. Its structure is as Fig. 1.

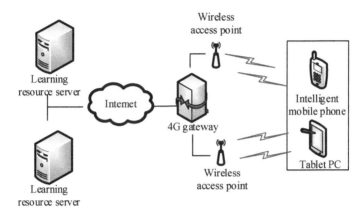

Fig. 1. Structure of mobile internet distance education teaching platform

In Fig. 1, in the Internet distance education platform, online audition and online course teaching are the most frequently used, so the selection of courses must be high-quality. That is to say, on the basis of digital transformation of existing teaching resources, multi-channel development and construction of teaching resources will be carried out. The best of the best courses will be selected, and the targeted, classic and representative courses will be provided to learners for multi-channel download and learning [5]. The construction of teaching management mode makes the course teaching quality, teaching design process, management order, service individualization, learning autonomy, teachers' high level and teaching evaluation socialization, which constitute the teaching management mode of modern distance education mobile network. No matter what kind of learning platform learners use, they can complete the learning task more easily and smoothly, and regard learning as their own interest [6]. In the Internet distance education platform, the function design of mobile learning is in the main position in the whole platform design, including mobile learning course module, mobile course test module, mobile course question answering module and Download Zone module. The framework of distance teaching data processing platform is as Fig. 2.

In Fig. 2, in the process of platform design, the requirement of device configuration is supported by ESB bus, which is the abbreviation of enterprise service bus and the product of the combination of Web services, XML and middleware technology [7]. In the overall framework of Internet distance education platform, ESB provides the most basic connection center in the network to eliminate the technical differences between different applications. Coordinated operation of different servers, played a compatible role, and realized the integration and communication between different services. This is to meet the functional requirements of multi service in the Internet distance education platform. Among them, the server needs to build five server clusters: basic platform server cluster, application server cluster, data storage server cluster, data exchange server cluster, basic

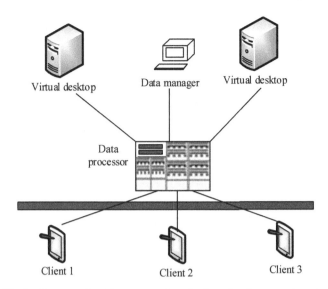

Fig. 2. Framework of data processing platform for distance education

application network and security [8]. Furthermore, three simulation warehouses are established in the platform to effectively manage massive teaching resources, including user data warehouse, application data warehouse and resource data warehouse. The structural framework of Internet distance education platform is as Fig. 3.

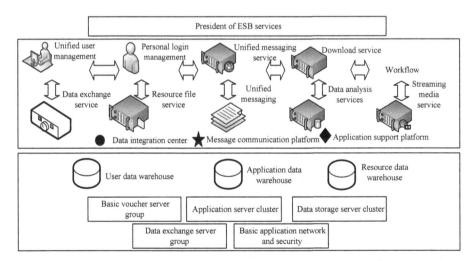

Fig. 3. Structure framework of internet distance education platform

In Fig. 3, add educational information query module, mobile microblog module and database module in the platform to facilitate the management of teachers, and form the whole Internet distance education platform.

2.2 Function Optimization of Mobile Internet Distance Education Teaching Platform

In order to ensure the quality of teaching, this paper further describes the functional modules, contents and user objects of the Internet distance education platform as Table 1.

Table 1. Function modules and objects of internet distance education platform

Functional module	Content	User object
Mobile learning course	1. Select courses by major: check the course introduction, key and difficult points, course specific content and review guidance	Learner
	2. Deal with students' course selection, add and modify course introduction, key and difficult points, course specific content and review guidance	Teacher
	3. Delete and modify the course introduction, key and difficult points, specific contents and review guidance	Administrators
Mobile classroom test	1. Select test course, grade test, check answers and evaluation	Learner
	2. Write test content and view student test statistics	Teacher

According to Table 1, the construction goal of the Internet distance education platform is to build a comprehensive information platform, which can centrally manage the people, money and materials in the Open University. Through a series of information application platforms, it supports the school's teaching, management, service and other activities, so as to realize the goal of "everyone can register, learn and test all the time and everywhere" proposed by the Open University [9]. The construction of network platform should be "unified standards, data concentration, application integration, hardware cluster", integrate and optimize the educational resources, and build an open university education platform with the goal of unified user management, unified authority control, unified resource sharing service, and personalized information service. The network platform should realize educational informationization, scientific decision-making and standardized management. The network platform should build an advanced virtual open campus, provide all kinds of resources and services for all kinds of learning users, provide all kinds of teaching, scientific research, learning and other interactive activities for educators and the educated, and improve the level of running a school and the quality of Teaching [10]. Through the mobile learning network platform, educational administrators at all levels can grasp the status of schools, enrollment, teachers, examinations and graduation in real time. Through standardized management, they can provide support for decision-making and ensure the teaching quality of open education. The construction goal of Internet distance education platform is as Table 2.

Table 2. Construction objectives of internet distance education platform

Integrated information platform	
The construction goal of mobile learning system	Everyone can register, study and test all the time and everywhere
	Unified standard, data centralization, application integration, hardware cluster
	Unified user management, unified authority control, unified resource sharing service and personalized information service
	Educational informationization, scientific decision-making and standardized management

According to Table 2 the goal and content of constructing the Internet distance education platform, using the modern, holistic, platform and network thinking mode, and based on the digital transformation of the existing teaching resources, the platform of

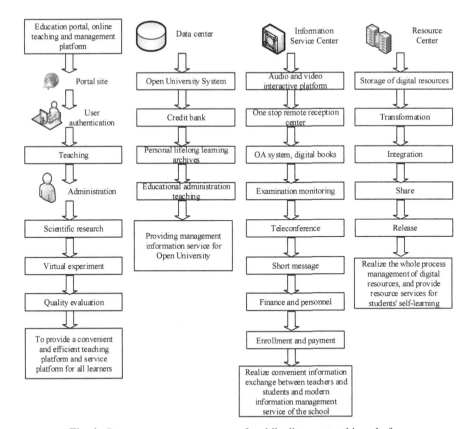

Fig. 4. Data management structure of mobile distance teaching platform

Internet distance education platform is gradually established by creating, introducing, developing and constructing teaching resources through multiple channels [11]. Through the construction of a portal platform and three centers, an information service system serving open education and lifelong education is built, and teaching resources are managed and optimized. Based on this, the data management function structure of mobile distance education platform is improved as Fig. 4.

As shown in the Fig. 4, through the mobile learning network platform, learners can learn from time to time, participate in online communication, manage personal information, meet the personalized requirements of different learners, and meet the learning needs of learners on the job, at school, at home anytime and anywhere [12]. Through the mobile learning network platform, teachers and educational administrators can carry out teaching research, teaching, educational administration management, examination evaluation management, financial charges, textbook subscription and other educational administration management. Through such a platform, teachers can timely understand students' learning situation, interactive communication and learning feedback needs.

2.3 Operation Method of Internet Distance Education Platform

Mobile learning network platform fully embodies the education concept of Open University, adapts to the trend of diversification and diversification, provides different forms, different levels and different types of education services for different learning subjects, and meets the learning needs of learners on the job, at school and at home anytime and anywhere. The Internet distance education platform adopts the service-oriented software architecture, realizes the service of the application platform and the integrated application among various services, quickly integrates the business application platform, and easily realizes the transformation of various functions [13]. The Internet Distance Education Platform relies on the rich educational resources in the existing distance higher education platform to complete its own educational functions, so the Internet distance education platform has the structure of general distance education websites, which inevitably shows the similarity and inheritance with general distance education websites. The structure and function framework of Internet distance education platform is as Fig. 5.

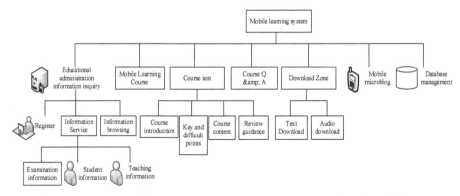

Fig. 5. Structure and function framework of internet distance education platform

In Fig. 5, in the framework of Internet distance education platform, learners can query information anytime and anywhere, use mobile phones to learn and test courses, ask questions online whenever they encounter problems, Download Text and audio courseware, listen to lectures anytime and anywhere, and communicate with other learners and express their opinions through mobile microblog. Teachers can carry out mobile teaching through the Internet distance education platform, answer questions online anytime and anywhere, view learners' self-test results online, communicate with learners, publish announcements and so on, so as to improve the practical application effect of education and teaching platform and better ensure the quality of teaching.

3 Analysis of Experimental Results

In order to ensure the rationality of the Internet distance teaching platform designed in this paper, the use process of the Internet distance teaching platform is simulated under the web network environment. Observe the use process of the platform, use SF test software to test the fluency of the platform, and use FPS index to measure the stability of the platform. The security level index is used to evaluate the security of the platform, the THU index is used to measure the operation effect of the platform, and the satisfaction index is used to evaluate the satisfaction of students and teachers to the platform. In order to ensure the validity and accuracy of the experimental verification, comparative test is used. Compare the traditional teaching platform with the internet teaching platform designed in this paper, analyze the experimental results and draw a conclusion. The experimental parameters are as Table 3.

Table 3. Experimental data

Number of tests	Types of uploaded data	Client call rate %
1	3.0	0.1
2	4.0	0.2
3	5.0	0.3
4	6.0	0.4
5	7.0	0.5
6	8.0	0.6

According to the experimental data in Table 3, using SF test software to load on the platform for use, the loaded SF software will not affect the normal use of the platform, and has little impact on its network data transmission. In order to ensure the accuracy of detection, this paper compares the traditional security evaluation methods of the information management and control platform with the research methods in this paper, and designs the operation evaluation parameters. The evaluation and record of information management and control platform running under different platforms are as Table 4.

Table 4. Comparison of platform operability evaluation

Data volume	Traditional method		Method of this paper	
	Information content	Safety degree	Information content	Safety degree
	500	74%	500	96%
2000	1600	68%	1600	95%
20000	17000	72%	17000	93%
50000	40000	65%	40000	91%

Analysis of Table 4 shows that when the negative average information is 40000, the average security level of the traditional method is 69.8%, and the average security level of this method is 93.8%. Therefore, the security level of text method is high, which can effectively improve the security of the platform. Further testing software uses FPS indicators to measure platform fluency, and thu parameters to measure platform operation effect. The FPS indicators and thu parameters of the Internet distance teaching platform designed in this paper from the perspective of industry education integration are compared, and the specific test results are as Fig. 6.

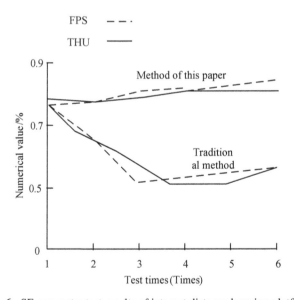

Fig. 6. SF parameter test results of internet distance learning platform

According to Fig. 6, the average values of FPs and THU parameters are 81% and 78% respectively, which can be maintained at more than 70% during operation, while the average values of FPS and THU parameters are 62% and 58% respectively. Which proves that the mobile Internet of things distance teaching platform in the perspective of industry education integration has good stability in the practical application process, and can ensure the effect of multiple people using at the same time. The satisfaction of teachers and students in the process of using the platform is further investigated, and the statistical results are as Fig. 7.

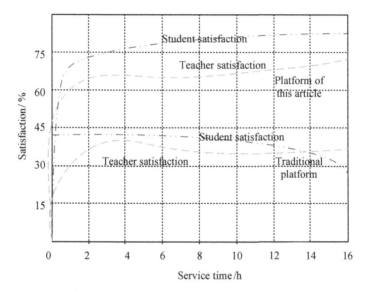

Fig. 7. Comparison results of platform use satisfaction

According to Fig. 7, when the service time is 16h, the average satisfaction of students and teachers in the traditional platform is 35% and 32% respectively, while the average satisfaction of students and teachers in this platform is 77% and 63% respectively. Therefore, in the actual use of the platform, the satisfaction of teachers and students is significantly higher than that of the traditional platform, which has a good use effect. To sum up, the Internet distance teaching platform designed in this paper can maintain a good running state in the network environment, and can maintain high fluency and upload efficiency. The satisfaction of teachers and students in the use process is also significantly higher than that of the traditional teaching platform, which proves that the platform has high use value.

4   Conclusion

In order to improve the use satisfaction and security of traditional education and teaching platform, and ensure the stability of operation, the mobile Internet Distance Education and teaching platform is constructed from the perspective of industry education integration. Through the construction of distance education teaching platform structure and distance education data processing platform framework, optimize the function of mobile Internet distance education teaching platform, realize the operation of mobile Internet distance education teaching platform. The experimental results show that:

(1) The average security level of the platform is 93.8%, which can effectively improve the security of the platform.
(2) The average values of FPS and THU parameters are 81% and 78% respectively, which proves that the distance teaching platform of mobile Internet of things has good stability in the practical application process and can ensure the effect of multiple people using at the same time.
(3) In this paper, the average satisfaction of students and teachers is 77% and 63% respectively. In practical use, it can effectively improve the satisfaction of teachers and students, and has a good effect.

References

1. Ai, F., Wang, N.: Integration of urban-rural planning and human geography for online education under the impact of COVID-19. J. Intell. Fuzzy Syst. **39**(6), 8847–8855 (2020)
2. Lenihan, S., Foley, R., Carey, W.A., Duffy, N.B.: Developing engineering competencies in industry for chemical engineering undergraduates through the integration of professional work placement and engineering research project - ScienceDirect. Educ. Chem. Engineers, **32**(5), 82–94 (2020)
3. Li, H.: Construction of distance teaching platform based on mobile communication technology. Int. J. Netw. Virtual Organ. **20**(1), 35–43 (2019)
4. Babu, M., Suman, K., Rao, P.S.: Drafting software as a practicing tool for engineering drawing-based courses: content planning to its evaluation in client-server environment. Int. J. Mech. Eng. Educ. **47**(2), 118–134 (2019)
5. Costa, R.D.D., Souza, G.F.D., Castro, T.B.D., et al.: identification of learning styles in distance education through the interaction of the student with a learning management system. Revista Iberoamericana de Tecnologias del Aprendizaje, PP(99), 1–1 (2020)
6. Wen, J., Zhang, W., Shu, W.: A cognitive learning model in distance education of higher education institutions based on chaos optimization in big data environment. J. Supercomput. **75**(2), 719–731 (2018). https://doi.org/10.1007/s11227-018-2256-2
7. Alshehri, A., Rutter, M.J., Smith, S.: An implementation of the UTAUT model for understanding students' perceptions of learning management systems: a study within tertiary institutions in Saudi Arabia. Int. J. Dist. Educ. Technol. **17**(3), 1–24 (2019)
8. Ghadirian, H., Salehi, K., Ayub, A.F.M.: Assessing the effectiveness of role assignment on improving students' asynchronous online discussion participation. Int. J. Dist. Educ. Technol. **17**(1), 31–51 (2019)
9. Gezgin, D.M.: The effect of mobile learning approach on university students' academic success for database management systems course. Int. J. Dist. Educ. Technol. **17**(1), 15–30 (2019)
10. Jing, L., Bo, Z., Tian, Q., et al.: Network education platform in flipped classroom based on improved cloud computing and support vector machine. J. Intell. Fuzzy Syst. **39**(99), 1–11 (2020)
11. Fu, W., Liu, S., Srivastava, G.: Optimization of big data scheduling in social networks. Entropy **21**(9), 902 (2019)
12. Liu, S., Li, Z., Zhang, Y., et al.: Introduction of key problems in long-distance learning and training. Mobile Netw. Appl. **24**(1), 1–4 (2019)
13. Liu, S., Sun, G., Fu, W. (eds.): e-Learning, e-Education, and Online Training. LNICSSITE, vol. 339. Springer, Cham (2020). https://doi.org/10.1007/978-3-030-63952-5

Design of International Teaching Quality Evaluation System for Railway Locomotive Specialty Based on Mobile Terminal

Wen-hua Deng$^{(\boxtimes)}$

Wuhan Railway Vocational College of Technology, Wuhan 430205, China

Abstract. Domestic railway locomotive colleges attach great importance to the supervision, evaluation and guidance of international teaching. It is necessary to build a set of teaching evaluation system suitable for the development of railway locomotive colleges. The current teaching quality evaluation system still has the problem of poor evaluation accuracy. Therefore, this paper designs an international teaching quality evaluation system for railway locomotive specialty based on mobile terminal. The system is divided into PC Web terminal and mobile terminal, and the appropriate chip and network equipment are selected to complete the selection of central integrated chip and the design process of network interface. According to the results of hardware optimization, a new evaluation index system is designed, and the teaching quality evaluation model is constructed by using analytic hierarchy process. By combining hardware with software, the design of international teaching quality evaluation system for railway locomotive specialty based on mobile terminal has been completed. The test shows that the application effect of the system is better than that of the current system, which has a certain role in promoting the international teaching of railway locomotive specialty.

Keywords: Mobile terminal · Teaching quality evaluation · Analytic hierarchy process · Scaling method

1 Introduction

The international teaching quality of railway locomotive specialty is not only the lifeline of the survival and development of railway locomotive specialty, but also the foundation of railway locomotive specialty. Continuously improving the teaching quality of railway locomotive specialty is the eternal theme of education and teaching. Teaching quality evaluation of railway locomotive professional teachers is one of the key links of teaching management, which plays a positive role in promoting and promoting the teaching quality of teachers [1, 2]. Research shows that good teaching qualities can not only improve students' academic performance in school, but also improve students' salaries after graduation. In recent years, important education documents at home and abroad reveal the important position of teaching quality evaluation in the whole higher education system, and indicate the transformation orientation of teaching quality evaluation in the

S. Liu and X. Ma (Eds.): ADHIP 2021, LNICST 417, pp. 419–429, 2022.
https://doi.org/10.1007/978-3-030-94554-1_33

aspects of content, mode and method. On the basis of the diversification of modern higher education, the education evaluation mode is facing the practical need of comprehensive transformation to informatization.

In reference [3], a series of teaching activities of "I ask you to answer" were designed to improve the deficiency of the homework form of students' self-made questions. The activity design included the design of roles and tasks, links and processes, evaluation mechanism and activity work page. The links of question writing, question answering, evaluation and discussion correction were integrated, and the quality blind evaluation was embedded in each key link. Through the practice of circuit course at Tianjin University, it shows that this method can make students fully integrate and deepen their knowledge, and the teaching effect is very significant while reducing the workload of teachers. In reference [4], a teaching evaluation system based on B/S mode is designed, in order to meet the needs of teaching quality evaluation of university teachers, UML technology designed and developed by MyEclipse and webstorm is used as the back-end tool and front-end tool. Taking tomcat 7.0.54 as the server and MySQL as the database, the system is standard, efficient, practical and easy to expand. Mobile Internet is regarded as the core and one of the most important trends of the future network development. The combination of education evaluation and the mobile Internet has become the inevitable trend of the development of education, and the informatization of education evaluation is the inevitable result of the development of the times and technology. Therefore, fully relying on the mobile Internet platform to achieve continuous, comprehensive and real evaluation results of teachers' teaching quality, and constantly using the evaluation results to improve the teaching quality of colleges and universities is the current trend [5].

According to the above analysis, the focus of international teaching quality evaluation of railway locomotive specialty should shift from infrastructure construction to application level. Therefore, a mobile terminal based international teaching quality evaluation system for railway locomotive specialty is designed, and a reasonable evaluation index is established in order to improve students' participation.

2   Hardware Design of International Teaching Quality Evaluation System for Railway Locomotive Specialty

Through the research and analysis of the shortcomings of the current international teaching quality evaluation system of railway locomotive specialty, the optimization design is based on the mobile terminal, in order to optimize the data processing speed and reliability of the current international teaching quality evaluation system of railway locomotive specialty more effectively. On the premise of optimizing the hardware and software of the whole system, the hardware structure of the system is designed. The hardware structure of the system after design is shown as follows.

According to the content in Fig. 1, the software part of the system is designed based on the construction of the hardware part of the system. The optimized hardware structure of the system is used as the basis of software development, and the system hardware structure in Fig. 1 is optimized to complete the selection of each device. In order to flexibly adapt to the integration of all kinds of terminals, this topic is based

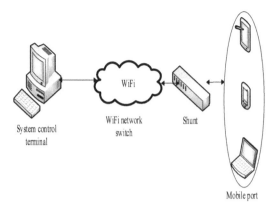

Fig. 1. Hardware structure of the system

on the theory of service computing, on the basis of in-depth study of service-oriented architecture (SOA), using web services technology, the interface of system integration is developed, so that the system can integrate all kinds of terminal programs, and achieve good scalability and cross terminal adaptability. Specifically, at the bottom of the system architecture, the web service is used to develop the system interface. On the basis of the web service interface, the mobile end and PC end of the system are developed. Therefore, the mobile end and PC end can adopt different programming technologies for development.

2.1 Central Integrated Chip Selection

The intelligent control of mobile terminal is the development trend of computer technology control. Using single chip microcomputer to control the mobile terminal is the most commonly used means to realize the intelligent control of mobile digital terminal. Through literature research, it can be seen that MCU integrates CPU, ram, ROM (EPROM or EEPROM), clock, timer/counter, serial and parallel I/O ports with multiple functions on one chip. In addition to the above basic functions, some also integrate A/D, D/A and other modules, such as Intel's 8098 series. It has the characteristics of processing capacity, low price, complete development environment and complete development tools. The system with single chip microcomputer as the digital control core not only has simple circuit, but also can realize more complex control with strong flexibility and adaptability. Therefore, this study will use a single-chip microcomputer as the main part of the central integrated controller. According to the system design requirements, inter series single-chip microcomputer is selected as the minimum control system. The schematic diagram of the chip and its peripheral circuit is shown in Fig. 2. The main frequency of the system is 8 m. There is no need to install additional program memory. The main components of the minimum system are described in detail as follows.

In order to ensure that the MCU can be effectively connected with other devices in the mobile terminal, the D/A conversion interface is set in it. D/A conversion is a typical interface technology of application measurement and control system. The main content of its design is the rotation of D/A integrated chip, configuration of peripheral circuits

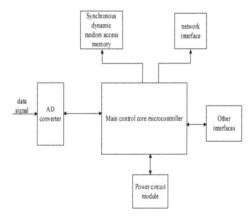

Fig. 2. Structure chart of central integrated control chip

and devices. Considering the performance, structure and application characteristics of the system, the D/A conversion chip selected in this study is set as multiple interfaces to meet the interface requirements of the system and reduce the development cost of the system.

2.2 Network Communication Interface Design

In this study, the wireless network terminal as the main platform of system design, to achieve the management of wireless network, must have the corresponding hardware support, in order to cooperate with the management end, to achieve the use process of the whole mobile terminal.

The hardware architecture is the bottom module of the whole network communication interface, and the hardware must support openway system [6, 7]. At the same time, the signal conversion chip is added in this interface to improve the communication ability of the system. At present, many companies in the market provide a variety of lUDs signal conversion chips. The obvious advantage of this design is that the form of front-end signal level is not limited. At the same time, there are many kinds of chips to choose from. According to the requirements, the chips with excellent performance can be selected in different transmission application fields, and different conversion chip characteristics can be used to meet different design requirements. This topic selects the lUDs signal conversion chip from Texas Instruments Company, which can not only meet the high-speed, multi-channel transmission characteristics, but also has strong anti electromagnetic interference ability, and improves the data transmission ability of the system to a certain extent.

The central integrated control chip and network communication interface designed in this study are introduced into the original hardware framework of the system. So far, the hardware design of international teaching quality evaluation system for railway locomotive specialty is completed.

3 Software Design of International Teaching Quality Evaluation System for Railway Locomotive Specialty

3.1 Construction of Evaluation Index System

This paper attempts to use the method of literature analysis to summarize the existing teaching quality evaluation related indicators, and systematically sort out, sort out the existing construction of the teaching quality evaluation index system, and sort out the key indicators. With the help of China Wanfang, CNKI and other paper websites, this paper collects journals, monographs and academic papers related to teaching quality, teaching quality evaluation, university evaluation index, education informatization, mobile teaching evaluation in five years, so as to screen, sort out and summarize the existing research on teaching quality evaluation index in China. 14 indicators are listed by exhaustive method, as shown in Table 1.

Table 1. Evaluation index system of international teaching quality of railway locomotive

Serial number	Index content	Direction
1	Excellent scientific research ability	Teachers' teaching direction
2	With noble sentiments	Teachers' teaching direction
3	Have the spirit of dedication	Teachers' teaching direction
4	Teachers' professional skills	Teachers' teaching direction
5	Teachers' professional conscience	Teachers' teaching direction
6	Teachers' professional style	Teachers' teaching direction
7	Professional honor of Teachers	Teachers' teaching direction
8	Internationalization of students' learning content	Students' learning direction
9	Students have international thinking	Students' learning direction
10	Students pay attention to international current affairs	Students' learning direction
11	Uphold academic spirit and abide by academic norms	Students' learning direction
12	To undertake social responsibility for national prosperity and national rejuvenation	Students' learning direction
13	Application of advanced technology	Students' learning direction
14	Good learning attitude	Students' learning direction

At present, most of the research uses the analytic hierarchy process, the Delphi method, the entropy weight method, the average method to determine the weight of indicators, and several methods have their own advantages and disadvantages [8–10]. In this study, AHP is used to assign the weight of indicators. AHP can determine the weight value of each indicator by comparing different indicators. It is a method of multiple objectives decision-making combining qualitative analysis and quantitative analysis. It is systematic, easy to operate and needs less quantitative data. This method provides a good solution for multi index and multi solution [11]. In this study, AHP is used to deal with the index system, which provides the basis for the follow-up evaluation.

3.2 Building Evaluation Model

According to the evaluation index system, the scale method is used to build the evaluation model in this study. After comparing the evaluation indexes at the same level, the judgment matrix is obtained.

$$A = \begin{bmatrix} a_{11} & a_{12} & ... & a_{1n} \\ a_{21} & a_{22} & ... & a_{2n} \\ ... & ... & ... & ... \\ a_{m1} & a_{m2} & ... & a_{mn} \end{bmatrix} \tag{1}$$

The judgment matrix A should satisfy the following two conditions:

$$a_{ii} = 1(i = 1, 2, ..., n) \tag{2}$$

$$a_{ij} = a_{ji}(i, j = 1, 2, ..., n) \tag{3}$$

According to the above formula, the judgment matrix A is calculated to obtain the weight coefficient. A new matrix is obtained by column normalization of the judgment matrix.

$$B = (b_{ij})_{n*n} \tag{4}$$

$$b_{ij} = \frac{a_{ij}}{\sum_{i=1}^{n} a_{ij}} \tag{5}$$

The matrix C is obtained by summing the matrix B according to the row.

$$C = (c_i)_{n*1} \tag{6}$$

$$c_i = \sum_{j=1}^{n} a_{ij} \tag{7}$$

Once again, the matrix is normalized [12, 13], and the eigenvector is obtained.

$$D = (d_i)_{n*1} \tag{8}$$

$$d_{ij} = \frac{c_i}{\sum_{i=1}^{n} c_i} \tag{9}$$

According to the above formula, complete the evaluation process of international teaching. In order to ensure the reliability of the evaluation results, the consistency of the results is tested. Set the judgment matrix as order $m * m$ and the maximum eigenvalue as α_{max}. when $\alpha_{max} = m$, the matrix is said to have consistency, but in general case $\alpha_{max} \neq m$, then it is necessary to use the consistency ratio to test it

$$CR = \frac{CI}{RI} \tag{10}$$

CR is the consistency ratio; CI is the consistency test index, and:

$$CI = \frac{\alpha_{max} - 1}{n - 1} \tag{11}$$

$$\alpha_{max} = \frac{1}{n} \sum_{i=1}^{n} \frac{(R * C)_i}{c_i} \tag{12}$$

In the above formula, RI is the average random consistency index. Use the above formula to complete the evaluation process of international teaching quality, and output the evaluation results to the image system. So far, the design of railway locomotive professional international teaching quality evaluation system based on mobile terminal has been completed.

4   System Test Analysis

This test task is to find as many errors as possible before the software is put into operation. There are two goals of testing: the first is the process of executing a program to find errors in the program. The second point is that a good test plan is one that is likely to find errors that have not been found so far.

4.1   System Test Environment

In this study, the design process of the international teaching quality evaluation system for railway locomotive specialty based on mobile terminal is completed. In order to analyze the effect of its use, they built the system test link, compared it with the original system, and determined the difference between them. The test environment of the system is described in Table 2. The network environment is campus network, with a bandwidth of 100 m, the server used is Lenovo, and the database environment is SQL server2016.

Using the above system test environment as the comparison platform between the mobile terminal system and the original system, the differences between the two systems are studied.

Table 2. System test environment

Experimental network	Campus laboratory network		
The server	Lenovo architecture server		
Database	SQL Server2016		
Client	Model	ThinkPad X 1	
	Number	2	
	PC mobile terminal	Hardware configuration	Software configuration
		CPU: Intel(R)Core(TM)i5-2430 M 2.4 GHz	Operating system: Windows 10
		Memory: 16 GB	Internet Explorer 8
		Hard disk: 1 TB	–

4.2 System Testing Programme

Generally speaking, when the international teaching quality evaluation system of railway locomotive specialty is adopted, there are two kinds of errors: output error and data consistency error. The reason of output error is the defect of program design or the slow network speed. The so-called consistency error is caused by the user's error when uploading information. The above two different errors can be tested respectively. In the process of system testing, six basic principles are strictly observed to ensure the smooth completion of system testing. The specific function test contents and results are shown in Table 3.

Table 3. System function test results

Function test content	Test results	Is this correct
Sign in	Login successful	1
User query	Information query successful	2
Increase in evaluation content	Success	3
Modification of evaluation content	Success	4
Deletion of evaluation content	Success	5
User information correction	User information corrected	6

From the above experimental results, it can be seen that the function of mobile terminal system meets the current functional requirements, and the subsequent performance test can be carried out. In this test, the system indicators are set as follows: teaching quality evaluation accuracy, teaching quality grade division error value and teaching quality evaluation running time. According to the above indicators, the application effects of the two systems are compared and analyzed.

4.3 System Test Results

As can be seen from Fig. 3, in the process of multi system simultaneous testing, the accuracy of teaching quality evaluation of mobile terminal system is significantly better than that of existing systems, and the highest accuracy rate can reach 87.5%. After many system tests, the accuracy of teaching quality evaluation of mobile terminal system is relatively stable, there are few problems of large fluctuations in the accuracy of evaluation, and it has a high and stable evaluation ability. Compared with the mobile terminal system, the evaluation accuracy of the comparison system is relatively low, the evaluation accuracy changes greatly and is unstable. Using the existing system to analyze the international teaching quality will lead to the teaching design and the current situation is not consistent. According to the above experimental results, the mobile terminal system is better than the current system.

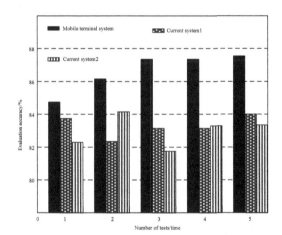

Fig. 3. Accuracy of teaching quality evaluation

In the process of this experiment, the accuracy of risk classification results is reflected as the data difference between each level after risk classification. In order to improve the representativeness of this test index, the above images are used to display and analyze it. As can be seen from Fig. 4, the result of teaching quality grading is more accurate and accurate, with lower error value and in line with the current teaching evaluation requirements. The results of risk assessment are divided into four levels according to the classification of teaching quality, and there are obvious differences between each level. The current system divides the teaching quality into three levels, and the differences between the levels are small, so it is unable to carry out a comprehensive analysis and timely teaching optimization on the international teaching quality.

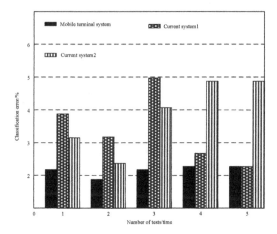

Fig. 4. Error value of teaching quality grading

Table 4. Length of evaluation of teaching quality

Test object	Test index	Test result/s
Mobile terminal system	Maximum running time	3.15
	Minimum running time	2.27
	Average running time	2.74
Original system 1	Maximum running time	4.13
	Minimum running time	3.85
	Average running time	4.06
Original system 2	Maximum running time	4.37
	Minimum running time	4.20
	Average running time	4.33

According to the test results in Table 4, the evaluation speed of the mobile terminal system is fast and stable. Although the current system also has high evaluation efficiency, it is still different from the mobile terminal system. In the follow-up research, this performance should be optimized in order to improve the use effect of the system. According to the above test results, the use effect of the mobile terminal system is better than the current system.

5   Conclusion

This paper studies and discusses the evaluation of international teaching quality of railway locomotive specialty under the network environment, and brings teachers into the overall environment of the teaching platform. Through the intelligent interaction among students, teachers, students and teachers, students and systems, teachers and systems, through the multi-user interaction mode and multi-agent cooperation mechanism, an all-round teaching quality evaluation under the network environment is realized Intelligent education services. The research on international teaching quality evaluation of railway locomotive specialty has always been concerned by China, and the field of thesis and creation is constantly emerging. However, in the development of Internet technology, how to combine technology with research theory in-depth research. As countries, schools and parents pay more and more attention to education, people from all walks of life are paying more and more attention to teachers' teaching qualities. Due to the limited ability and knowledge of the author, the author only makes an in-depth analysis on the teaching quality of a single railway locomotive major, and does not explore the education problems in a wider range of majors. But it is these problems that can be further studied that have become a new starting point for my future research. There is no end to finding problems and constantly summarizing and exploring them.

Fund Projects. Scientific research project of Hubei Vocational and Technical Education Association, development and Research on international teaching standards of railway locomotive specialty under the background of "double high plan" (ZJGB2020003).

References

1. Jiang, L., Xie, F.: Design of effectiveness evaluation system of foreign language assistance teaching on mobile terminal. Mod. Electron. Techniq. **43**(18), 132–134 (2020)
2. Liu, Q., Chen, C., Liu, C., et al.: Content design and teaching effect evaluation of Microbiology course based on the construction of network teaching resources. Microbiology **47**(04), 1117–1125 (2020)
3. Yu, X., Jia, H., Li, G.: Design for teaching project of "answer my question" with process evaluation and double-blind peer review. Proc. CSU-EPSA **31**(02), 144–150 (2019)
4. Yan, H.: Construction of diversified quality evaluation system for the whole process of graduation design of environmental design major. J. Anshan Norm. Univ. **21**(04), 81–84 (2019)
5. Zhang, J., Hu, Y., An, Z.: Design and implementation of web-based teaching evaluation system. J. Yuxi Norm. Univ. **35**(03), 118–123 (2019)
6. Wang, R.: Design of home-school connection and student evaluation system based on mobile terminal. J. Shaoguan Univ. **40**(03), 25–28 (2019)
7. Liu, S., Liu, G., Zhou, H.: A robust parallel object tracking method for illumination variations. Mob. Netw. Appl. **24**(1), 5–17 (2018)
8. Yue, Q., Wen, X.: Application of improved GA-BP neural network in teaching quality evaluation. J. Nat. Sci. Heilongjiang Univ. **36**(03), 353–358 (2019)
9. Luo, Y., Deng, K., Tian, G., et al.: Immersion VR teaching system "VisAll." Comput. Simulat. **37**(11), 194–198+303 (2020)
10. Xu, F., Chen, Y., Huang, Z., et al.: An empirical study on teaching quality evaluation of pharmaceutical administration. J. Shenyang Pharmaceut. Univ. **36**(12), 1119–1126 (2019)
11. Wang, Y., Ma, Y., Yu, S.: Reform of computer course scoring model for sergeant cadet education and its effect on teaching quality. Comput. Eng. Sci. **41**(S1), 229–233 (2019)
12. Liu, S., He, T., Dai, J.: A survey of CRF algorithm based knowledge extraction of elementary mathematics in Chinese. Mob. Netw. Appl. **26**(5), 1891–1903 (2021)
13. Liu, S., Fu, W., He, L., Zhou, J., Ma, M.: Distribution of primary additional errors in fractal encoding method. Multim. Tools Appl. **76**(4), 5787–5802 (2014)

Design of Mobile Teaching Platform for Vocal Piano Accompaniment Course Based on Feature Comparison

Ying Chen[1](✉), Jia-yin Chen[2], and Ai-ping Zhang[3]

[1] College of Art, Xinyu University, Xinyu 338000, China
[2] College of Music, Ningbo University, Ningbo 315211, China
[3] School of Software, East China JiaoTong University, Nanchang 330013, China

Abstract. With the rapid development of the Internet, the current education system courses often use mobile teaching platforms to improve teaching effects. However, due to the large and unclear feature extraction range, the established mobile teaching platform database information is confused, and the stored data resources are not comprehensive. Poor operation stability, etc. In order to improve the mobile teaching effect of vocal piano accompaniment courses, the text designs a mobile teaching platform for vocal piano accompaniment courses based on feature comparison. By constructing a mobile teaching platform framework for vocal piano accompaniment courses, we will improve and update the mobile teaching platform resource database and expand the content of resources. Design the course teaching evaluation module, use feature comparison technology to extract the fundamental frequency parameters of vocal piano accompaniment and Mel cepstrum parameters, create a scoring mechanism to compare vocal piano accompaniment, to evaluate the teaching effect of vocal piano accompaniment course, and complete the course teaching evaluation module design. Experimental results show that in the actual application process of the mobile teaching platform in this paper, the delay time is 1.9 s, the response speed is fast, the memory occupancy rate is only 35%, and the stability is high, which can effectively improve the actual teaching effect.

Keywords: Feature comparison · Vocal music · Piano accompaniment course · Mobile teaching platform

1 Introduction

Along with technological innovation, various mobile devices have quietly entered our lives and changed the way we learn. As the most representative electronic product in the contemporary era, smart phones have undoubtedly become the most widely used electronic tools in students' lives [1]. The development of the mobile Internet and new media technology has made the speed of information dissemination faster and wider,

S. Liu and X. Ma (Eds.): ADHIP 2021, LNICST 417, pp. 430–442, 2022.
https://doi.org/10.1007/978-3-030-94554-1_34

and has provided a lot of convenience for teachers' teaching and students' learning. For the teaching of vocal piano accompaniment courses in colleges and universities, it can be said that opportunities and challenges coexist in the era of new media. On the one hand, the educational content of paper textbooks has the characteristics of authority, authenticity and rigor, which are the main reference basis for teachers' teaching. On the other hand, there are disadvantages in terms of interactivity, poor image quality, and slow update of educational content [2]. Therefore, with the advantages of new media and mobile Internet, the presentation forms and teaching methods of piano knowledge are enriched.

After years of development, the country's mobile teaching has been greatly developed, and various mobile teaching platforms have also been established, such as Chaoxing Xuetong, Little Ant Teaching Platform and so on. Literature [3] adopts research methods such as literature research method, questionnaire survey method, experimental research, interview method, and constructivist learning theory, informal learning theory, cognitive load theory, and Maslow's demand theory as the theoretical basis. It also analyzed the feasibility of vocal piano accompaniment course teaching based on the WeChat public platform, and built a vocal piano accompaniment course teaching platform "Mousse Music Class". On the basis of modern information technology, Liu Jia [4] deeply studied the design of the cloud platform for learning online music courses on mobile terminals. Through the cloud platform as the carrier of the online music course learning environment atmosphere, users can access the network, servers, applications and other resources in real time, so that students can enjoy online and offline course on-demand, online live broadcast, course content push and other related service functions. It helps to meet the students' individualized and diverse music course learning needs, and also helps music teachers to successfully complete music teaching tasks. Chen Hao and Wu Yuqi [5] designed a music feature recognition system design based on the Internet of Things technology in order to improve the ability of music feature recognition. The sound sensor is used to collect the original music signal, the digital signal processor is used to analyze the music signal, and the music signal is processed through the network transmission layer. Finally, the data result is transmitted to the music signal database in the system. Use the internal music feature analysis module of the system to realize the feature recognition of music signal, and identify the music form and the corresponding music feature content of the music emotion according to the recognition result. However, these system platforms are usually more complicated, the design is not flexible enough, and data acquisition is time-consuming. It usually takes more time for students to find the resources they need. Although the above-mentioned cloud platform design software can meet some music teaching courses, it is not suitable for vocal piano accompaniment courses. For this reason, feature comparison technology is introduced, and a mobile teaching platform design for vocal piano accompaniment courses based on feature comparison is proposed.

2   A Mobile Teaching Platform for Vocal Piano Accompaniment Courses Based on Feature Comparison

2.1   Designing a Mobile Teaching Platform Architecture for Vocal Piano Accompaniment Courses

The design of a mobile teaching platform for vocal piano accompaniment courses divides the overall framework of the mobile teaching platform into two parts, the client and the server. This research mainly designed two main functions of the web page adaptive matching platform and the video transcoding module on the server side of the mobile teaching platform of the vocal piano accompaniment course. The overall structure is shown in Fig. 1.

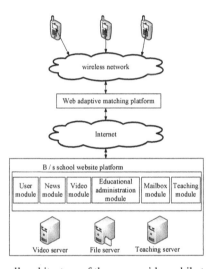

Fig. 1. The overall architecture of the server-side mobile teaching platform

Computer users and mobile phone users share all the information resources and services in the website database through the mobile teaching platform [6]. The web adaptive matching service platform is a core of the entire mobile teaching platform. The platform can automatically convert the web pages in the school website based on the B/S model into Wap pages. So that mobile phone users can browse the same content as when browsing on a computer anytime and anywhere.

The school's Web site itself can implement functions such as news release, educational administration platform login, learning of teaching resources, email service, and viewing of teaching videos. At present, there is still a gap in the processing capabilities of smartphones for video reception and online viewing compared with computers. In addition, when there are more users accessing video resources, it will bring a certain burden to the original server and database containing data and video [7]. Therefore, the relevant information of the teaching video is separated from the original file database.

There are certain restrictions on the format, flow and speed when watching videos on mobile phones. Therefore, video transcoding technology has been researched. On the server side of the B/S teaching website, FFmpeg is used to re-encode and compress the videos and package them to adapt to wireless network transmission and mobile phones. The format of the video to watch. Therefore, a video transcoding module is added to the original video module, with the purpose of re-encoding and compressing existing and future videos that need to be added to meet the needs of mobile terminals.

Based on the overall structure of the server-side mobile teaching module designed above, it is also necessary to design a web page adaptive matching module to realize web page crawling and perform complex page format analysis. Then use the data extraction component to convert it into a standard XML data stream format. By analyzing the screen resolution of the mobile phone terminal, the browser version and other information, the corresponding template information in the data conversion module in the extraction module is determined, and the corresponding data is loaded. Then display the web page information on the mobile terminal to realize the conversion from the web page to the Wap page [8].

The web pages in the school web site are dynamic web pages. You can build a DOM tree by extracting tags from the web pages, remove tags that are not easily recognized by mobile terminals, extract web tags and attributes that are applicable to smart phone browsers, and generate Wap templates. And write the ID number of the generated template and save it in the web page adaptive matching module. When the user visits a certain ID web page, the Wap template can be called, and at the same time, the relevant data in the database can be called and displayed on the smart phone terminal [9]. Therefore, the web page adaptive matching module needs to construct five sub-modules: data extraction sub-module, data conversion denoising sub-module, thumbnail generation sub-module, template extraction sub-module, and rendering and generation sub-module. The function of this module is mainly to extract data from existing HTML documents in the original Web site through web page denoising technology (JTidy) and XSLT technology, and to normalize the extracted data into XML documents. Then through the XSLT technology, the content of the XML document is reorganized into a Wap document, so as to realize the purpose of logging onto the teaching website anytime and anywhere with the smart phone terminal. The establishment of the web page adaptive matching module is based on the existing web site, maintaining the relevant structure and related data of the original site, and improving work efficiency while meeting the accuracy of information. And targeted removal of some information that wasted traffic due to loading in the original Web site, thus reducing the waste of traffic in wireless network transmission. The specific web page adaptive matching platform module is shown in Fig. 2.

2.2 Establish a Mobile Teaching Platform Resource Library for Vocal Piano Accompaniment Courses

The mobile teaching platform database for vocal piano accompaniment courses is designed this time. The mobile teaching platform database is divided into three levels: basic resources, teaching unit resources and courseware resources, as shown in Fig. 3.

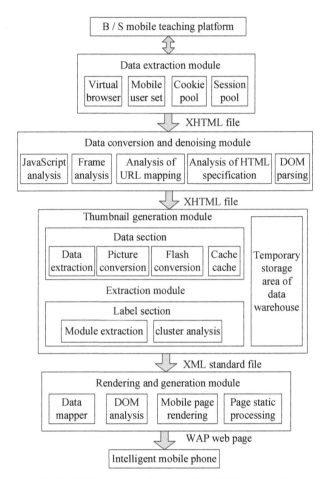

Fig. 2. Web page adaptive matching platform module

It can be seen from Fig. 3 that the bottom layer is the basic resources, including text, graphics and images, animation, audio, video, literature, teaching materials, words, vocabulary, grammar, accompaniment, example sentences, application tools, etc. Used in multimedia course teaching. According to the definition of multimedia materials, media materials are divided into five categories: text, graphic images, animation, audio, and video. These materials are the most basic elements of the course. These materials need to be classified, indexed and stored in accordance with the standards of each sub-library. These basic materials can be used by teachers to construct individualized teaching integration units and different forms of courseware. The basic materials have the widest shareability and are most suitable for mass production in a pipeline. The acquisition of basic resources can also be obtained from public resources, such as Internet resources.

The middle layer is the teaching unit resources used in different teaching links for a certain teaching content, including cases, test questions, test papers, homework, common problems, terminology, reference materials, etc. Its content can be a certain knowledge

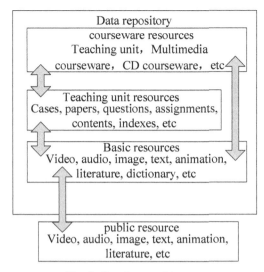

Fig. 3. Database architecture

point, a certain chapter or a certain comprehensive topic, etc., and the teaching function is relatively single. Teaching unit resources can be used directly for teaching, and can also be used to construct complete courseware flexibly and quickly.

The highest level is courseware resources, which use basic resources and teaching unit resources to construct different forms of courseware. Multimedia courseware is to edit and process the video of the teacher's lecture process or the video of some teaching content in the form of video recording, and then put it on the Internet in the form of CD, TV program or digitized for the learner's reference [10]. Its advantage is that the display method is vivid and vivid, especially the lecture process video of excellent teachers has the value of learning, reference and preservation.

Different forms of courseware can be constructed using basic resources or teaching unit resources. Because different teachers face different objects, achieved teaching goals, teaching ideas, teaching design, teaching methods, etc., there are differences. Therefore, integrated teaching resources must have individual characteristics, but a courseware can completely reflect the teaching content and teaching links of course teaching, and is suitable for students to learn independently. The above three levels of resources together constitute the teaching curriculum library.

2.3 Design Course Teaching Evaluation Module

Extracting Characteristic Parameters of Vocal Piano Accompaniment Based on Feature Comparison
The vocal piano accompaniment signal is a time-varying signal. The short-term stability of the vocal piano accompaniment signal can be used to observe the vocal piano accompaniment signal over a short period of time to track the change of the signal in time [11]. This study uses the fundamental frequency change trajectory and Mel cepstrum

parameters to evaluate the input vocal piano accompaniment. The Mel cepstrum parameter characterizes the content of the sound. The fundamental frequency trace shows the change of sound pitch, reflecting the fluctuation of pitch and the change of rhythm.

(1) Extraction of fundamental frequency parameters. The pitch period is calculated using AMDF with center clipping. Compared with commonly used pitch estimators based on correlation, homomorphic signal processing, and linear predictive coding, AMDF does not involve multiplication and division, and is more suitable for applications on platforms such as embedded platforms. The valley point at the pitch period point is sharper than the peak point of the short-term autocorrelation function. Therefore, the estimation accuracy is higher and more stable. For the n-th frame vocal piano accompaniment signal S_n.

$$AMDF_n(t) = \frac{1}{M} \sum_{M=0}^{M-t-1} |S_n(M) - S_n(M+t)| \tag{1}$$

In formula (1), M represents the frame length; t represents the candidate pitch period. When $S_n(M)$ is periodic and the period is T, then there is a valley near the $AMDF_n(t)$ pitch period T or its integer multiples. Therefore, the first valley point can be selected as the pitch period of the vocal piano accompaniment. The estimated fundamental frequency $F_0(n)$ is.

$$F_0(n) = \frac{1}{\min\limits_{t>0}\left(\arg \; localMin\limits_{t>0}(AMDF_n(t))\right)} \tag{2}$$

In formula (2), $localMin$ represents the function to obtain the local minimum [12]. In actual calculations, the position of the first minimum valley point sometimes does not coincide with the actual pitch period, which is mainly caused by the interference caused by the formant characteristics of the vocal tract. Therefore, before using AMDF to estimate the pitch period, it is necessary to preprocess the vocal piano accompaniment signal to eliminate the influence of formants and improve the estimation effect. The preprocessing method uses the center clipping nonlinear transformation. After the clipped signal passes through AMDF, the trough at the pitch period point becomes more obvious, and the effect of pitch period estimation can be improved to a certain extent. In addition, linear smoothing must be performed on the obtained fundamental frequency curve to eliminate wild points with larger errors.

The acquired fundamental frequency parameter is an important parameter in the score of vocal piano accompaniment, and its level changes reflect the intonation and rhythm characteristics of vocal piano accompaniment. The intonation that imitates the standard pronunciation is a basic requirement in the learning of vocal piano accompaniment. The use of pitch changes for scoring is a feature of the scoring algorithm. It is different from the previous scoring algorithms for vocal piano accompaniment courses based on content accuracy. It evaluates the quality of a vocal piano accompaniment from a higher reading level.

(2) Extraction of Mel cepstrum parameters. After the vocal piano accompaniment signal is pre-emphasized, the suppressed high frequency part of the vocal piano accompaniment signal is compensated [13, 14]. In order to reduce the Gibbs effect, a Hamming window is added to each frame of the vocal piano accompaniment signal. After fast Fourier transform, the frequency spectrum of each frame can be obtained, and the output logarithmic spectrum $m_j(j = 1, 2, \cdots, 20)$ of each frequency band can be obtained through a set of 20 triangular filters. After discrete cosine transform, the 12-dimensional Mel cepstrum parameter c_j can be obtained.

$$c_j = \sum_{j=1}^{p} m_j \cos\left(\frac{\pi k}{p}(j - 0.5)\right), k = 1, 2, \cdots, 12 \tag{3}$$

In formula (3), $p = 20$ represents the number of triangular bandpass filters. j represents the number of logarithmic spectrums. k represents the number of frequency bands.

Because the pitch frequency of the piano accompaniment signal depicts the vocal characteristics of the speaker, there are some pitch differences in the piano accompaniment music. The score of piano accompaniment in this study focuses on the change of tone and rhythm. In order to better compare different vocal piano accompaniment, we need to regularize the feature parameters, so that the feature parameters with individual differences can be compared under the same benchmark.

Comparison of Piano Accompaniment in Vocal Music

In order to compare the piano accompaniment of vocal music to be graded with the reference standard piano accompaniment, we can estimate the difference between the two characteristic parameters to reflect the similarity between them. Because they are different in reading speed and pause time, it is impossible to compare them directly. Therefore, the dynamic time warping method is used to find the closest comparison path between the two. In general, DTW correction is applied to each feature parameter, which not only increases the computational complexity of the algorithm, but also cuts off the correlation between feature parameters. Firstly, the WFCC parameter is used to correct the DTW nonlinearity of the two pieces of vocal piano accompaniment, so that the input vocal piano accompaniment and the standard vocal piano accompaniment are corresponding to each other in the similar position. At this time, we can get a path with the least error and the corresponding DTW distance. The distance is the result of the comparison of MFCC characteristics of two pieces of vocal piano accompaniment, which reflects the pronunciation difference in the content of two pieces of vocal piano accompaniment. Based on this correction path, the fundamental frequency change trajectories $f_1(n)$ and $f_2(m)$ can be compared at the corresponding positions of similar contents. The object of comparison is the difference between the fundamental frequency point $|f_1(n) - f_2(m)|$ and its variation $|\Delta f_1(n) - \Delta f_2(m)|$. Among them, $\Delta f_1(n) = |f_i(n) - f_i(n - 1)|$. The smaller the difference is, the more similar the tone is. It can be seen that only one DTW operation is needed to compare the two feature parameters.

It is assumed that the MFCC eigenvector of reference standard vocal piano accompaniment is $M_1 = [m_1(1), m_1(2), \cdots, m_1(T)]$. The eigenvector of fundamental frequency is $P_1 = [p_1(1), p_1(2), \cdots, p_1(T)]$. Where T is the length of piano accompaniment of reference vocal music. The MFCC feature vector of piano accompaniment to be evaluated is $M_2 = [m_2(1), m_2(2), \cdots, m_1(S)]$. The fundamental frequency eigenvector is g. Where $P_2 = [p_2(1), p_2(2), \cdots, p_2(S)]$ is the length of piano accompaniment to be evaluated, there is the following formula.

$$C = \begin{pmatrix} C_1 \\ C_2 \end{pmatrix} = DTW(M_1, M_2)$$

$$\begin{pmatrix} P \\ M \end{pmatrix} = \begin{pmatrix} P_1 & P_2 \\ M_1 & M_2 \end{pmatrix} \begin{pmatrix} C_1 \\ C_2 \end{pmatrix}$$

(4)

In Eq. (4), C is the feature comparison matrix, which is obtained by DTW using MFCC feature vector of vocal piano accompaniment. Using the feature comparison matrix, the similarity of fundamental frequency change P and MFCC feature similarity M can be obtained.

Scoring Mechanism

The purpose of piano accompaniment score is to show whether the tone of the accompaniment is correct and standard, and whether it meets the requirements. The higher the score is, the higher the satisfaction of the accompaniment is. On the contrary, the lower the score is, the less accurate the tone of the accompaniment is. Therefore, the evaluation score can be defined as.

$$(P, M) = k_1 P + k_2 M + k_3 PM$$

(5)

In formula (5), k_1, k_2 and k_3 represent the weight of each scoring parameter in the scoring, P for the similarity of fundamental frequency change, and M for MFCC feature similarity. The selection of weight can be different according to different requirements or the focus of scoring. In order to make the computer better simulate the score of accompaniment experts, we can train the weights and find out the best mapping relationship between computer score and manual score.

3 Experiment

In order to verify the design of the mobile teaching platform for vocal piano accompaniment course, two groups of commonly used mobile teaching platforms are selected for comparative experiment. Choose the vocal music piano accompaniment course as the comparative experimental object of this design, verify the design of vocal music piano accompaniment course mobile teaching platform. From two aspects of platform response delay and memory occupancy, the running stability of three groups of mobile teaching platform for vocal piano accompaniment course was compared.

3.1 Experimental Preparation

This test vocal piano accompaniment course mobile teaching platform, will choose PHP server and Android client-based related equipment as the platform test necessary tools. Based on the actual situation, the platform will be specially configured with relevant staff for unified testing and installation of the system, and there are strict requirements for basic configuration of mobile phones, mainly reflected in Android 2.2, which is the minimum configuration. The server side shall carry out relevant tests with the help of the server of business application. In the actual test process, the server shall be replaced by the same PC if necessary. The specific configuration is shown in Table 1.

Table 1. Server hardware and software configuration

	Name	Specifications
Software	Compiling software	Eclipse + ADT
	Operating system	Windows 7 + Windows Server 2003
	Database	MySQL 5.6
Hardware	Graphics memory	1 GB
	Hard disk capacity	500 GB
	Memory capacity	2 GB
	CPU brand	Intel E5.2620
	Inspiron model	620S–356
	Product name	Dell 620S–356

Choose the Monkey platform testing tool to test the response delay and memory occupancy of the mobile teaching platform, and let three groups of mobile teaching platforms. Run on the server hardware and software configuration shown in Table 1, compare the response delay and memory occupation rate of the three groups of mobile teaching platforms, and verify the stability of the three groups of mobile teaching platforms.

3.2 Experimental Result

The First Group of Experimental Results

Based on the experimental parameters set in this experiment, the monkey platform test tool is selected, and a total of 300 people are selected to test the platform concurrency. Set the initial concurrency of the three groups of mobile teaching platforms to 50 people, and increase the concurrency of 50 people every 5 min until the total concurrency of the platform reaches 300 people. Verify the response delay time of three groups of platforms facing different concurrency, and the experimental results are shown in Fig. 4.

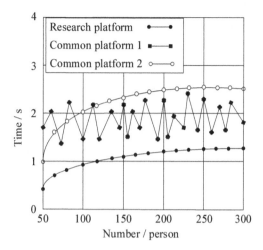

Fig. 4. Comparison of platform response delay test

As can be seen from Fig. 4, with the increase in the number of visitors, the response delay time of the three groups of mobile teaching platforms increases, which is affected by the running cycle to a certain extent. Among them, the common platform 1 faces the concurrency of 300 people. The delay time is basically maintained at 1.9 s, without large delay, which can ensure that 300 people visit the mobile teaching platform at the same time. However, its balance is poor. Whenever the concurrency of the platform changes, the delay time of the mobile teaching platform will change, which can only meet the basic needs of the platform. Common platform 2 faces the concurrency of 300 people, and the delay time increases with the increase of concurrency. However, when the number of people who log on to the mobile teaching platform at the same time reaches 250, the platform delay time shows a downward trend, which is not different from the actual cognition. It shows that the common platform 2 crashes, only 250 people can access the mobile teaching platform at the same time, but it has a good balance. The research platform faces the concurrency of 300 people, and the delay time increases with the increase of concurrency. However, when the number of people who log in to the mobile teaching platform at the same time reaches 250, the delay time of the platform is basically unchanged, and it is difficult to ensure the normal operation of the platform. It shows that the research platform can only guarantee 250 people to visit the mobile teaching platform at the same time, and the curve is smooth, and it also has a good balance. However, compared with common platform 2, with the increase of concurrency, the latency of the research platform is significantly less than that of common platform 2, which shows that the research platform has short response latency and fast response speed.

The Second Group of Experimental Results
On the basis of the first group of experiments, the second group of experiments was carried out. Using the monkey platform test tool, test three groups of mobile teaching

platform memory occupancy. In this group of experiments, 150 users are selected to test the memory occupancy rate, and the test results are shown in Fig. 5.

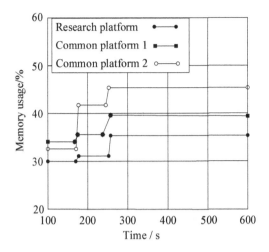

Fig. 5. Memory usage

As can be seen from Fig. 5, under the same amount of concurrency, the change of test time will also change the actual utilization of existing resources to a certain extent. Among them, the memory occupancy rate of the platform is faster than that of the two groups of platforms, and the memory occupancy rate is only 35%.

4 Conclusion

The design of a mobile teaching platform can play a guiding role in the planning of vocal piano accompaniment courses. This article builds a mobile teaching platform architecture, designs a mobile teaching platform database, uses feature comparison technology to extract vocal piano accompaniment parameters, and according to the actual parameter results obtained, The teaching effect of the vocal piano accompaniment course is optimized and upgraded, and the course teaching evaluation model is improved. The design of the mobile teaching platform for vocal piano accompaniment courses based on the comparison of features in this study is beneficial to optimize the update and upgrade of the mobile teaching platform, effectively improve the response speed of the system, reduce the system memory usage, increase the system load capacity, and expand the vocal piano accompaniment content The search range improves the overall teaching effect of the vocal piano accompaniment course system. In the actual application process of this article, the mobile teaching platform has always improved the teaching efficiency of vocal piano accompaniment and ensured the quality of teaching. At the same time, it has also improved the classroom teaching efficiency of teachers, and provided more reliable teaching methods for the learning of vocal piano accompaniment courses. The design of multimedia teaching system adds theoretical support.

References

1. Wenbo, Z.E.N.G.: Research on construction of interactive teaching platform for design majors supported by mobile technology. Ind. Design **10**, 76–77 (2020)
2. Zhang, Y.: Mobile live broadcasting platform for aerobics video teaching based on Android. Inf. Technol., **44**(5), 41–53 (2020)
3. Yang, Y., Lin, S., Zhou, Yuan, C., et al.: The design and practice of the micro-course mobile teaching system based on the smartphone and the WEB platform. Tech. Autom. Appl., **39**(4), 182–185 (2020)
4. Jia, L.I.U.: Design and practice of network music course learning cloud platform based on mobile terminal. Tech. Autom. Appl. **39**(7), 148–150 (2020)
5. Chen, H., Wu, Y.: Design of music feature recognition system based on Internet of Things technology. Modern Electron. Techn., **43**(10), 43–50 (2020)
6. Zhou, H., Li, Z., Han, B., et al.: Teaching platform of identification of chinese materia medica based on mobile internet. Autom. Inf. Eng., **41**(6), 41–48 (2020)
7. Chen, H., Zhang, F.: Construction and application of mobile teaching platform based on "Super Star Learning" from the perspective of big data. wuxian hulian keji, **17**(1), 38–39 (2020)
8. Zhao, L.: Design and implementation of architectural design teaching platform based on mobile cloud technology. J. Jiamusi Univ. Nat. Sci. Edn., **38**(2), 140–142 (2020)
9. Dongyi, C.H.E.N., Jianguo, C.H.E.N., Yurong, L.I.: Design of experimental teaching platform for infrared detection combined with new method for extracting characteristic frequency of human body movement. Experim. Technol. Manage. **37**(6), 107–111 (2020)
10. Hui, W.A.N.G., Lijun, Y.U., Jian, H.U., et al.: Research on intelligent mobile service robot experiment teaching platform. Laborat. Sci. **23**(6), 98–101 (2020)
11. Zhang, X., Sun, L., Tang, T.: Research on mobile learning platform for blended learning. Microcomput. Appl., **36**(5), 28–31 (2020)
12. Liu, S., He, T., Dai, J.: A survey of CRF algorithm based knowledge extraction of elementary mathematics in Chinese. Mobile Netw. Appl. (2021). https://doi.org/10.1007/s11036-020-01725-x
13. Liu, S., Fu, W., He, L., Zhou, J., Ma, M.: Distribution of primary additional errors in fractal encoding method. Multimed. Tools Appl. **76**(4), 5787–5802 (2014). https://doi.org/10.1007/s11042-014-2408-1
14. Liu, S., Bai, W., Liu, G., et al.: Parallel fractal compression method for big video data. Complexity **2018**, 2016976 (2018)

The Mixed Teaching Quality Evaluation Model of Applied Mathematics Courses in Higher Vocational Education Based on Artificial Intelligence

Hai-ying Chen[✉]

Department of Public Courses, Tianmen Vocational College, Tianmen 431700, China

Abstract. With the deepening of research on the mixed teaching of Applied Mathematics Course in higher vocational colleges, and the requirements for teaching quality are constantly improved, so it is necessary to conduct a comprehensive evaluation on it. However, due to the backward processing technology of evaluation indicators in traditional evaluation model, the calculation error of evaluation results is relatively large. Therefore, this paper designs the evaluation model of mixed teaching quality of Applied Mathematics Course in Higher Vocational Colleges Based on artificial intelligence. This paper constructs the evaluation index system of mixed teaching quality of Applied Mathematics Course in higher vocational colleges, and completes the consistency calculation. According to the results of the index selection, the neural network of artificial intelligence technology combined with fuzzy evaluation method is used to construct the hybrid teaching quality evaluation model. After the evaluation results are obtained, the evaluation results are classified. At this point, the design of the mixed teaching quality evaluation model for higher vocational applied mathematics courses based on artificial intelligence is completed. Constructing the experimental results, by comparing the experimental indicators, we can see that the evaluation error of this model is lower than that of the traditional model. It can be seen that this model works best.

Keywords: Artificial intelligence · Neural network · Teaching quality evaluation · Mixed teaching

1 Introduction

Since the 20th century, mixed teaching quality evaluation has been systematically studied as an independent research field. In particular, the "eight-year study" conducted by the American Association for Progressive Education from 1934 to 1942 has made the field of curriculum quality evaluation mature, and the most The focus of attention is that the mixed teaching quality assessment of applied mathematics courses is more extensive [1, 2]. In recent years, my country's higher education has developed rapidly, and the number of higher vocational teachers has increased significantly. The effect of higher vocational education has also improved the country's economy. The development has

S. Liu and X. Ma (Eds.): ADHIP 2021, LNICST 417, pp. 443–454, 2022.
https://doi.org/10.1007/978-3-030-94554-1_35

had a significant impact, among which the education of applied mathematics is one of the fundamentals [3]. Higher vocational applied mathematics teachers are both educators and researchers. All kinds of requirements for teachers have reached a new height. It is the fundamental plan of educational reform and development to establish a teacher team with good political professional quality, reasonable structure and relatively stable. Establishing an objective, scientific and intelligent teacher management system is of great significance to strengthening teacher management, promoting the development of teachers, schools and society. In order to make the teaching quality evaluation management systematization, office automation and informatization of Applied Mathematics Course in higher vocational colleges, it is necessary to carry out research and discussion in various aspects. The content and system of mixed teaching quality evaluation, as well as the processing and analysis of the data of mixed teaching quality evaluation need further research. The purpose of mixed teaching quality evaluation is to promote curriculum construction, promote curriculum reform, improve teaching work, and improve curriculum teaching quality.

The application of mathematics curriculum mixed teaching quality evaluation model in China's education evaluation is carried out earlier, more and has obvious effect, and is generally welcomed. In Ref. [4], in order to improve the effectiveness and accuracy of teaching quality evaluation, a teaching quality evaluation model based on hybrid intelligent optimization algorithm is proposed. Entropy weight method is introduced to objectively determine the index weight and initial evaluation results of teaching quality evaluation system. The parameters of BP neural network are optimized by genetic algorithm based on adaptive mutation, and the teaching quality evaluation model is established. Compared with BP neural network and GA-BPNN (genetic algorithm back propagation neural network) model, the prediction accuracy is improved by 15.04% and 5.41% respectively. At the same time, the convergence speed of the algorithm is improved. Reference [5] first proposed a general mathematical model for data quality detection and evaluation. On the basis of this model, ontology technology was used to define the mapping of conversion rules from general mathematical model to ontology model. Considering that most of the data is stored in relational database, taking relational database as an example, according to the proposed mathematical model and conversion rules, The extraction and construction of data quality evaluation ontology and the definition of complex quality rules are realized. Finally, an application system is implemented to verify the correctness and scientificity of the system. The model is reasonable and extensible, It is universal. However, the research and practice of the mixed teaching quality evaluation model system based on artificial intelligence technology for applied mathematics course has been from the 21st century. With the popularization of artificial intelligence applications, the network has entered into thousands of households and penetrated into people's life, study, work and other aspects. The mixed teaching quality evaluation model of Applied Mathematics Course in Higher Vocational Colleges Based on artificial intelligence has attracted much attention because of its strong interactivity, wide spread range, open space-time, convenient data collection and management, personalized information exchange, and fast data statistics and analysis functions. The evaluation content should be practical. The mixed teaching quality evaluation method is practical, technical, advanced, diversified, flexible and so on. The evaluation results are scientific, objective and fair.

2   Design of Mixed Teaching Quality Evaluation Model for Higher Vocational Applied Mathematics Course Based on Artificial Intelligence

Based on the artificial intelligence environment, the system process of setting the mixed teaching quality evaluation model of higher vocational applied mathematics is analyzed, and the teaching quality evaluation model is completed according to the following process (Fig. 1).

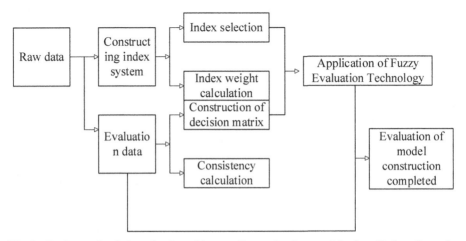

Fig. 1. Design path of the mixed teaching quality evaluation model of applied mathematics courses in higher vocational education based on artificial intelligence

According to the above-mentioned process, the design process of the mixed teaching quality evaluation model of Higher Vocational Applied Mathematics Course Based on artificial intelligence is completed, and the corresponding experimental links are set to determine the use effect of the design model in this paper.

2.1   Constructing the Mixed Teaching Quality Evaluation Index System of Applied Mathematics Course in Higher Vocational Education

This paper collects a lot of literature related to the research, and looks through the teaching evaluation table of teaching quality in many colleges and universities. In addition, it also makes a comprehensive analysis of the existing research results. According to the purpose of classroom teaching and the basic requirements of teaching, combined with the situation of mixed teaching of Applied Mathematics in higher vocational colleges, the evaluation index system of mixed teaching of Applied Mathematics Course in higher vocational colleges is preliminarily established.

In this study, the importance of each index in the evaluation index system is divided into 1–5 levels. Calculate the score of each first-level indicator, the calculation formula of the indicator score is as follows:

$$Q = \frac{1}{n} \sum_{i=1}^{n} c_i d_i \tag{1}$$

Among them: Q is the index score; c is the score of the evaluation grade; d_i is the number of people scored at the i level; n is the total number of participants. In this paper, the 9-scale method [6–8] proposed by Sardi is used to construct pairwise judgment matrix. Through the judgment and decision-making of experts in related fields according to their work and practical experience, the judgment matrix of all levels of indicators is obtained, and it is verified by the consistency operation of hierarchical single ranking.

Level single sorting is to calculate the maximum value of the eigenvalues of each judgment matrix and the corresponding eigenvectors to obtain the level single sorting, and obtain the importance data sequence of the index layer to the target layer. In order to obtain the optimal decision. The specific steps are first to solve the maximum eigenvalue α_{max} of the judgment matrix W, and then use the formula $Q\beta = \alpha_{max}\beta$ to solve the corresponding eigenvector β of α_{max}. After standardization, β is the ranking weight of the relative importance of corresponding elements in the same level to a factor in the previous level. To test the consistency of the matrix. Although it is impossible to require all judgments to be completely consistent, the judgments should be made roughly consistent. Therefore, it is necessary to check the consistency of the judgment matrix. First calculate the consistency index P of matrix W, the specific formula is shown below.

$$P = \frac{\alpha_{max} - n}{n - 1} \tag{2}$$

In the formula, n is the order of the judgment matrix. When W has complete consistency, $P = 0$. The larger P, the worse the consistency of matrix W. In order to evaluate the consistency test of hierarchical total ranking, we need to calculate its consistency test index. P is the consistency index; H is the average random consistency index; G is the random consistency proportion.

$$P = \sum_{i=1}^{n} c_i P_i \tag{3}$$

In the formula, P_i is the consistency index of the evaluation layer corresponding to c_i.

$$H = \sum_{i=1}^{n} c_i H_i \tag{4}$$

In the formula, H_i represents the average random consistency test index of the evaluation layer corresponding to c_i. Thus, the consistency verification evaluation of the overall evaluation index system can be completed, and the specific formula is as follows.

$$G = P/H \tag{5}$$

When the calculation result of this formula is $G \leq 0.1$, it is considered that the calculation result of the total ranking has satisfactory consistency, then the index construction result conforms to the teaching quality evaluation standard.

2.2 Constructing a Mixed Teaching Quality Evaluation Model

In view of the multi-objective, multi-level and complex non-linear problem of the quality evaluation of the mixed teaching of Applied Mathematics in higher vocational colleges, and the existing evaluation methods and models of the mixed teaching quality of the applied mathematics course in higher vocational colleges have difficulties in determining the standard weight, too strong subjectivity and randomness, easy to appear over fitting, slow speed of optimization and slow convergence speed of standard BP neural network In this study, artificial intelligence neural network is used to put forward the fuzzy evaluation model of hybrid teaching quality. The main idea of the model is to introduce artificial intelligence technology to improve the gradient descent method of BP neural network to improve the convergence rate, and optimize the network structure to ensure the stability of the model. In addition, new evaluation indexes are added to the traditional evaluation indexes to construct the teaching quality evaluation index system used in this study, so as to ensure that the model comprehensively evaluates teaching activities. The normalized data set of evaluation index samples is used as the input feature vector of the model to improve the calculation efficiency of the model. The neural network structure used in this design is as follows (Fig. 2).

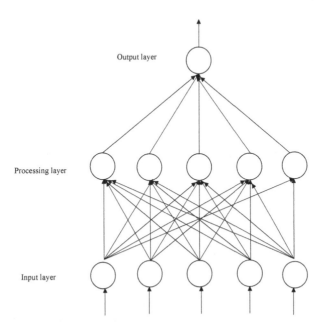

Fig. 2. Neural network structure

According to the above network organization, the mathematical model of fuzzy comprehensive evaluation [9, 10] is composed of four factors (R, T, V, X). R is the evaluation factor set $R = \{R_1, R_2, ..., R_n\}$; T is the comment set $T = \{T_1, T_2, ..., T_n\}$; X is the weight distribution vector of each index in the R index set; $X = \{X_1, X_2, ..., X_n\}$, $\sum_{i=1}^{n} X_i = 1$; V is the evaluation matrix $V = \{V_1, V_2, ..., V_n\}$, V_i is the single factor evaluation result $V_i = \{V_{i1}, V_{i2}, ..., V_{in}\}$ of the evaluation factor R_i, where V_{in} is the membership degree of the i th evaluation factor r_i to the n comment Grade t_i, and the evaluation matrix is formed by each factor evaluation vector as the row, and the fuzzy evaluation matrix is constructed through the above process.

Suppose there are n evaluation grades T and m low-level evaluation indicators R. There is a fuzzy relationship Z. between the two, expressed as a matrix:

$$Z = \begin{cases} R_1 \\ R_2 \\ ... \\ R_a \end{cases} \begin{matrix} m_1 \ m_2 \ ... \ m_n \\ \begin{bmatrix} Z_{11} \ z_{12} \ z_{1n} \\ Z_{21} \ z_{22} \ z_{2n} \\ \\ Z_{a1} \ z_{a2} \ z_{an} \end{bmatrix} \end{matrix} = \left(z_{ij} \right) m \times n \tag{6}$$

In the formula, z_{ij} represents the membership degree of the i evaluation index to the j grade. Assuming that the weight of the evaluation index is $R = \{R_1, R_2, ..., R_n\}$, the comprehensive evaluation result K is:

$$K = Z \times R \tag{7}$$

In practical evaluation work, the personnel involved in teaching evaluation are often composed of many different types of personnel, and the importance of the evaluation results of various evaluators is different. At this time, hierarchical clustering can be used to aggregate the evaluation data of the same degree of importance. At the same time, in the evaluation, the more important data set is directly selected as the evaluation data to obtain the comprehensive higher vocational applied mathematics curriculum mixed teaching evaluation result, and finally the weighted average is used to obtain the general higher vocational applied mathematics curriculum mixed teaching evaluation result.

2.3 Grading of Assessment

According to the evaluation model constructed above, complete the evaluation process of mixed teaching quality evaluation of applied mathematics courses in higher vocational education. In this part, the evaluation results are graded. In order to ensure the effectiveness of the classification of the evaluation results, some indicators are selected for scoring and combined with the evaluation results to obtain comprehensive and highly accurate evaluation results. The specific indicators and scoring results are as follows (Table 1).

Table 1. Scoring results of key indicators

Index	Bottom indicators	Evaluation criteria	Score
Teaching team	Applied mathematics teacher	Curriculum teachers have strong teaching ability	10 branch
		Distinctive teaching characteristics	
	Overall quality of teaching team	Strong sense of responsibility, team spirit	10 branch
	Research on teaching reform	Active teaching thinking and innovation in reform	20 branch
Content of courses	Course content	The contents are novel and informative	10 branch
	Reform in education	Combine inside and outside class	10 branch
Teaching effectiveness	Peer review	Excellent evaluation and good reputation	20 branch
	Student evaluation	The content of the course is authentic and the evaluation is good	20 branch

According to the contents of the above table and the evaluation results, the teaching quality evaluation is completed. Because the calculation results of the evaluation model designed in this paper are more complex, therefore, in the process of grading evaluation, the formula of grading is set as follows:

$$e_i = \frac{y_i - y_{min}}{y_{max} - y_{min}} \tag{8}$$

$$e_i = \frac{y_i - y_{mid}}{\frac{1}{2}(y_{max} - y_{min})} \tag{9}$$

$$y_{mid} = \frac{y_{max} + y_{min}}{2} \tag{10}$$

In the above formula, y_{max} represents the maximum value in the scoring result, y_{min} represents the minimum value in the evaluation result, and e_i and y_i represent the pre-processing scoring result and the post-processing scoring result. The y_{mid} in the formula represents the intermediate value of the change in the score result. The above-mentioned formula controls the results of the evaluation results to complete the overall process of teaching evaluation. At this point, the design of the mixed teaching quality evaluation model for higher vocational applied mathematics courses based on artificial intelligence is completed.

3 Analysis of Experimental Demonstration

3.1 Experimental Environment Setting

In order to verify the effectiveness of the artificial intelligence-based hybrid teaching quality evaluation model for higher vocational applied mathematics courses designed above, an experimental link is constructed and the design model in the article is compared with the traditional model.

The small-scale evaluation data used in this experiment is the data set collected before the experiment, while the large-scale data set uses the short-term traffic flow data set. Because the large-scale evaluation of education data involves a lot of privacy, it is difficult to collect, so large-scale short-term traffic flow data is used instead of verification, and the structure and dimension of teaching quality evaluation data are similar to those of short-term traffic flow data. Set the activation function of the deep noise reduction encoder as the Sigmoid function, the learning rate is generally set to 0.02, the unsupervised training target accuracy is 0.005, the maximum number of training times is 3000, the weight is randomly assigned, and the threshold is set to 0. The deep noise reduction autoencoder introduces the Adam algorithm, and optimizes the number of hidden layers and the number of neurons to minimize the error between the output data and the original data to obtain the essential characteristics of the original data. The important parameters of support vector regression are adjusted in the supervised output layer to improve the prediction accuracy of the model. Using average absolute percentage error, mean square error, etc. as performance comparison indicators to compare with other teaching quality evaluation models to verify the effectiveness of the design evaluation model in this article.

3.2 Experimental Comparison Index Setting

In order to verify that the evaluation model designed in the article has more advantages than other models in the teaching quality evaluation of applied mathematics courses in higher vocational education, this section introduces the average absolute percentage error, the mean square error, the symmetric average absolute percentage error, and the root mean square error as the models. The performance comparison index for evaluating prediction accuracy, the specific formula is as follows.

$$A = \frac{1}{n} \sum_{i=1}^{n} \left| \frac{a_i - b_i}{a_i} \right| \tag{11}$$

$$B = \frac{1}{n} \sum_{i=1}^{n} (b_i - a_i)^2 \tag{12}$$

$$C = \frac{100\%}{n} \sum_{i=1}^{n} \frac{|a_i - b_i|}{(|a_i| - |b_i|)/2} \tag{13}$$

$$D = \sqrt{\frac{\sum_{i=1}^{n} (a_i - b_i)^2}{n}} \tag{14}$$

In the above formula, A represents the average absolute percentage error of the model, B represents the average absolute percentage error of the model, C represents the symmetric average absolute percentage error of the model, D represents the root mean square error of the model n represents the sample size of the test data, b_i represents the actual truth of the test data Value, a_i represents the predicted value of the model on the test data. Using the above formula, the use process of the design model to the traditional model is calculated, and the comparison between the design model and the traditional model is completed. In order to obtain the experimental data effectively, five experiments were carried out and the corresponding data results were recorded.

3.3 Analysis of Results

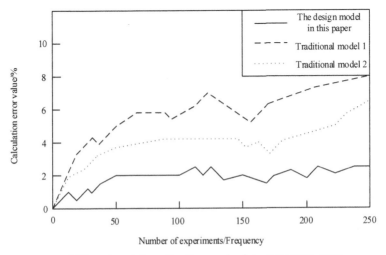

(a)Experimental results of mean absolute percentage error

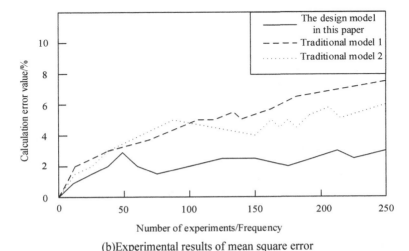

(b)Experimental results of mean square error

Fig. 3. Analysis of experimental results

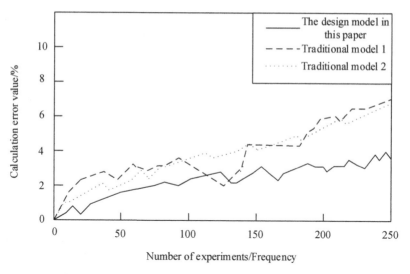

(c)Experimental results of symmetric mean absolute percentage error

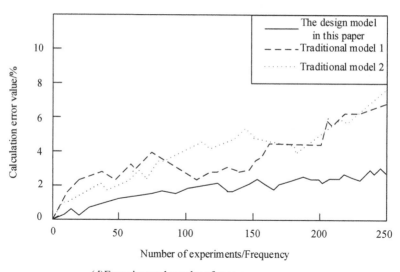

(d)Experimental results of root mean square error

Fig. 3. continued

According to the above experimental results, when the small-scale evaluation sample data set is input to train and verify the model, and the prediction evaluation results are inverse normalized, the experimental results of the design model in various models are the best. At the same time, compared with the traditional model, the training time of the design model is higher than that of the traditional model, but the other four performance indicators are better than the traditional evaluation model. Compared with the traditional model 1, although the design model in the article has no big difference in training time and average absolute percentage error, it is better than it in other performance indicators, especially the mean square error and symmetric average absolute percentage error. Important indicators to verify the effectiveness of the design model in the article. On the other hand, although the design model variance error and the symmetric mean absolute percentage error in the article are better than the traditional model 2, the difference between them is not very large, and the time performance of the traditional model 2 is much lower than the model in this chapter, thus verifying The design model in the article is more suitable for processing small-scale data sets. From the above comparison, we can see that both the design model and the traditional model can complete the teaching quality evaluation and each has its own advantages, but the effect of the design model in this paper is obviously better than that of the traditional system (Fig. 3).

4 Conclusion

Improving the teaching quality of applied mathematics courses in higher vocational education has become the focus of current higher education. With the continuous deepening of teaching reform in colleges and universities, how to improve teaching quality and train more talents has become the core issue in college teaching reform. The establishment of a teaching quality evaluation system that focuses on improving the teaching quality of applied mathematics courses in higher vocational education also highlights its Significance. Teaching quality evaluation refers to the process of using effective technical means to comprehensively collect, sort out and analyze the teaching situation and make value judgments to improve teaching activities and teaching quality. Some of the classification rules generated in this study have low adaptability to the actual situation. The reason for this phenomenon may be some shortcomings in the selection of data set attributes. Therefore, in future research, it is necessary to increase the capacity of sample data to improve the representativeness of sample data. In this way, the final construction of the mixed teaching quality evaluation model of applied mathematics courses in higher vocational education has more application value and becomes a more practical analysis model.

References

1. Li, X.: Wireless mobile communication system risk assessment based on artificial intelligence algorithm. Mod. Electron. Tech. **43**(01), 12–15 (2020)
2. Qiu, D., Li, Z., Yu, Z.: Curriculum and teaching paradigm shift in the context of artificial intelligence. Contemp. Educ. Cult. **12**(02), 48–55 (2020)
3. Liao, H., Zhang, Q.: Construction and evaluation of blended teaching readiness index in colleges and universities. E-educ. Res. **40**(03), 59–67 (2019)
4. Yue, Q., Wen, X.: Application of improved GA-BP neural network in teaching quality evaluation. J. Nat. Sci. Heilongjiang Univ. **36**(03), 353–358 (2019)
5. Zhang, X., Yuan, M.: General data quality assessment model and ontological implementation. J. Comput. Res. Dev. **55**(06), 1333–1344 (2018)
6. Wu, M., Liu, Z., Wang, Y., et al.: Quality evaluation model of network operation and maintenance based on correlation analysis. J. Comput. Appl. **38**(09), 2535–2542 (2018)

7. Xu, F., Chen, Y., Huang, Z., et al.: An empirical study on teaching quality evaluation of pharmaceutical administration. J. Shenyang Pharm. Univ. **36**(12), 1119–1126 (2019)
8. Zhuo, L., Zhang, M., Wang, G., et al.: No-reference MS-SSIM video quality assessment model based on support vector regression. J. Beijing Univ. Technol. **44**(12), 1486–1493 (2018)
9. Du, S., Liang, Y.: Establishment of drug substance evaluation model based on analytic hierarchy process. Chin. J. Pharm. **50**(01), 108–112 (2019)
10. Liu, G., Liu, S., Muhammad, K., et al.: Object tracking in vary lighting conditions for fog based intelligent surveillance of public spaces. IEEE Access **6**, 29283–29296 (2018)

Computer Aided Online Translation Teaching System

Jie Cheng[1] and Gang Qiu[1,2(✉)]

[1] Changji University, Changji 831100, China
[2] School of Software, Shandong University, Jinan 250101, China

Abstract. In the process of machine translation, similarity threshold is an important factor restricting retrieval and translation. When the fuzzy interval similarity threshold is low, the translation effect of the traditional online translation teaching system is poor, but if the threshold is too high, it will lead to retrieval difficulties and affect the translation progress. In view of this situation, a computer-aided online translation teaching system is designed. In the hardware design of the system, the content addressed memory is mainly studied to provide space for data storage and facilitate address search for the partition of translation cache package; In the software design, firstly, the neural loop network is established for deep learning, the natural language feature vector is extracted, and the source sentence is encoded to obtain the sentence vector with a certain length. After adding some algorithms and constraints, the decoding is completed with the help of neural network to eliminate the gradient imbalance in the process of translation ambiguity; Optimize the translation training model, reorder the extracted feature vectors, and optimize the translation results. In order to verify the effectiveness of the designed system, experiments are designed. The results show that when the fuzzy interval is less than 0.5, the performance of the designed system is significantly better than the traditional system. With the increase of similarity threshold, the performance gap between the two methods gradually narrows.

Keywords: Computer-aided · Online translation · Teaching system

1 Introduction

With the continuous progress of science and technology and Internet technology, more and more computers have entered people's life, especially in the field of higher education. Computer teaching has become an important symbol of the modernization of higher education and the trend of educational development. As an extension of traditional teaching, the role of computer information teaching has been gradually valued by people. More and more schools have begun to build campus computers and digital campuses. On this basis, the computer-aided teaching system has the possibility of development and implementation, which better makes up for the shortcomings of classroom teaching [1, 2]. With the development of computer and multimedia technology, computer information teaching system will have more development space.

© ICST Institute for Computer Sciences, Social Informatics and Telecommunications Engineering 2022
Published by Springer Nature Switzerland AG 2022. All Rights Reserved
S. Liu and X. Ma (Eds.): ADHIP 2021, LNICST 417, pp. 455–466, 2022.
https://doi.org/10.1007/978-3-030-94554-1_36

In the network environment, the teaching assistance platform can not only "enrich both voice and emotion", but also make full use of and share teaching resources, realize the communication between teachers and students, and turn the traditional one-way teaching mode into a multi-directional communication and interactive virtual learning community. It has been scientifically proved that timely and high-quality information interactive transmission can effectively improve learning efficiency. After entering the system, teachers can upload their class videos, lesson plans, courseware, reference materials and assignments to the database, and students can download them at any time to facilitate the further development of teaching activities [3, 4]. In addition to the functions of random questions, online Q & A, online test and online review, the teaching auxiliary platform can also communicate and interact with students, understand the teaching effect and improve teaching methods in time. Students can learn in a differentiated environment through the online translation function provided by the system, make full use of the resources in the database and enhance their learning consciousness.

Reference [5] believes that under the background of the wide popularization of smart phones, more and more people in teachers and students begin to use smart phones for mobile teaching or learning. With the continuous development of hardware conditions, many PC translation systems begin to support smart phones. Users can easily translate and learn foreign languages on mobile phones and other mobile devices. Based on the existing mobile translation software, this paper expounds the implementation of English translation on smart phones, and puts forward corresponding solutions to the problems existing in the current translation software, which provides a new idea for the construction of English learning system. Reference [6] constructs the structural level of English translation system, including translation data collection module, information feature extraction module and analysis model construction module. The language model of the English translation system is established, the probability distribution of the specific sentence sequence or word sequence of the translation is counted by using the model, and the information features of the user's English translation documents and the information features of the translation training set are extracted. According to the feature extraction results, the similarity of feature keywords is calculated, and the BP network optimized by particle swarm optimization is used for fitting calculation. Finally, the teaching of English translation is realized. However, the performance of the above traditional system is poor when the fuzzy interval similarity threshold is low. However, in practical applications, the threshold of similarity is too high, which will lead to difficulties in retrieval and affect the progress of translation.

In view of this situation, this paper designs a computer-aided online translation teaching system. Computer aided translation technology can help translators complete translation tasks quickly and efficiently. This technology is based on database driver. In a broad sense, computer tools that can assist translation can be classified as computer-aided translation tools, such as the integrated use of computers and related application software, grammar checking tools and network resources; In the narrow sense, computer-aided translation tools refer to special software and related technologies developed to improve the translation process. The core technology of computer-aided translation software is translation memory technology and database technology. Its core modules are translation memory system, corpus management system, translation alignment tool, translation project management tool and so on. It is a bold attempt to apply it to the online translation teaching system.

2   Computer-Aided Online Translation Teaching System

2.1   Hardware Design

In the online translation teaching system designed in this paper, the memory module is the basis of computer-aided translation. In the hardware design, content addressable memory is first introduced, which is a memory to realize associative translation on the basis of traditional memory technology. Content addressable memory has a large number of storage units [7, 8], which can provide space for the storage of translation data. The data is stored in a specific location, and the current location can be searched through the stored data content to get the storage address of the data. The memory is equipped with a hardware clock cycle. In this cycle, the keyword matching can be completed once (Fig. 1):

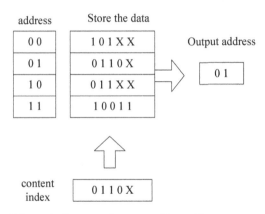

Fig. 1. Schematic diagram of content-addressable memory operation

After inputting the key words in the memory, the input key words are compared with all the table items in the memory, and finally all the matching table items are returned in the corresponding address in the memory. In the returned address, it will be used in an associated table to complete addressing and partitioning, and return the corresponding content. In content addressable memory, there are only two display states of bits in each unit, namely "0" and "1", which are used to reflect the two states of "match" and "mismatch". Therefore, content addressable memory can realize accurate matching search. The translation cache address is generally used as a subnet address or a prefix address, and the length of the prefix address is also called the bit length. In the classification application subset, the matched prefix has the maximum prefix length. If there are redundant translation cache packages, they need to be implemented by classification. In order to meet the longest prefix matching rule, to ensure that each possible address prefix length in the network transmission process can be equipped with a separate search matching chip, so a uniform length translation cache prefix set needs to be reserved in the chip. There are as many kinds of prefix lengths as there are as many memory chips as needed. So far, the system hardware design is completed.

2.2 Software Design

Establish a Recurrent Neural Network

In the online translation teaching system, the main idea is to rely on computer to complete the online translation conversion of the teaching system. In the software design, it mainly relies on the deep learning of recurrent neural network to complete the extraction of natural language feature vector. The neural network is composed of a large number of neurons. The neuron structure in the artificial cyclic neural network is similar to that of biological neurons. When the neuron is stimulated by the external environment, the artificial cyclic neural network will behave abstractly after being stimulated. As a mathematical language, the structure of neurons when they are stimulated is shown in the figure below (Fig. 2):

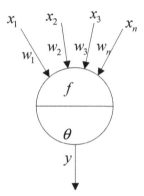

Fig. 2. Schematic diagram of external stimulation of neurons

In the above figure, $X = (x_1, x_2..., x_n)$ represents the external input received by the neuron, $W = (w_1, w_2..., w_n)$ represents the weight corresponding to the external input, θ represents the threshold in the neuron, which can be set manually, and y represents the output of the neuron. The above relationship can be expressed for:

$$y = f(\sum_{i=1}^{n} w_i x_i - \theta)x_1 \tag{1}$$

In the structure of artificial circulation neural network composed of neurons, it is mainly divided into three layers: input layer, hidden layer and output layer. In the traditional machine translation process, the whole sentence is decomposed into separate words and phrases for translation, but this translation method will lead to the lack of context semantic relevance. After the cyclic neural network is established, the source sentence can be encoded, processed to obtain a sentence vector of a certain length, and certain algorithms and constraints are added to complete the decoding with the aid of the neural network to achieve accurate online translation. And training is performed on the established recurrent neural network. The schematic diagram of translation training is as follows (Fig. 3):

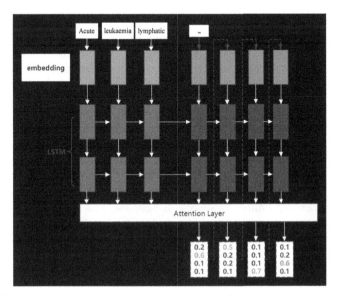

Fig. 3. Schematic diagram of translation training based on neural network

In the process of translation training as shown in the figure above, it can effectively reduce the engineering design involved in the translation work, and encode and decode through the encoder and decoder in the recurrent neural network. The encoder can process the word vector and the hidden unit in the upper layer to get the hidden unit at a certain time by the following formula:

$$h_t = f(x_t, h_{t-1}) \tag{2}$$

In the above formula, h_t represents the hidden unit, f represents the non-linear activation function, t represents a certain time, and x_t represents the input sequence vector at time t. As the length of sentences and the number of iterations increase, the established artificial recurrent neural network can solve the shortcomings of the original method and eliminate the "gradient imbalance" problem that occurs in the online translation process.

Optimizing Translation Training Model

Although the phrase-based machine translation system does not require the characteristics of the language in the bilingual corpus, any two languages can be trained as a machine translation system [9, 10]. However, due to the characteristics of the language itself, specific analysis needs to be carried out for specific language types when building language models and translation models. Therefore, in the process of building machine translation system software, the main goal of model training is to complete simple sorting between words. In the process of translation, for a given source statement e, there is $e = e_1, ..., e_n$. In simple sorting, for each unit that needs to be reordered, there is the following formula:

$$scroe_L(e_i) = \theta \cdot \phi(e, i) \tag{3}$$

In the above formula, ϕ represents the feature vector, θ represents the corresponding weight, and $scroe_L$ represents the scoring function of the re-ranking model. Simple reordering will sort all the basic units that need to be reordered according to $scroe_L$:

$$\pi^* = sort(e1, ..., en, scroe_L) \tag{4}$$

At present, some machine translation systems can only receive one sentence or one article, that is, pure text data. However, in the modern information processing industry, the demand for machine translation system as a component of a large-scale information processing system is growing [11, 12]. The information obtained by this large-scale system is no longer pure text data, but may be any other kind of data, such as the output of character recognition or speech recognition. This output result is generally data that has not been proofread, and it is no longer a single data, but an output with multiple errors or multiple possibilities. Hybrid network decoding can receive this multi-possibility input, and obtain better translation results through its algorithm. Under the above theoretical guidance, the training and reordering process of the obtained model is shown in the following figure (Fig. 4):

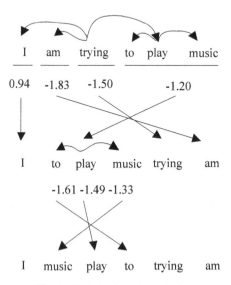

Fig. 4. Training based reordering

This simple sorting method can complete inference in a certain time complexity during model training, which has certain advantages for improving system performance. So far, the design of computer-aided online translation teaching system has been completed.

3   Experiment

3.1   Experiment Preparation

In order to verify the effectiveness of the designed system, the universal verification methodology (UVM) is used as the development library of verification environment, and the designed system is carried on the verification platform. The framework of UVM verification platform is shown in the following figure (Fig. 5):

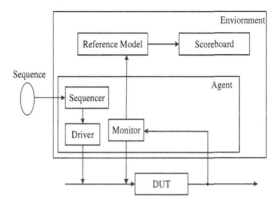

Fig. 5. Typical UVM verification platform structure

For the management partition system of translation cache, its performance can be judged by the miss rate of online translation block and the efficiency of partition exception instruction location. In the above verification platform, the cache partition based on fuzzy clustering analysis is verified. According to the generation frequency of the target code block, the transition of translation set partition management is detected. The following figure is the state chart of the algorithm used to detect the partition management transition of online translation set in this system performance test (Fig. 6):

According to the generation frequency characteristics of the target translation block, the threshold value 1 is set to 60% and the threshold value 2 is set to 40% during the performance test. The profile is used to count the frequency of generation, and a statement is added to the target code block to count the frequency of generation. The program just started running at S1, once the generation frequency is greater than the threshold 1, enter S2, which means that the working set is being constructed. At this

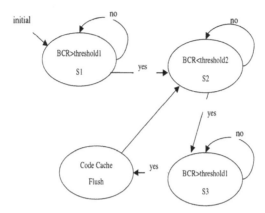

Fig. 6. Schematic diagram of the algorithm for detecting changes in the translation set

time, if it is detected that the generation frequency is less than the threshold 2, then enter S3, indicating that the working set has been constructed. When S3 detects that the generation frequency is greater than threshold 1, it means that the new working set is started to be built, the translation cache will be emptied immediately and the previous working set will be emptied from the translation cache.

The performance test of this system simultaneously compares the BLUE value of the traditional online translation teaching system and the system designed in this paper under the similarity threshold in different fuzzy intervals, so as to compare and analyze the performance.

3.2 Parameters and Experimental Process Settings

In order to make the results more stable without oscillation, the experimental system needs to perform three trainings to obtain three different sets of n-best translation results, and conduct experiments on the basis of these three sets of results and take the average value. In order to prevent over-fitting, this article uses the CNTK training ranking model, taking the default average distribution parameters as the initialization parameters, the range of the random number is $[-0.5, 0.5]$, and the activation function is as follows:

$$\alpha(n) = \begin{cases} 0 \ \ if \ n < 0.03 \\ 1 \ \ else \end{cases} \tag{5}$$

The state of hidden layer in deep neural network determines the number of parameters needed in the experiment. The larger the hidden layer is, the more complex the task can be solved. Therefore, in this paper, according to the data of the development set, the training set and the test set, we set the relevant super parameters of the system (Table 1).

Table 1. Hyperparameter settings

Parameter	Symbol	Numerical value
Feature vector	d	50
Hidden layer	d_h	800
Regularization rate	λ	10^{-8}
Adagrad's initial learning rate	α	0.01
Mini-batch size	b	10000
Word vector size	d_{rnn}^w	25
Embedded distance size	d_{rnn}^d	25
Regularization parameter	λ_{rnn}	10^{-4}
Margin loss rate	κ	0.2
Correction factor	f	8

In the online translation teaching system of this article, a random reordering correction result is selected for analysis. The correct word order of the sentence is "French leaders have made speeches". In the reordering system, the top 10 candidate sentences of the original recognition are as follows (Table 2):

Table 2. Translation candidates

Ranking	Result	Score
1	The eight leaders spoke in succession	-275.415154
2	The German leaders have delivered speeches one after another	-225.454152
3	The heads of state of the Republic of China made speeches one after another	-283.141212
4	Leaders of major powers have made speeches one after another	-293.450231
5	The leaders of all countries have made speeches	-284.115451
6	The heads of state of Pakistan made speeches one after another	-305.417521
7	Leaders of major powers have made speeches one after another	-2645.145541
8	Foreign heads of state have made speeches one after another	-281.521252
9	French leaders have made speeches one after another	-263.545841
10	The heads of state of Pakistan made speeches one after another	-284.504241

By dividing the results in the above table, the candidate sentences in the above table can be realized in the form of lattice, as shown in the following figure (Fig. 7):

As can be seen from the figure above, there are many candidates for each tone in lattice structure, and the recognition result is affected not only by the language model,

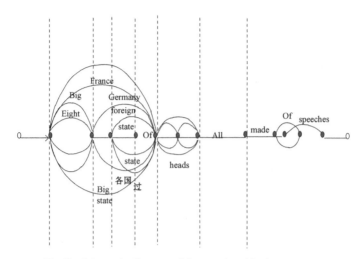

Fig. 7. Schematic diagram of the translated lattice structure

but also by decoding algorithm and other factors. Therefore, in the process of system performance test, the system designed in this paper and the traditional system are used for performance test, and the performance test results are counted and analyzed, and the blue value method is used for performance judgment.

3.3 Experimental Results and Comparison

In this paper, the performance of the two methods is compared by setting different fuzzy interval similarity thresholds. 100 experiments are carried out under each fuzzy interval similarity threshold, and the average experimental results under each threshold are taken to show the performance. The experimental results are shown in the table below (Table 3):

Table 3. Performance comparison under different fuzzy interval similarity thresholds

Fuzzy interval similarity threshold	Traditional system	Text system
[0.0,0.1)	0.221	53.41
[0.1,0.2)	4.85	55.38
[0.2,0.3)	6.13	56.81
[0.3,0.4)	8.37	59.31
[0.4,0.5)	13.94	62.84
[0.5,0.6)	24.87	63.97
[0.6,0.7)	42.64	66.31
[0.7,0.8)	57.31	67.29
[0.8,0.9)	65.87	68.34
[0.9,1.0]	70.35	69.51

It can be seen from the experimental results in the above table that the higher the threshold of fuzzy interval similarity, the higher the similarity between the test set data and the training set data, and the closer to the correct answer, whether in this system or in the traditional system. This paper designs a computer-aided online translation teaching system. In the previous translation fuzzy retrieval process, only when the threshold similarity threshold between the dialogue to be translated and the reference translation segment is the closest, the final translation result will be affected by the similarity threshold of the translation memory retrieval module. Therefore, the results from the above table are analyzed. When the output unit in the retrieval process is only the same as the sentence to be translated, the translation unit in the traditional method will participate in the translation process of the translator, and it is not output as the final translation result. Therefore, the traditional system performance is slightly lower among the fuzzy regions. The data in the table above can be seen that when the fuzzy interval is below 0.5, the performance of the system designed in this paper is obviously better than that of the traditional method. With the increase of similarity threshold, the performance gap between the two methods is gradually reduced. However, it is important to note that in practical applications, too high threshold of similarity will cause difficulty in retrieval and affect the progress of translation. In the case of reasonable setting of similarity threshold, the system designed in this paper has more advantages than traditional systems, which can improve the translation results.

4   Conclusion and Outlook

This topic is based on the actual needs of computer-assisted translation courses, combined with the current situation of course teaching and the current research status of computer-assisted teaching systems at home and abroad, using existing campus network resources, and using neural networks and translation memory technology in accordance with software engineering development methods. And computer-aided technology developed and designed a computer-assisted online translation teaching system suitable for the teaching and practice of computer-assisted translation courses in local colleges and universities. However, the time-consuming translation teaching of this article system still needs to be further optimized. In the follow-up research work, in-depth research will be conducted on this issue.

References

1. Cao, L.: Interactive computer aided teaching system based on .NET platform. Mod. Electron. Techniq. **43**(03), 134–137+141 (2020)
2. Li, S., Qi, C.: Computer aided translation model simulation based on bilingual E-chunk. Comput. Simulat. **36**(12), 345–348+352 (2019)
3. Lin, H.: Design of English translation online assistant system based on multi-language interaction. Mod. Electron. Techniq. **42**(06), 30–33 (2019)
4. Yisa, W., Alifu, K., Hao, Z.: Research on Chinese language assistant teaching system in deaf-mute schools. Comput. Eng. Appl. **56**(11), 225–229 (2020)
5. Liang, Y., Liang, L.: Construction of an English assisted English translation learning system based on smart phones. Automat. Instrument. **25**(08), 142–144 (2018)

6. Guo, L.: Design of intelligent computer scoring system based on natural language processing for English translation. Mod. Electron. Techniq. **42**(04), 158–160 (2019)

7. Fu, R., Hu, D.: On cultural translation competence improvement in translation teaching of Chinese Universities from the perspective of the going global strategy of Chinese culture. J. Mudanjiang Coll. Educ. **19**(02), 42–44 (2019)

8. Liu, S., Li, Z., Zhang, Y., Cheng, X.: Introduction of key problems in long-distance learning and training. Mob. Netw. Appl. **24**(1), 1–4 (2018)

9. Wei, X.: The application of contrastive English-Chinese studies of reference in translation and teaching. J. Hezhou Univ. **35**(01), 126–131 (2019)

10. Liu, S., Liu, D., Srivastava, G., Połap, D., Woźniak, M.: Overview and methods of correlation filter algorithms in object tracking. Compl. Intell. Syst. **7**(4), 1895–1917 (2020)

11. Yu, X., Cheng, L., Zhang, H., et al.: Conception of improving the international display of Chinese scientific journals based on online translation system. Chinese J. Sci. Tech. Period. **30**(02), 173–178 (2019)

12. Liu, S., Liu, X., Wang, S., Muhammad, K.: Fuzzy-aided solution for out-of-view challenge in visual tracking under IoT assisted complex environment. Neural Comput. Appl. **33**(4), 1055–1065 (2021)

Online and Offline Hybrid Learning System of Ideological and Political Course Based on Mobile Terminal

Chi-ping Li[1](✉), Ling-li Mao[2], and Xian-bin Xie[3]

[1] Guangzhou Maritime University, Guangzhou 510725, China
[2] Guangzhou Huali Science and Technology Vocational College, Guangzhou 511325, China
[3] School of Economics and Management, Hunan Software Vocational College,
Xiangtan 411100, China

Abstract. Aiming at the shortcomings of low coverage and weak interaction of traditional ideological and political courses online and offline blended learning systems, to improve the adaptability of the online and offline blended learning systems for ideological and political courses, a mobile terminal-based ideological and political course is designed. On-line and off-line blended learning system. Based on the mobile terminal, the hybrid learning server is designed using the three-tier structure of the B/S model to complete the overall system architecture design. Construct a learner model based on the characteristics of a blended learning model, design a learning process, and test the effect of software operation. The test results show that the online and offline hybrid learning system for ideological and political courses based on mobile terminals can effectively improve online and offline interactivity and expand coverage.

Keywords: Mobile terminal · Ideological and political courses · Online and offline · Blended learning

1 Introduction

With the continuous development and integration of modern multimedia technology and computer network technology, various mobile terminals, such as tablet computers and smart phones, have been rapidly popularized in people's daily production and life. Nowadays, smart phones are no longer just communication tools, but portable mobile terminals with independent operating systems and can install various applications and software. They can be used to complete corresponding tasks, and can also realize wireless network connection through mobile communication and other networks, so as to gradually develop into popular and popular electronic products. Affected by the rapid development of modern mobile devices, we should give full play to their advantages, innovate and optimize modern lifestyles and learning and working methods. With the wide application of WiFi technology and streaming media technology, the number of

© ICST Institute for Computer Sciences, Social Informatics and Telecommunications Engineering 2022
Published by Springer Nature Switzerland AG 2022. All Rights Reserved
S. Liu and X. Ma (Eds.): ADHIP 2021, LNICST 417, pp. 467–479, 2022.
https://doi.org/10.1007/978-3-030-94554-1_37

people who use mobile devices to watch video in real time is gradually increasing, and the role of mobile terminals in teaching is irreplaceable.

With the continuous advancement of mobile terminal technology, the application of mobile terminals in all walks of life is increasingly inseparable from it. For example, in the fields of information management and online learning, computers can be seen everywhere. With the development of mobile networks and intelligent terminal technology, mobile phones and other mobile terminals have been widely used in learning and office, so that their learning is no longer limited by time and space, and they can use mobile devices to acquire the required knowledge at any time. The connection of the mobile terminal needs to use the information it transmits during the entire process, which is also an important aspect of system design, which will directly affect the user's learning experience [1]. With the gradual development of online courses, students pay more and more attention to the interactivity of online learning. With the development of mobile terminals, the implementation conditions of online learning systems are more perfect, the cost is lower, the service quality is higher, and the stability and flexibility are higher. All of these have laid a good foundation for the development of online learning systems for college students, provide convenient learning channels for different types of students, promote communication between students, and facilitate the evaluation of learning effect so as to ensure the significant improvement of learning effect [2, 3].

Based on the mobile terminal technology, this paper extends the hybrid learning system in the ideological and political online learning system, and designs the software and hardware of the ideological and political online hybrid learning system. Multimedia information elements such as language, sound and image are displayed on the off-line classroom screen by mobile terminal, and learning is completed through the interactive operation between students and mobile terminal. The research on the online and offline learning system of Ideological and political course based on mobile terminal can not only stimulate students' learning interest and enthusiasm, but also shorten the whole cognitive process, so as to optimize and improve the teaching process, and significantly improve the teaching efficiency and teaching quality.

2 Hardware Design of Online and Offline Hybrid Learning System in Ideological and Political Courses

2.1 Overall Architecture Design

The popularization and innovation of online and offline mixed learning of political theory courses need to rely on the technological means of mobile terminals [4], optimize and improve the overall structure of the original online learning system, gradually form teaching features and enrich the service mode of mixed learning [5]. At present, most online learning systems adopt the B/S three-tier structure mode, and the overall system architecture is shown in Fig. 1.

Among them, the three-tier structure of B/S pattern includes data service layer, Web service layer and application service layer. It can provide an adequate data services layer for the system to run. The Web Services layer can use the Internet browser to provide system users with an online learning interface to optimize system services. And

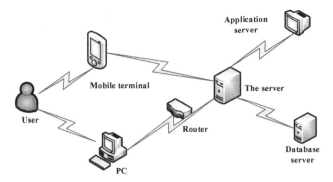

Fig. 1. Three-tier B/S architecture online learning platform

the application service layer can provide data service layer operation permissions for various types of users [6]. After the mobile device enters the online learning system page, it can learn courses, download materials, and modify user information. Web services layer can be used to analyze and standardize the user demand information, and transfer the data information to the middle layer. After receiving these data information, the application layer server will make corresponding feedback according to the user's request and provide the user with appropriate online services.

2.2 Hybrid Learning Server Design

The design of hybrid learning server is inseparable from the background data service. The whole mobile learning system runs from the background server [7]. The server-side mixed background learning in this paper is developed and implemented under the Android system, the development tool used is MyEclipse, and the programming language is Java. The hybrid learning server of the system adopts the classic SSH architecture. According to the analysis of system requirements, it is necessary to design the corresponding API interface on the background learning server for the mobile client to use, and the mobile client implements related functions by calling the corresponding API interface, its processing flow is shown in Fig. 2.

The client calls its corresponding API interface and sends a network request to the background through the HTTP protocol. The hybrid learning server first verifies whether the network request is valid, and then verifies whether the signature parameter in the network request is valid. If the two judgment methods are both valid, mixed The learning server will execute the corresponding request task, and then encapsulate the processed data into JSON data through the Struts2 framework and return it to the client. The client parses the JSON data returned in the background to obtain the corresponding information [8]. If the two judgments are not completely correct, the background will send the error reason code to the client. The common response status codes of the server are shown in Table 1.

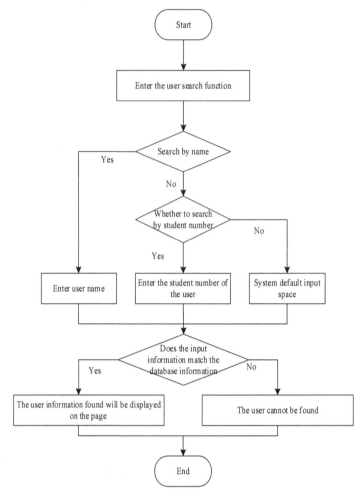

Fig. 2. Server side processing flow

The hybrid learning server returns the status code to the client, and the client needs to find the cause of the error and find the error according to the status code. Based on the first digit of the status code, you can determine whether the response is successful, whether the client or server is wrong [9]. If the first number is "2", it is judged that the response is successful; if the first number is "4", it is judged as a client-side error; if the first number is "5", it is judged as a server-side error. Based on the first number, developers can judge the basic problems. The background hybrid learning interface provided by the server mainly includes the IOS mobile learning client interface and the Android mobile client interface.

Table 1. Common corresponding status codes of servers

Status code	Explain	Details
200	Success	The server has successfully processed the request
202	Accepted	The server has accepted the request but has not yet processed it
400	Wrong request	The server could not parse the request
403	No Access	The server rejected the request
404	Not found	The server cannot find the requested page
408	Request timed out	Server request timed out
500	Server internal error	The server encountered an error and could not complete the request
503	Service is not available	The server is currently unavailable
504	Gateway timeout	The server acts as a gateway or proxy, but did not receive a request from the upstream server in time

3 Software Design of Online and Offline Hybrid Learning System in Ideological and Political Courses

3.1 Building a Learner Model

On the basis of the existing student model, the basic information data items and knowledge structure data items of the traditional middle school students model are retained, and the learner behavior data and cognitive level data items are introduced to construct a learner model which is conducive to adaptive learning path recommendation. The core data items are shown in Fig. 3.

To establish a personalized learner model, first of all, we should fully respect the differences between learners' cognitive levels; second, the data items of the student model need to dynamically express the learning tendencies and learning styles of different learners in learning activities [10]. Due to the particularity of the field of cyber security, professionals are required to have strong practical abilities, adaptability, and comprehensive knowledge analysis capabilities. This article proposes "understanding, application, the cognitive level standard of six basic aspects: analysis, synthesis, evaluation, and creation, and revised it into five basic aspects of "strategic analysis, management skills, technical research, engineering practice, and special breakthroughs", and the five basic aspects of learners The basic cognitive level score is used as an important basis for evaluating and recommending learning paths.

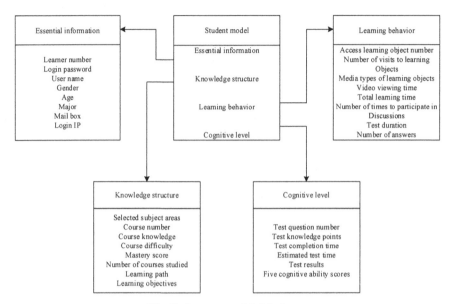

Fig. 3. Learner model data items

3.2 Design of the Online and Offline Hybrid Learning Process for Ideological and Political Courses

In the learning system designed for the online and offline hybrid learning process of ideological and political courses [11], users can be divided into two categories: learners and teachers. The operations performed by the two in the system are not the same [12]. Figure 4 shows the workflow of the online and offline blended learning system for ideological and political courses.

From Fig. 4, we can see that in the operation and use process of the two types of users in the system, the learner is the main body using the system and the main target of the adaptive learning system. When logging into the system for the first time, learners need to fill in some basic information to establish the initial exclusive model. Learners select learning courses in the system according to their own learning objectives, and then customize the learning path according to their current knowledge situation and knowledge map. After each class, the system detects the students' mastery of learning content in the form of adaptive evaluation [13]. Through the establishment of knowledge model and cognitive level, the system will dynamically recommend the most suitable course entity to complete the learning goal by calculating the skill completion index of each class. Users can choose the follow-up course according to the recommended entity. Compared with student users, the process of teacher/manager user using the system is relatively simple.

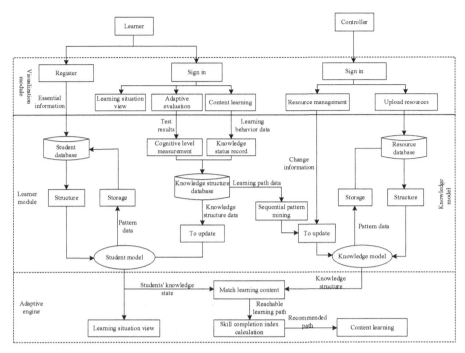

Fig. 4. System flow chart

4 Test and Analysis

System testing is an important process to ensure the reliability of the system. This process is an important part of the product's life cycle. This process provides important feedback information for the developers of the system to detect the use and operation of the system for development the user makes timely adjustments to the system to provide users with a good sense of experience.

4.1 System Test Environment

Test server: Dell OptiPlex 5040, Intel Core i3-6100, CPU is quad-core 3.70 GHz, operating system is 64-bit Windows7 Ultimate, RAM is 8 GB; Java language JDK version is 1.6.0, Tomcat server version is 6.0, The MySQL database version is 8.0.13. Web client: Google Chrome browser.

The test environment of IOS mobile phone client mainly uses the iPhone with different models and screen sizes loaded with IOS system as the test environment. The test process not only needs to test the running effect of the software on the mobile phone, but also tests the software for differences. Size of the mobile phone screen adaptation. The test environment of IOS mobile client is shown in Table 2.

Table 2. IOS mobile client test environment

Mobile phone model	Operating system	CPU	Memory	Screen size
iPhone SE	iOS 9.2	A9	2G	4 in.
iPhone 6 Plus	iOS 11.0	A8	1G	5.5 in.
iPhone 6s	iOS 12.0	A9	2G	4.7 in.

4.2 Test Process

Step 1: Open Tomcat server and enter IOS mobile phone client;

Step 2: The iOS mobile client enters the registration interface, fills in the relevant information and sends the mobile verification code, fill in the SMS verification code in the corresponding position, and click the registration button;

Step 3: After completion, enter the login interface, input the previous registration information, and click the login button to enter the main interface of mobile phone client;

Step 4: Click the "Discussion Area" control on the homepage of the IOS mobile client to enter the interactive discussion function. In the discussion area, click the post button, type questions about the article in the input box, and then click the post button. Please check the Discovery area to see if the article can be displayed successfully. Then click on the question post sent by other users, reply to the question in the post, and check whether the content of the reply is displayed in the post;

Step 5: Go back to the home page, click the circle control below to enter the assignment submission function. Student users need to choose whether to take photos to upload homework or upload homework from album;

Step 6: If you choose to take a photo to upload the job, enter the photo mode, point the camera at the place where you need to take a photo, click to take a photo, and then click the upload button after the preview is completed; if you choose to upload the job from the album, enter your mobile phone album, select the job picture that needs to be uploaded, whether the picture is complete after the preview is completed, and finally click the upload button;

Step 7: The teacher user enters the "view student homework" module to check the situation of students' homework, click the student user's homework view, and give the score;

Step 8: Click to enter the "Curriculum Learning" module, select the subject that needs to be studied. After entering, there are two options of "Content Learning" and "Question Practice", enter the content learning function, display the theoretical knowledge of the subject for users to learn. The user can complete the correct answer to the question on the interface. According to whether the user answers correctly or not, the interface will have different colors;

Step 9: Enter the "Personal Information Management" module, edit your own information, click the save button after completion, and the edited information will be displayed on the interface. Then click to enter the system setting interface, click to exit, and observe whether the user has completed this operation;

Step 10: Open the Google Chrome browser on the computer with Windows system, and enter "Google Chrome" in the URL bar http://localhost:8080/jwlearning/admin/ to enter the login page of web client;

Step 11: Enter relevant information, click the login icon, and enter the homepage of the web client after the operation is successful;

Step 12: first click the system administrator management function under the administrator management directory to check whether the page displays relevant information. Then click modify information to change the administrator's user name. Then click Modify administrator password, enter the original password and new password of the administrator, you can modify the administrator's password. Then click Delete User to move an administrator directly. Finally click Add Administrator, enter its account and password, click Confirm, and check whether the page is displayed successfully;

Step 13: First click the user management function under the user information management directory, and the user information of all registered IOS mobile learning clients will be displayed on this page. Click a user's detail icon on the page, and the page will display the user's details. Then click the task query function, you can preview the task submitted by the user, and you can delete the task. Finally, click "Delete" the user to remove it;

Step 14: Click the user search function under the user information management to search the user after entering the student number or name. When the input is finished, click the OK button to check whether the user information is displayed on the page.

4.3 Mobile Phone Client Test Results

The first is the matching test of mobile phone screen. Three iPhone phones with different models and sizes are installed on an app. Through the test, the software can well match different iPhone models, and different mobile phones can also display the app UI interface normally. This shows that the IOS mobile phone client has achieved the initial UI design goal. After completing the screen matching test, the function test is carried out according to different modules. The test results are shown in Table 3 and Table 4.

Table 3. Mobile client homepage function test results

Functional module	Test case	Expected results	Conclusion
Home	1. Click the "Discussion" button	Enter the discussion area selection interface	Adopt
	2. Click the "job upload" button	Enter the job upload selection interface	Adopt
	3. Click the "Learn" button	Enter the learning course selection interface	Adopt
	4. Click the "I" button	Enter the personal center function interface	Adopt
	5. Click the "Home" button	Return to home page	Adopt

Table 4. Mobile client personal center function test results table

Functional module	Test case	Expected results	In conclusion
Personal center	1. Click the "I" module on the main interface	Switch to the "I" module interface	Adopt
	2. Click the "View Homepage" button	Jump to the personal details interface	Adopt
	3. Click the "edit data" button	Jump to edit list	Adopt
	4. Enter the edit profile and modify the personal letter Click the "save" button	Prompt to modify the status information and jump to the interface	Adopt
	5. Click the "my concerns" button	Show my watch list	Adopt
	6. Click the "my fans" button	Show my fan list	Adopt
	7. Click the "my homework" button	Show jobs submitted by users	Adopt
	8. Click the "Settings" button	Enter the setting interface	Adopt
	9. Click the "clear cache" button	Clean up the memory garbage in the application	Adopt
	10. Click the help and feedback button	View help and feedback	Adopt
	11. Click the "about software" button	View the information of this software	Adopt
	12. Click the "log out" button	The user logs out and returns to the login interface	Adopt

4.4 Web Client Test Results

The test results of each functional module of web client are shown in Table 5.

Table 5. Test results of web client user information management function

Functional module	Test case	Desired result	Conclusion
User information management	1. Click the "user information management" button in the main menu	Query data, display the information of all registered users	Adopt

(continued)

Table 5. (*continued*)

Functional module	Test case	Desired result	Conclusion
	2. Click on a user's information	Jump to user details page	Adopt
	3. Click the "Homework" button	Jump to the page of all jobs submitted by the user	Adopt
	4. Click the "delete" button	Delete the administrator information	Adopt

The above test results can show that the performance of the online and offline hybrid learning system for ideological and political courses based on mobile terminals is relatively good in all aspects.

Running time is an index to directly judge the advantages and disadvantages of hybrid learning system. Under the same external environment, the proposed method, reference [4] method and reference [5] method are selected for the comparative experiment of running time.

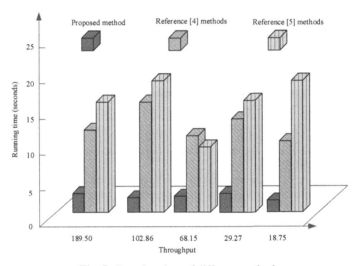

Fig. 5. Running time of different methods

The running time results of the three methods are shown in Fig. 5. This test is completed under different throughput. According to statistics, the running speed of the proposed method is much higher than that of the other two methods. The proposed method has the shortest time and the highest efficiency. Although the operation time of the method in reference [4] is stronger than that of the method in reference [5], the method in reference [4] still has large operation errors due to too long operation time and too slow speed. The method in reference [5] has the highest operation time in each operation experiment. To sum up, the proposed method has the best operation effect, the highest efficiency and the shortest time. It is the system with the best operation performance among the three methods.

5 Conclusion

This paper proposes an online and offline hybrid learning system for ideological and political lessons based on mobile terminals. The real goal of online and offline hybrid learning system is to break through the "rigidity" of traditional online ideological education, combined with the flexibility of online ideological education, so that students can complete the learning of knowledge according to their own goals, interests and rhythm. In the teaching process, according to the characteristics of teaching objectives and objects, teachers scientifically and reasonably select and use modern teaching media, actively participate in the teaching process, apply various multimedia information to students, form the overall structure of the teaching process and achieve the best teaching effect. The test results show that this system can effectively improve the online and offline interaction and expand the coverage. As the whole system is independently developed by the author, due to the limited time and number of people, the whole system still needs to optimize the interface and improve the function in the later stage.

References

1. Huang, Y., Kechadi, T.: An effective hybrid learning system for telecommunication churn prediction. Expert Syst. Appl. **40**(14), 5635–5647 (2013)
2. Bhaskaran, S., Santhi, B.: An efficient personalized trust based hybrid recommendation (TBHR) strategy for e-learning system in cloud computing. Clust. Comput. **22**(1), 1137–1149 (2019)
3. Liu, S., Liu, D., Muhammad, K., Ding, W.: Effective template update mechanism in visual tracking with background clutter. Neurocomputing **458**, 615–625 (2021). https://doi.org/10.1016/j.neucom.2019.12.143
4. Gao, P., Li, J., Liu, S.: An introduction to key technology in artificial intelligence and big data driven e-learning and e-education. Mob. Netw. Appl. (2021). https://doi.org/10.1007/s11036-021-01777-7
5. Zhao, X., Cen, L., Long, S., et al.: Abstract: study on and realization of hybrid recommendation-based adaptive learning system. J. Softw. Eng. **9**(4), 886–894 (2015)
6. Meryem, A., Ouahidi, B.E.: Hybrid intrusion detection system using machine learning. Netw. Secur. **2020**(5), 8–19 (2020)
7. Cocana-Fernandez, A., Ranilla, J., Sanchez, L.: Energy-efficient allocation of computing node slots in HPC clusters through parameter learning and hybrid genetic fuzzy system modeling. J. Supercomput. **71**(3), 1–12 (2015)
8. Abidi, M.H., Alkhalefah, H., Mohammed, M.K., et al.: Optimal scheduling of flexible manufacturing system using improved lion-based hybrid machine learning approach. IEEE Access **8**, 96088–96114 (2020)
9. Tao, H., Chen, D., Yang, H.: Iterative learning fault diagnosis algorithm for non-uniform sampling hybrid system. IEEE/CAA J. Automat. Sin. **4**(03), 148–156 (2017)
10. Tao, H., Chen, D., Yang, H.: Iterative learning fault diagnosis algorithm for non-uniform sampling hybrid system. IEEE/CAA J. Automat. Sin. **4**(3), 534–542 (2017)

11. Liu, S., Li, Z., Zhang, Y., Cheng, X.: Introduction of key problems in long-distance learning and training. Mob. Netw. Appl. **24**(1), 1–4 (2018)
12. Gil, A., de la Prieta, F., López, V.F.: Hybrid multiagent system for automatic object learning classification. In: Corchado, E., Romay, M.G., Savio, A.M. (eds.) Hybrid Artificial Intelligence Systems, pp. 61–68. Springer, Heidelberg (2010). https://doi.org/10.1007/978-3-642-13803-4_8
13. Wei, H., Hongxuan, Z., Yu, D., et al.: Short-term optimal operation of hydro-wind-solar hybrid system with improved generative adversarial networks. Appl. Energy **250**(PT.1), 389–403 (2019)

Mobile Cloud Teaching System for Ideological and Political Network Courses Based on P2P Technology

Yan-ming Zhan[1][(✉)] and Lin Chen[2]

[1] Jiangxi Normal University, NanChang 330000, China
[2] NanChang University, NanChang 330000, China

Abstract. Due to the unreasonable network structure of the traditional online course mobile cloud teaching system, the download rate of the system under complex requests is slow. Therefore, a mobile cloud teaching system for ideological and political network courses based on P2P technology is designed. The system is mainly divided into software design and hardware design. In the hardware design, the P2P topology network structure is mainly designed, and the internal structure of the development board of the mobile terminal and the wireless communication chip are designed. The network structure adopts the P2P hybrid topology structure to realize the complementarity between the advantages of different structures. In the software design, the system adopts haystac's cloud storage solution. Design the access process of database information according to the actual situation. In addition, the joining and exiting of user nodes in the system are managed, and the design of the system is completed so far. The system performance test results show that the download rate of the designed system under simple request is not much different from that of the traditional system. Under complex requests, the download rate of the design system is 3.9 MB/s higher than that of the traditional system.

Keywords: P2P technology · Ideological and political network courses · Mobile cloud teaching · Cloud storage

1 Introduction

Nowadays, electronic information technology is developing at an unprecedented speed. Many traditional fields have been impacted, resulting in various changes. Based on the advanced network communication technology, the distance teaching mode came into being. With the development of distance education, the current teaching mode has diversified. Both teaching mode and teaching content have undergone great changes [1, 2]. Accordingly, the learning mode of students is also changing. In today's classroom, students are no longer simply listening as the main way of learning. They began to turn into the main body of the classroom, with questions purposefully seeking teaching resources. Therefore, the long-distance teaching mode with large teaching materials,

© ICST Institute for Computer Sciences, Social Informatics and Telecommunications Engineering 2022
Published by Springer Nature Switzerland AG 2022. All Rights Reserved
S. Liu and X. Ma (Eds.): ADHIP 2021, LNICST 417, pp. 480–490, 2022.
https://doi.org/10.1007/978-3-030-94554-1_38

convenient communication channels and good interaction mechanisms has become the mainstream teaching mode in the field of education. In the study of Ideological and political courses, we also need to adapt to the current mainstream teaching mode. Since the learning resources are provided on the server side, once there is a problem on the server side, the system cannot run. Second, the cost is high. In order to better meet the needs of the system, it is necessary to construct a variety of different teaching services, and at the same time as the number of clients increases. Higher requirements are put forward for both the server and the data. Third, poor scalability. When the number of users increases, the number of expensive servers needs to be increased to provide more stable services.

Currently, the network teaching system is divided into three generations. The first generation is to provide students with teaching materials and related materials through web pages, and to connect with other related education networks. In addition to providing learning materials online, the second generation also requires students to conduct asynchronous two-way communication through e-mails, electronic bulletin boards, online exercises and measurements. In addition to the first and second generations, the third generation requires simultaneous two-way communication through online chat rooms, telephone conferences, video conferences or MUD systems. The current world network courses are developing towards the third generation. In related research, reference [3] designed a wireless network-based ideological and political multimedia network teaching resource integration system. This method clarifies the theoretical elements and the evaluation process of teaching resource management, and uses XML as the data exchange carrier to realize the automatic integration of teaching resources. Call the subscription function of the dSPACE framework to store the subscription information in the database. Reference [4] studied the artificial intelligence model of the real-time monitoring of the ideological and political course teaching system, constructed a real-time monitoring system of the ideological and political classroom based on artificial intelligence algorithms, and constructed the model function module according to the actual needs of the ideological and political course.

During the use of the traditional online course mobile cloud teaching system, due to the inadequate network structure optimization, the system downloads course files at a slower rate under complex requests. Therefore, this paper designs a mobile cloud teaching system for ideological and political network courses based on P2P technology. This article uses P2P technology to design the internal structure of the development board of the mobile terminal and the wireless communication chip. In the software design, the system adopts haystac's cloud storage solution. Design the access process of database information according to the actual situation. In order to improve the quality of mobile teaching of ideological and political network courses.

2   Design of Mobile Cloud Teaching System for Ideological and Political Network Courses Based on P2P Technology

2.1   Hardware Design

In the ideological and political network course mobile cloud teaching system designed in this paper, the hardware is mainly designed for the P2P topology network structure.

Among them, the development board of the mobile terminal is a very important part. The mobile terminal development board is installed on the student's personal host to complete the auxiliary functions of the online course mobile cloud teaching system [5, 6]. Among them, the embedded development board uses stm81151k4, which can be programmed with one key, has stable performance, and has more expansion interfaces. It has a USART configuration jumper, which can be connected to a five-wire asynchronous serial port. The structure is shown in Fig. 1.

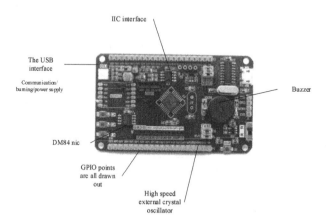

Fig. 1. stm81151k4 development board structure

In the development board, the external five wire asynchronous serial port is connected to the GPS module, which is responsible for receiving the operation information in the network system. IIC interface is connected with GPRS module, which is mainly responsible for information exchange with management system center. Students or teachers can contact the management system center through the interactive function to report the unexpected situation or the completion of learning tasks in the use process. During this period, it mainly relies on Internet communication to realize the interaction between information. The key hardware used in Internet communication is the wireless communication chip: fibocom, which allows a node to communicate with multiple nodes at the same time, and its wireless communication speed can reach 2 m (BPS). When the wireless communication chip turns on the receiving mode and uses the same channel, it can receive six channels of data that are not transmitted through the data channel. The receiving diagram is shown in Fig. 2.

In Fig. 2, when six different wireless modules appear and are in the system enabled state, they can communicate with a chip in the receiving state. And the receiving chip can accurately identify the signals transmitted by multiple different terminals. After receiving the data, the chip transmits the response signal at the same time after recording the IP address of the transmission. The transmitting IP address and the receiving address need to be consistent to receive the response signal. The internal structure of the wireless communication chip is shown in Fig. 3.

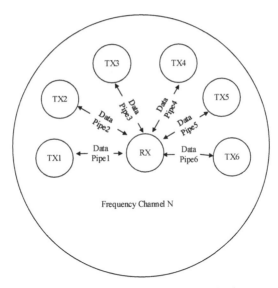

Fig. 2. Schematic diagram of wireless communication reception

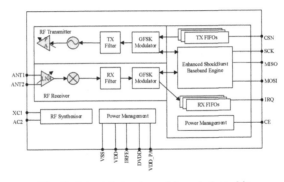

Fig. 3. Internal structure of the wireless chip

From the perspective of micro processing, we only need to design the pins of Ant1, xc1, VSS and IRQ. Ant1 means that the chip is in working state when the chip selection line Ant1 is set to low capacitance. Xc1: SCK clock line controlled by chip. VSS means: the chip controls the power line. IRQ means: when the terminal signal and data are transmitted, the microprocessor will interrupt the communication of the system signal control chip.

In the design of this paper, the main purpose is to optimize the network structure of the system. So we need to introduce P2P technology. P2P is a distributed network architecture, which can divide tasks or workload among peers. In applications, peer nodes are equally important and effective participants. In the traditional mode, users can't even play 50% of the performance of a computer. In this case, the waste of hardware resources is conceivable. After using P2P technology, users can share their hardware resources, including CPU, memory, hard disk, network bandwidth, etc., to other users. On the contrary, he himself can get the hardware resources shared by other users. In this way, we can allocate resources more reasonably, make full use of resources, and do the most at the least cost [7, 8]. In the process of sharing, it does not need the intervention of the central server or other hosts, only through each participating node to complete the corresponding sharing operation process. The P2P topology established in this paper is shown in Fig. 4.

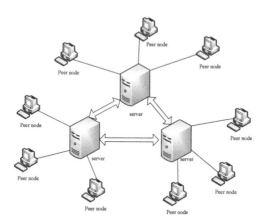

Fig. 4. Mixed topology

The mixed topology is a combination of centralized topology and all-distributed non-structured topologies. By combining their advantages, its maximum potential can be played. A hybrid topology selects some performance nodes as a super node, which manages a part of the node, and the super node is found through a certain discovery algorithm in the entire topology. Then forward the various nodes under each node by the super node. In such a structure, efficiency can be improved. The hardware design of the system is completed.

2.2 Software Design

Design Database Information Access Process

The main feature of mobile cloud teaching in this paper is cloud storage, which has PB level scalability, can share PC computing resources, open, programmable. Given that the cloud storage framework is relatively mature, stable. In the technique selection, the system takes the Haystac's cloud storage scheme, as shown in Fig. 5.

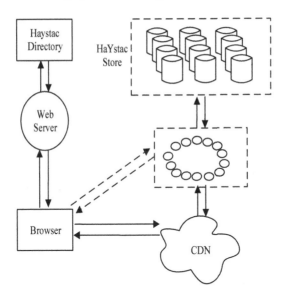

Fig. 5. Haystack cloud storage solution

The platform system involves the cloud storage of file formats including audio and video, compressed text, pictures and other unstructured files. The database module of web teaching website is the most important module in the platform system. All JSP pages are written around the operation of database. The following will briefly introduce the design process of the system database process. The user registration module stores the personal information submitted by the user into the database, and can enter the information display module, information operation module and user management module through the user login module. The data access flow between these modules is shown in Fig. 6.

The operation of each module for the database is usually not directly acting on the underlying physical database, but is connected to the database through the connection pool. The connection pool is connected to the current partial connection. The registration module operates the same database table with the login module. When registering, the system is connected through the connection in the pool (if there is no need to connect, you need to access the database to establish the connection. At the same time, the connection to the connection pool will get the data of the data, write the information to the database. The login module is just the contrary. After the connection pool is connected, the system reads the data from the database, verifying the user's correctness of the login information by the user login module [9–11]. When the user logs in success, you can enter the information display, information operation, information management module, and obtain the corresponding service. The display, operation, and management of information are also connected to the database through the connection pool. The design of the data access process is completed.

Manage User Nodes

The user node management module includes two functions of the user node and the exit function of the user node. We set a configuration file for each user node, which is

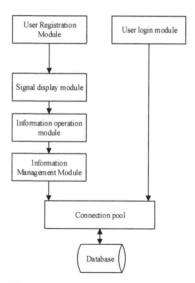

Fig. 6. Database module data process

specifically used to store user node information. When the user node is added to the system, we need to connect to the system server for registration of the relevant node, so you need to know the IP address of the system server. There are two kinds of dynamic states of user nodes. One is the joining of user nodes. User nodes should be added according to the type of file nodes in the user node. Each user node may be a file of Mn, SCN or WCN. Therefore, the user node needs to distinguish between different file nodes. User nodes also need to determine if the user node is a new user node during the joining process, which has never been incorporated out of the system. Because the files when the system is preserved in the old user node, it is necessary to perform different processing according to the various file types in the user node during the user node. By viewing the user_type field in the user node information profile, we can determine the new and old situations of the user node.

Another dynamic state of the user node is an exit of the user node. The exit requirement of the user node is discussed. For the normal exit of the user, in the source node information table, the new joined node authentication and allocate the teaching group, and update node information. When the node is properly exiting, a request is issued to the source node, and the source node deletes the node information table corresponding information and feeds back to the corresponding teaching group. The group leader stops the request to exit the group of node services to delete the corresponding group information.

When the user uses the system, if a network failure or power off is encountered, the system will determine whether the user is unusually exiting, and the group length node does not receive an exit request from the user node. Therefore, the transmission system needs to be judged through the heartbeat mechanism to survive. The normal node sends a status information to the source node to the source node to survive. If the source node does not receive a node status information in N specific time t, the node has been exited. Feedback to the corresponding group length node, stopping the Node topology to the unresponsive node, updating the node topology. If the group leader node is not responded, the group length node is deleted, and the replacement team leads to the new node length and configure the corresponding group of node information to the new group.

3   System Test

This paper implements the design of the mobile cloud teaching system under the P2P technology. Based on the traditional network architecture, the user node expansion layer is added, mainly implementing the coexistence of different modules of the system. In the performance test of the system, it is mainly to download the download rate between the two by comparison in the system, and download the download rate between the two in the traditional system, indicating the advantages of the system.

3.1   Construction System Test Environment

The entire system test environment deployment is divided into a central cloud storage server and user node expansion cloud, where the central cloud storage server uses an open source Moosefs distributed file system. Due to the limited conditions, the server test environment is deployed on the laboratory server, and the Windows Server 2012 operating system and VMware Workstation Pro are installed in this server, 5 virtual machines via VMware virtual. One of them as MASTER, one as a MetaLogger, three as Chunk Servers, each virtual machine configuration is basically the same, hardware is 2.5 GHz CPU, 2 GB memory, operating system is Debian8, and deployment is shown in Table 1.

Table 1. Test environment deployment

Server	Name	IP
Metadata server	mfs-master	172.20.156.120
Backup server	mfs-metalogger	172.20.156.121
Data storage server	mfs-chunkserver-1	172.20.156.122
Data storage server	mfs-chunkserver-2	172.20.156.123
Data storage server	mfs-chunkserver-3	172.20.156.124
Mobile terminal	Kyd-user-1	172.20.156.130
Mobile terminal	Kyd-user-2	172.20.156.131
Mobile terminal	Kyd-user-3	172.20.156.132
Mobile terminal	Kyd-user-4	172.20.156.147
Mobile terminal	Kyd-user-5	172.20.156.148
Mobile terminal	Kyd-user-6	172.20.156.149

User node expansion cloud test requires a large number of terminals to test. Due to the limited conditions, there are not enough personal terminals. So in the lab server, 20 virtual machines are virtualized for testing. Each virtual machine is configured substantially the same, the hardware is 2.5 GHz CPU, 1 g memory, 40 G disk space. The operating system is a text of Windows 10. Its IP address is 172.20.156.130-172.20.156.149. The entire test environment architecture is shown in Fig. 7.

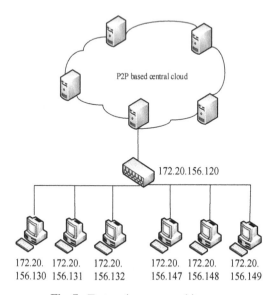

Fig. 7. Test environment architecture

The performance test of this article is to test a variety of normal, peaks, and abnormal load conditions by automated test tools to test system download rates. By load testing, it is determined that the download performance of the system under various workloads is to test the change of the download rate when the load is gradually increased. The system uses Jmeter to simulate the actual situation for pressure testing. Simply use two test cases: simple requests and complex requests. In a simple request, set up 30 concurrent users every 20 s to request five simple download services. Repeat 30 times, a total of 4,500 requests.

In complex requests, 30 complicated users are requested for five complex download services every 20 s. Repeat 30 times, a total of 4,500 requests. It is equivalent to adding 90 online users per minute, and accesses 10 min according to each person, which is equivalent to 900 at the same time online processing power. Under the same experimental conditions, the system and conventional systems designed in this paper are tested and the test results are compared and analyzed.

3.2 Comparative Analysis of Experimental Results

In the above system test environment, the two systems are obtained under a simple request, and the test results of 4000 times are tested, as shown in FIG.

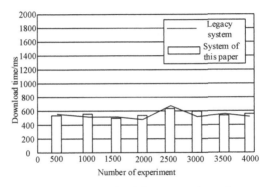

Fig. 8. Simply request download time comparison of two systems

It can be seen from the results in Fig. 8 that under a simple request, the test download time of the two systems is not much different. After calculation, the average download rate of the traditional system is 5.3 MB/s, and the average download rate of this article is 5.5 MB/s. The above test results show that under the simple request test, the performance difference between the two systems in terms of downloading is relatively small. In order to further clarify the advantages of the method in this paper, the download time of the two systems under complex requests was tested, and the test results of 4000 tests were obtained, as shown in Fig. 9.

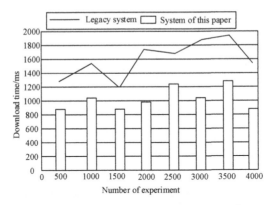

Fig. 9. Comparative download time comparison of two systems under comprehensive request

It can be seen from the results in Fig. 9 that under complex requests, the test download time of the two systems has a significant gap. The average download rate of the traditional system is 7.2 MB/s, and the average download rate of this article is 11.1 MB/s. The above test results show that in the case of complex request testing, the download rate of the system designed in this paper is significantly improved compared with the traditional system. Because the network structure of the method in this paper adopts the P2P hybrid topology structure to realize the complementarity between the advantages of different structures, thereby effectively increasing the download rate, and verifying the advantages of the system designed in this paper.

4 Conclusion

This article is designed and described with a P2P technology-based network course mobile cloud teaching system, and is designed in detail from both hardware and software. In the experiment, the performance and advantages of the system were verified. The design of the P2P technology-based thinking network curriculum mobile cloud teaching system is far better than the traditional system under complex request, which will bring a new experience to the distance education model. Solve problems that are difficult to control in network teaching. With the improvement and wide application of the network, the mobile cloud teaching system based on P2P technology will provide a better platform for distance teaching.

References

1. Lyu, H.: Research on the practical application of education management system based on mobile cloud platform in colleges and universities. wuxian hulian keji **16**(20), 43–44 (2019)
2. Cao, L.Z., Zhou, H.: Design and application of a mobile multimedia teaching system based on cloud computation with the course of computer network technology as an example. J. Jilin Teach. Inst. Eng. Technol. **36**(3), 87–89 (2020)
3. Liu, R.: Design of ideological and political multimedia network teaching resources integration system based on wireless network. Sci. Program. **2021**(3), 1–15 (2021)
4. Luo, Y.: Artificial intelligence model for real-time monitoring of ideological and political teaching system. J. Intell. Fuzzy Syst. **40**(1), 1–10 (2020)
5. Tang, T.: Design and application of the mobile teaching system based on blended learning. Softw. Eng. **22**(12), 47–51 (2019)
6. Guo, J., Wang, L.L.: On the construction of blended college English mobile teaching mode under the background of "Internet+" era. J. Hubei Correspond. Univ. **32**(4), 148–149 (2019)
7. Xu, D.H., Luo, Z.B., Xu, X.H., et al.: Research on internet plus medical image teaching model based on the online interactive cloud teaching platform. Med. Educ. Res. Pract. **27**(3), 400–403 (2019)
8. Liu, S., Glowatz, M., Zappatore, M., Gao, H., Jia, B., Bucciero, A. (eds.): eLEOT 2018. LNICSSITE, vol. 243. Springer, Cham (2018). https://doi.org/10.1007/978-3-319-93719-9
9. Liu, S., Liu, D., Srivastava, G., Połap, D., Woźniak, M.: Overview and methods of correlation filter algorithms in object tracking. Compl. Intell. Syst. **7**(4), 1895–1917 (2020). https://doi.org/10.1007/s40747-020-00161-4
10. Jiang, K.W., Liu, G.J.: Design of a P2P-based shared book resource mode. J. Jilin Teach. Inst. Eng. Technol. **35**(9), 85–88 (2019)
11. Liu, S., Li, Z., Zhang, Y., et al.: Introduction of key problems in long-distance learning and training. Mob. Netw. Appl. **24**(1), 1–4 (2019)

Mobile Multimedia Teaching System for Ideological and Political Theory Courses Based on Smart Phones

Lin Chen[1] and Yan-ming Zhan[2(✉)]

[1] NanChang University, NanChang 330000, China
chenlin454@tom.com
[2] Jiangxi Normal University, NanChang 330000, China

Abstract. The existing mobile multimedia teaching system for ideological and political theory courses has poor data balance, which leads to excessive communication overhead and affects the overall teaching effect of the system. The mobile multimedia teaching system for ideological and political theory courses is designed based on smartphones. In terms of system hardware design, the Zynq-7000 chip is used as the main control unit to design the ZYNQ core processing module; the high-speed storage circuit is designed to match the acquisition speed and storage speed. In terms of system software design, calculate the load capacity of the gateway, design the teaching system to share the communication network; set the data balance mode of the ideological and political theory course based on the smart phone to shorten the task request delay. Experimental test results show that the communication overhead of the system designed in this paper is significantly less than that of the existing system, the transmission delay range is 30–50 ms, the packet loss rate is less than 20%, and the network load balance can reach 70%, so it is beneficial to balanced multimedia The load capacity of the teaching system reduces link congestion, makes the communication network data transmission more reliable, and is more conducive to the practical teaching system application of ideological and political theory courses.

Keywords: Smart phone · Ideological and political theory class · Mobile · Multimedia teaching system

1 Introduction

Since entering the new century, the Party Central Committee has successively issued a number of documents on further improving and strengthening the ideological and political education of college students. With the continuous renewal and improvement of the student teaching system in the new era, the school pays more and more attention to cultivating students' comprehensive quality capabilities, improving students' psychological qualities, and fully paying attention to the healthy development of students. Ideological and political education, as an important teaching content of students' moral education,

S. Liu and X. Ma (Eds.): ADHIP 2021, LNICST 417, pp. 491–503, 2022.
https://doi.org/10.1007/978-3-030-94554-1_39

can guarantee the comprehensive development of morality, intelligence, physical education, and art as students grow up. To a certain extent, ideological and political courses can be said to be one of the innovative results of the reform of ideological and political education courses in colleges and universities in recent years. In order to protect the students' ideological and political ability and improve the traditional ideological and political courses, such as the lack of teaching content and the single education angle, from the perspective of the healthy development of students, the mobile multimedia teaching system is used to combine the education value of ideological and political education for students. Establish correct values. Under the premise of continuous advancement of society, science and technology have also advanced. In this process, mobile smart terminals and mobile Internet usher in new development space. As a brand-new teaching method, multimedia teaching can effectively improve the classroom teaching effect of ideological and political theory in colleges and universities, and stimulate students' learning initiative. In the use of multimedia methods to teach ideological and political theory courses, it is necessary to combine the actual situation of students, grasp the good interest and appropriateness of the ideological and political teaching content, and combine the teaching requirements of ideological and political theory courses to make the multimedia teaching courseware present timeliness and simplicity [1]. In this context, a mobile multimedia teaching system for ideological and political theory courses has gradually been developed and put into use. At this stage, the academic community has carried out extensive and in-depth research on the teaching system, and has achieved certain research results. Literature [2] designed an effectiveness evaluation system for the shortcomings of low evaluation accuracy of the teaching system, which greatly improved the evaluation accuracy. Literature [3] designed and developed a university physics mobile learning platform based on the Android system to free learners from the constraints of time and space. As we all know, smart phones are one of the necessary tools in today's society. People need smart phones in their daily life, study and work. Smart phones have changed people's lifestyles. As long as they are connected to the mobile Internet, they can access information. Access makes people's lives more convenient. With the continuous development of the mobile Internet, people no longer only rely on the original fixed terminal equipment in the process of obtaining information, and are no longer restricted by places, and can consult information anytime and anywhere. Therefore, this paper designs a mobile multimedia teaching system for ideological and political theory courses based on smart phones to improve the teaching effect of ideological and political courses.

2 Hardware Design of Mobile Multimedia Teaching System for Ideological and Political Theory Course

2.1 Design ZYNQ Core Processing Module

The ZYNQ core processing module is the brain of the system. It needs to complete the management of the system's working status, the collection and storage of image data and other scheduling tasks, as well as the communication with external equipment. The main control module receives the control commands of the 422 serial port, sets the working mode of the camera and feeds its status back to the upper computer. The main control module can also switch between storage function mode and data export mode according to system instructions. In the storage mode, the main control unit can store the collected image data in a fixed format; in the data export mode, the main control unit can read the image data in the storage medium to the host computer through Gigabit Ethernet [4]. According to the current research status at home and abroad, combined with device resources and costs, and according to Xilinx related materials, it is finally determined to use Zynq-7000 series chips as the main control unit. In order to simplify the system code development cycle, the decoder chip is used to decode the LVDS signal of the Cameralink protocol input from the camera. The Cameralink interface can be divided into three types of signals, each of which has a different purpose. The image data signal is the most basic part of the interface. It consists of a set of differential clocks and 4 sets of differential data, and is used to transmit high-precision image data. The camera control signal consists of 4 groups of differential signals, which are mainly used to set the camera's external synchronization, pixel reset, etc. The serial communication signal is composed of 2 groups of differential signals, which can set the camera working mode and configure specific parameters [5]. The decoded data signal contains 3 clock signals, 4 synchronization signals and 80-bit data. A total of 87 I/Os are required for camera control pins (CC1-4). A total of 4 I/Os are required. A set of serial communication signals (SerTFG, SerTC) requires 3 I/Os, 4 serial ports use 8 I/Os and at least 48 I/Os are required for multiple storage solutions selected according to the storage array. In summary, ZYNQ's PL end I/O needs to use at least 150. Considering the quantity and cost of resources, ZYNQ plans to use XC7Z020-CLG484 of Xilix Company. The PL end of the Z7020 uses Artix7, which is highly integrated and can reduce the BOM cost. At the same time, the static power consumption is reduced by 65% compared with the 45 nm device, which has a very wide range of applications.

2.2 Design High-Speed Memory Circuits

According to system performance requirements and functional requirements, in order to ensure that the acquisition speed can match the storage speed, the system is designed to use multiple meters and multiple Micro SD cards to form a storage array. When the SD card uses SDR104 mode, it first needs to be initialized under 3.3 V level signal and switched to work under 1.8 V voltage. After entering the 1.8 V level standard, the mode selection is performed, so the voltage switching function should be provided in the schematic design. The system design uses NVT4857UK to realize the voltage switching function. NVT4857UK is a bidirectional two-level converter, which integrates an LDO module capable of outputting 1.8 V, and uses a 70 K resistor inside the chip to pull the signal line [6]. When the system is powered on, the B2 pin (SEL) is low by default, and the internal pull-up voltage is consistent with the A3 pin (VCC). When the PL terminal pulls the SEL pin high, the internal LDO module starts to work and outputs 1.8 V, which also makes the data pull up to 1.8 V. The system uses 8 Micro SD cards in total, and the circuit design is shown in Fig. 1.

The system design uses SDR104 mode, a single SD card uses 8 PL end I/Os, and 8 SD cards need to use up to 64 I/Os. In order to save system I/O, when all SD cards of the system are initialized, the 8 voltage switching control pins are controlled by the same

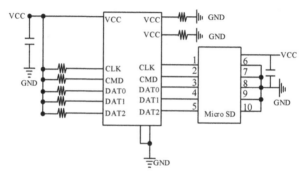

Fig. 1. High-speed memory circuit

I/O, and finally 57 I/Os on the PL end are used. In the core board I/O, the BANK34 and BANK35 pins can change the level standard and each BANK has 50 I/Os, in the design process. In order to ensure the signal quality and the normal and stable operation of the system, all pins of a single SD card are required to use the I/O of the same BANK. So far, the hardware design of the mobile multimedia teaching system for ideological and political theory courses is completed.

3 Software Design of Mobile Multimedia Teaching System for Ideological and Political Theory Course

3.1 Design Teaching System to Share Communication Network

The ideological and political theory course mobile multimedia teaching system shares the communication network. According to the node's video media data and the scheduling information list BM of other nodes, a reasonable client is selected, and the ideological and political theory teaching multimedia information is collected from it. When the user finds the teaching information he needs, he should look it up in the list of other nodes. The information management module mainly updates the management node scheduling list. When the same data module is provided by multiple providers, the provider with the least delay with the node should be selected. If there is no node providing data, the information management module updates the node list to ensure that each data has a provider. When the provider provides more information, the data should be transmitted first to improve the sharing speed of intelligent teaching multimedia resources. The campus network port in the school and other nodes form a network to ensure that users can use other node resources [7]. Different user nodes in the intelligent teaching multimedia sharing network have different functions. When all nodes in the network select one node as the gateway, the network may have a bottleneck. In most cases, the gateway node has the ability to connect to the external network. At this time, the most suitable gateway node should be selected as the gateway to speed up the sharing of intelligent teaching multimedia. Calculate the distance between each gateway node and the fixed node, and judge whether the gateway node is stable according to the distance, and then select the optimal gateway node as the gateway. The weighted average method can be used to measure the stability of the gateway node, and the formula (1) is as follows.

$$Y_i = \alpha D_{ij} + \beta L_{ij} + \chi P + \delta F + \phi W \tag{1}$$

In formula (1), Y_i represents the node selection factor, and i is used to describe the comprehensive metric that the selected node can be regarded as a gateway. D_{ij} represents the number of hops from gateway node i to node j, and the weight is α. L_{ij} represents the link quality from the gateway node i to the j node, and the weight is β. P_{ij} represents the energy of the gateway node, and the weight is χ. F_{ij} represents the load capacity of the gateway node, and the weight is δ. W_{ij} indicates that the stability weight of the gateway node is ϕ. By comparing Y_i, select the best node that is the gateway. The comprehensive performance of a node is inversely proportional to the selection factor, that is, the smaller the selection factor, the better the comprehensive performance. The load capacity of a node is related to the number of selected nodes as gateways. The greater the remaining load capacity of a multimedia information sharing network node for ideological and political theory teaching, the faster the sharing of multimedia teaching resources. In formula (1), the calculation formula of load capacity is:

$$F = K_i - \sum k_{i,t} \tag{2}$$

In formula (2), K_i represents the total bandwidth of the gateway node. $k_{i,t}$ represents the bandwidth when the gateway node i is selected at t. The calculation of the remaining load is related to time, and the remaining load calculated at different times is different. In order to get a suitable load, it should be smoothed.

$$F = \begin{cases} \lambda F_{i,t} + (1 - \lambda)\lambda F_{i,t-1}, & t > 0 \\ K_i, & t = 0 \end{cases} \tag{3}$$

In formula (3), λ represents the remaining load factor, and the value range is between 0 and 1. In order to calculate the stability of the gateway node, the distance from the node to the optimal node in a certain period of time is counted, and the stability of the gateway node is judged by the variance of the distance. The mean value of the distance between node j and gateway node i at time t is:

$$e\left(s_t^{i,j}\right) = \frac{1}{j}\sum_{i=0}^{j-1} s_{t-i}^{i,j} \tag{4}$$

Select the value with the smallest variance, which means that the gateway node at this time is the optimal gateway node, and the optimal gateway node is selected as the gateway exit. Using the node at this moment to access the external network can realize multi-node networked intelligent teaching multimedia resource sharing. And sharing multimedia The speed and precision are high. Integrating the above process, complete the design of the shared communication network of the teaching system.

3.2 Set Up Data Balance Mode for Ideological and Political Theory Courses Based on Smart Phones

Mobile learning based on the smart phone's own functions has limited learning software and learning resources, and has not played its own advantages. In recent years, with the development of various new technologies such as mobile network technology and wireless network connection technology. The mobile learning model based on external connections is more and more able to play its own advantages and is being widely used [8]. Therefore, smart phones are used as the key nodes of the teaching communication network to ensure the real-time performance of the system and the maximum utilization of resources. First, each smart phone is given the same initial PR value, and iteratively updated until the PR value stabilizes and the traversal ends. The process can be expressed as:

$$PR(G) = (1 - \alpha) + \alpha \left(\frac{PR(H_1)}{N(H_1)} + \cdots + \frac{PR(H_i)}{N(H_i)} \right) \tag{5}$$

In formula (5), $PR(G)$ represents the PageRank value of smart phone G; α represents the attenuation coefficient; H_i represents the web pages connected to G; and N represents the number of web pages connected to G. The Google matrix is constructed by the PR values of each webpage, and its iterative method can be expressed as:

$$GO = \omega S + \frac{1 - \omega}{n} EE^T \tag{6}$$

$$PR^{(k+1)} = G^T PR^k \tag{7}$$

In formula (6–7), GO is the n-order Google matrix; S is the adjacency matrix constructed by connection; ω is the damping coefficient; E is an n-bit column vector with all 1 elements; k is the number of iterations. The communication network node of the teaching system can be regarded as a web page, and the link can be regarded as a connection. Therefore, the PageRank algorithm can be used to rank the positioning nodes of the communication network of the teaching system and identify the key nodes with influence. The communication load reflects the operation mode of the communication network. Under the premise of the same topology, different load distributions result in different importance of nodes. Therefore, it is necessary to consider the influence of communication load on node identification. The greater the load capacity, the higher the importance of the node [9]. The load level also reflects the node's requirements for power communication. Under the same scale, the higher the level of load, the greater the communication failure and loss caused by the load. Considering the node load capacity and level comprehensively, the node importance index I_x is obtained, which can be expressed as:

$$\begin{cases} I_x = \theta_1 \bar{I}_{1,x} + \theta_2 \bar{I}_{2,x} \\ \bar{I}_{1,x} = \dfrac{I_{1,x} - \min(I_{1,x})}{\max(I_{1,x}) - \min(I_{1,x})} \\ \bar{I}_{2,x} = \dfrac{I_{2,x} - \min(I_{2,x})}{\max(I_{2,x}) - \min(I_{2,x})} \end{cases} \tag{8}$$

In formula (8), $I_{1,x}$ represents node load capacity, $I_{2,x}$ represents node load level, $\bar{I}_{1,x}$ and $\bar{I}_{2,x}$ distributions represent normalized load capacity and level results; θ_1 and θ_2 represent weighting factors. According to the communication network structure and operating state of the initial teaching system, calculate the node load, consider the importance of the nodes, and obtain a stable *PR* matrix through the power method iteration. The larger the element *PR* value in the matrix, the higher the importance of the corresponding node, and it can be identified It is a key node. By identifying key nodes, it can help adjust the load strategy of the entire teaching system communication network, and adjust and optimize the communication load flow [10]. Set the campus website as the main station and the smart phone as the sub-station. The two are abstracted together as node set *A*, the communication channel is abstracted as link set *B*, and the defined constraint condition is *C*, then the ideological and political course teaching system can be expressed as (*A*, *B*, *C*). The task request in the mobile multimedia teaching of the ideological and political theory course is set as *d*, including (source node, target node, bandwidth), and the transmission delay and link utilization rate of the teaching task request are comprehensively considered to construct the objective function. The routing objective function can be expressed as:

$$f(d) = \min \left(\sum_{d \in D} \alpha T_d + \beta \kappa \right) \tag{9}$$

In formula (9), α and β are constant coefficients; D represents the set of teaching task requests; T_d represents the communication delay; κ represents the maximum link utilization. Among them, the calculation formula of communication delay T_d is:

$$T_d = \frac{l}{v} a_{dl}^{xy} + bt_u + \Delta t \tag{10}$$

$$b = \sum_{(x,y) \in B} a_{dl}^{xy} \tag{11}$$

In formula (10–11), l represents the length of the path; v represents the transmission speed of the information channel; a_{dl}^{xy} represents the l-th service flow and the d-th path through the link l_{xy}, when it passes through a_{dl}^{xy}, it is 1, otherwise it is 0; b represents the a_{dl}^{xy} Set; t_u represents node switching delay; Δt represents delay jitter. The formula for calculating the maximum link utilization rate κ is:

$$\kappa = \max \left(\sum_{d=1}^{D} \frac{a_{dl}^{xy} Q_{dl}}{L_{xy}} \right) \tag{12}$$

In formula (12), Q_{dl} represents the path flow; L_{xy} represents the link capacity. The constraint conditions for the above-mentioned routing objective function are:

$$\begin{cases} \sum_{(x,y) \in B} a_{dl}^{xy} - \sum_{(y,x) \in B} a_{dl}^{yx} = 0 \\ \sum_{(y,x) \in B} a_{dl}^{xy} Q_{dl} \leq L_{xy} \cdot \kappa \end{cases} \tag{13}$$

The above conditions indicate that (1) the total inflow of nodes is equal to the total outflow; (2) the routing load needs to meet the maximum link load constraint. Using smart phone data balance adjustment to minimize the task request delay and link utilization additive measurement, to ensure the real-time performance of the teaching system and maximize the use of resources. Through the construction of the above hardware and software parts, the design of the mobile multimedia teaching system for ideological and political theory courses is completed.

4   Experimental Study

4.1   Teaching System Communication Function Debugging

Communication performance is an important part of the system. First of all, the communication image of the teaching system is debugged. If the debug result does not reach the set goal, the mobile multimedia teaching system cannot be applied to the teaching of ideological and political theory. This will not only improve teaching efficiency, but will reduce efficiency. Therefore, communication performance debugging must be carried out before deployment, and the mobile multimedia teaching system can be deployed to the school on the basis of performance debugging results in accordance with the requirements and goals. Select two indicators of transmission delay and packet loss rate to test the application effect of the communication function. The transmission delay test result is shown in Fig. 2, and the data packet loss rate test result is shown in Fig. 3.

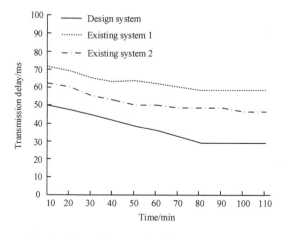

Fig. 2. Comparison results of transmission delay

According to the comparative test result in Fig. 2, the transmission delay of the designed system and the existing system will decrease with the increase in running time. And it converges when it runs for 80 min. The transmission delay starting point and convergence point of the system in this paper are lower than the existing system. The transmission delay is in the range of 30–50 ms, and the rate of decline is relatively fast, showing a rapid decline trend, indicating that the system designed in this paper is superior to the existing system in terms of data transmission time, and can quickly and effectively reduce the data transmission delay. Therefore, the identification of key nodes is accurate, which is beneficial to balance the load and reduce link congestion.

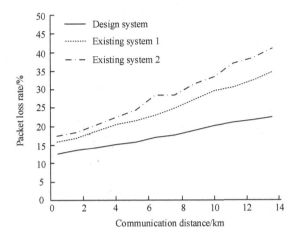

Fig. 3. Comparison results of packet loss rate

According to the comparative test results in Fig. 3, the data packet loss rate of the system designed in this paper and the existing system increase with the increase of communication distances. Compared with the existing system, the data packet loss rate of the system in this paper is relatively low, at 20%. Below, the growth rate is relatively slow. It shows that the method in this paper identifies key nodes, adds routing strategies for alternative paths, expands the search space, and helps to balance the load. Even if the logical link is interrupted in the dynamic changes of the network, the success rate of data packet delivery is relatively high, and the data transmission of the communication network is relatively more reliable.

4.2 Evaluation of Teaching System Data Balance

To test the data balance effect of the teaching system designed in this article, test the load balance of the communication network of the teaching system. Calculate the load of each path through the key node, set the load capacity of the node as the balance index of the node, and evaluate the balance index of the key node identified by the system and the existing system to detect the effect of data load balancing. The specific test results are shown in Table 1.

According to the test results in Table 1, the key node positions obtained by this system and the existing system are different. But in terms of load balancing index, the key node load index obtained by the system in this paper is higher than that of the existing system. Further test the overall load balance of the communication network of the teaching system, and the results are shown in Fig. 4.

Table 1. Key node load comparison test

Sort	Design system		Existing system 1		Existing system 2	
	Serial number	Equilibrium index	Serial number	Equilibrium index	Serial number	Equilibrium index
1	33	90.8	15	90.2	34	80.8
2	18	90.3	18	80.8	37	80.6
3	12	80.6	37	80.5	29	70.9
4	32	80.4	11	70.8	31	70.5
5	25	80.2	22	70.4	16	60.9
6	11	70.9	21	70.2	24	60.4
7	17	70.5	9	60.9	17	50.8
8	8	60.8	20	60.5	26	50.3
9	15	60.4	14	50.4	4	40.7
10	26	50.8	16	40.9	12	30.6

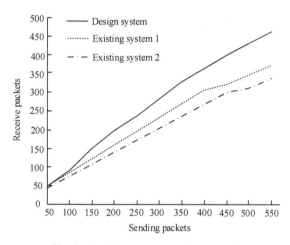

Fig. 4. Load balance comparison results

The communication status of the communication network link of the teaching system is compared, and the number of data packets sent and received is tested. According to the comparative test results in Fig. 4, at the stage when the network load is small, the load transmission and reception balance of the system in this paper and the existing system are relatively close; as the network load increases, the data reception of the existing system grows slowly. The system growth trend in this paper is fast, and the network load balance can reach 70%. It shows that the network load balance of the method in this paper is better than the traditional method. The data packet transmission and reception of the communication link are relatively flat, and the link congestion is less. In the teaching system.

4.3 Teaching System Communication Overhead Test

System communication overhead testing is an efficient distributed testing behavior. It is calculated that the teaching system changes as the number of concurrent users increases. In the network parallel computing system, the network communication overhead is an important factor that affects the operating efficiency of the network system. The running time and the communication overhead between computing nodes have an important impact on the design of the teaching system. However, network communication overhead contains many factors. This article uses a timing tool with an accuracy of 0.1 microseconds to measure and record the number of nodes that need to be updated when users participate and exit, compare and measure communication overhead, and compare the communication overhead of user participation under different systems. The value of the test result is shown in Tables 2 and 3.

Table 2. Comparison results of communication overhead when users participate

Number of concurrent users	Communication overhead (bits)		
	Design system	Existing system 1	Existing system 2
50	160	210	200
100	210	240	230
150	240	270	260
200	280	310	290
250	300	340	340
300	340	370	360
350	380	420	390
400	420	460	470
450	460	480	490
500	480	500	510
Average value	327	360	354

From the test results in Tables 2 and 3, it can be seen that the communication overhead of the system designed in this paper is significantly less than that of the existing system. The communication overhead is reduced by 33 bits and 27 bits when the user participates, and the communication overhead is reduced by 50 bits and 53 bits when the user exits. When a user participates or exits, different nodes may be affected multiple times, and as the number of users grows, the amount of communication increases accordingly. The data balance of the system designed in this paper is better, and the data redundancy is reduced, so the communication overhead is small and it has certain advantages.

Table 3. Comparison results of communication overhead when users log out

Number of concurrent users	Communication overhead (bits)		
	Design system	Existing system 1	Existing system 2
50	170	210	230
100	200	250	260
150	230	270	280
200	260	300	320
250	280	330	330
300	300	360	350
350	320	380	370
400	350	400	410
450	370	430	420
500	410	460	450
Average value	289	339	342

5 Concluding Remarks

The multimedia teaching system can play a guiding role in the mobile teaching planning of ideological and political theory classrooms. This article effectively improves the poor data balance of the traditional multimedia teaching system by constructing the system main control unit and the calculation gateway load capacity, resulting in excessive communication overhead. Big question. This research is based on a smart phone to design a mobile multimedia teaching system for ideological and political theory courses, which is conducive to the improvement of the system's balanced load capacity and reduces link congestion. At the same time, key nodes can be selected efficiently, alternative paths can be added, search space can be expanded, and the overall teaching effect of the ideological and political system can be improved. However, due to limited time and research conditions, the system designed this time still has shortcomings, and the results still have limitations. For example, only one school was selected as the experimental object in this study, and the data lacked universality. Therefore, in the subsequent experiment selection, we can go deeper and make multi-dimensional choices. At the same time, data security technology can be applied to the teaching system to further improve the privacy security of the system and provide more reliable teaching methods for the study of ideological and political theory courses. This will consolidate the results of this experiment and provide theoretical support for the design of multimedia teaching systems in the future.

References

1. Chen, J., Wu, Z.: Discussion on the effective application of multimedia means in vocational ideological and political theory course teaching. J. Nanchang Coll. Educ. **27**(6), 116–117 (2012)

2. Tong, Y., Tian, Z., Deng, H.: Design and application of the mobile learning platform of college physics based on Android. Coll. Phys. **39**(04), 48-52+72 (2020)
3. Jiang, L., Xie, F.: Design of effectiveness evaluation system of foreign language assistance teaching on mobile terminal. Mod. Electron. Techniq. **43**(18), 132–134 (2020)
4. Wu, T., Fu, H., Cai, J.: Experience of the application of 3D virtual simulation teaching system 3Dbody in the experimental teaching of system anatomy. Chinese J. Anat. **43**(01), 77–79 (2020)
5. Zhu, L., Sun, J., Zhou, M.: Design and implementation of dance video teaching system based on Spring MVC architecture. Mod. Electron. Techniq. **42**(07), 71-73+78 (2019)
6. Liu, S., Glowatz, M., Zappatore, M., Gao, H., Jia, B., Bucciero, A. (eds.): eLEOT 2018. LNICSSITE, vol. 243. Springer, Cham (2018). https://doi.org/10.1007/978-3-319-93719-9
7. Yu, Y., Zhu, P.: Mobile Learning Mode Based on Mobile Terminals in the "Internet +" Environment. Inf. Sci. **38**(02), 125-128+155 (2020)
8. Liu, S., Li, Z., Zhang, Y., et al.: Introduction of key problems in long-distance learning and training. Mob. Netw. Appl. **24**(1), 1–4 (2019)
9. Xie, H., Chen, X.-L., Chen, D., et al.: Research on quaternity and connotative form of microbiology and immunology based on "characteristic textbook + cloud platform + mobile terminal + virtual system" through blended teaching. Chinese J. Immunol. **35**(17), 2141–2146 (2019)
10. Fu, W., Liu, S., Srivastava, G.: Optimization of big data scheduling in social networks. Entropy **21**(9), 902 (2019)

Author Index